FLORA ZAMBESIACA

*Flora terrarum Zambesii aquis conjunctarum*

VOLUME SIX: PART FOUR

# FLORA ZAMBESIACA

## MOZAMBIQUE, MALAWI, ZAMBIA, ZIMBABWE, BOTSWANA

## VOLUME SIX: PART FOUR

**Authors**

D. J. NICHOLAS HIND

HENK J. BEENTJE

BENOIT F.P. LOEUILLE

**Edited by**

BENOIT F.P. LOEUILLE

on behalf of the Editorial Board:

FRANCES CHASE
*National Botanical Research Institute, Windhoek, Namibia*

MARTIN CHEEK
*Royal Botanic Gardens, Kew, U.K.*

DAVID CHUBA
*School of Natural Sciences, UNZA, Lusaka, Zambia*

IAIN DARBYSHIRE
*Royal Botanic Gardens, Kew, U.K.*

DAVID GOYDER
*Royal Botanic Gardens, Kew, U.K.*

SHAKKIE KATIVU
*University of Zimbabwe, Harare, Zimbabwe*

JAMESON SEYANI
*Herbarium and National Botanical Gardens, Malawi*

Published by the Royal Botanic Gardens, Kew
for the Flora Zambesiaca Managing Committee
2026

Citation: Hind, D.J.N. *et al.* (2026). Compositae (part 4).
In: B.F.P. Loeuille (ed), *Flora Zambesiaca*, Vol. 6(4). Royal Botanic Gardens, Kew.

First published in 2026 by
Royal Botanic Gardens, Kew,
Richmond, Surrey, TW9 3AE, UK
www.kew.org

Distributed on behalf of the Royal Botanic Gardens, Kew in North America by the University of Chicago Press, 1427 East 60th Street, Chicago, IL 60637, USA

ISBN 978-1-84246-848-7
eISBN 978-1-84246-849-4

British Library Cataloguing in Publication Data
A catalogue record for this book is available from the British Library

Design and page layout: Nicola Thompson, Culver Design
Production management: Georgina Hills

Printed in the UK by Halstan

EU Authorised Representative: Easy Access System Europe Oü, 16879218.
Mustamäe tee 50, 10621, Tallinn, Estonia (email: gpsr.requests@easproject.com).

For information or to purchase all Kew titles please visit shop.kew.org/kewbooksonline or email publishing@kew.org

Kew's mission is to understand and protect plants and fungi, for the wellbeing of people and the future of all life on Earth.

Kew receives approximately one third of its running costs from Government through the Department for Environment, Food and Rural Affairs (Defra). All other funding needed to support Kew's vital work comes from members, foundations, donors and commercial activities including book sales.

# CONTENTS

# FAMILY 97. **COMPOSITAE**

## LIST OF TRIBES INCLUDED IN VOLUME 6, PART 4

11. Senecioneae

## NEW TAXA AND COMBINATIONS PUBLISHED IN THIS VOLUME

# 97. COMPOSITAE

by D.J. Nicholas Hind

## Preface to Flora Zambesiaca Compositae parts 6(2–5)[1]

Since the publication of Flora Zambesiaca **6**(1) in 1992 (where Pope provided a general conspectus to the arrangement of tribes in the then remaining two parts, and Jeffrey & Pope provided a key to the tribes found in the Flora area), much has happened in Compositae research.

Bremer, Asterac. Cladist. Classif. (1994), provided the first of the modern treatments of the family, recognising 17 tribes in three subfamilies, with an estimate of 1535 genera and 23,000 known species ('excluding microspecies'). His treatment was an alphabetical coverage of genera within each of the infratribal groupings (each genus provided with a synoptic description highlighting 'unusual' or 'diagnostic' characters in italics), but lacked keys. With partially systematic partially alphabetical, tribal treatments, the Compositae volume in Kubitzki's Families and genera of vascular plants (FGVP) (Kadereit & Jeffrey, Fam. Gen. Vasc. Pl. **8**, [2006] 2007) provided the keys, but the tribal, infratribal, and infrafamilial taxa had dramatically changed since both Bremer's and Pope's volumes. In Jeffrey's opening summary the number of genera had increased to 'over 1,600' (1620 were treated), and subfamilies to four; treatments in the volume had increased the number of tribes to 30. Funk et al.'s 2009 'Compositae' volume (Funk *et al.*, Syst. Evol. Biogeogr. Compositae, 2009), following on from the conference on 'Systematics & Evolution of the Compositae' in Barcelona in 2006, summarized subsequent changes. Five subfamilies and 40+ tribes were recognized – too much to be able to communicate easily to students attempting to come to grips with such a large family: Jeffrey & Pope's tribal key in Flora Zambesiaca 6(1) was just under two pages (Royal Octavo); Jeffrey's global key to the family in the FGVP volume ran to eight pages, of two columns per page (of A4).

We are essentially following Pope's conspectus for tribal order (and ignoring subfamilial arrangements), but it is now convenient to recognize two additional tribes – the Platycarpheae and Athroismeae. The Platycarpheae is only represented by *Platycarphella*. The genera of the Athroismeae (*Anisopappus, Artemisiopsis, Athroisma, Blepharispermum*) are now placed after the Inuleae s.s. where part (*Anisopappus* and *Artemisiopsis*) would key out in Jeffrey & Pope's tribal key. Purely to maintain a manageable production line the coverage of tribes in remaining part are proposed as follows:

| *Volume 6 part 2* | *Volume 6 part 3* | *Volume 6 part 4* | *Volume 6 part 5* |
|---|---|---|---|
| 6. Platycarpheae | 8. Athroismeae | 11. Senecioneae | 12. Calenduleae |
| 7. Inuleae | 9. Astereae | | 13. Heliantheae |
| | 10. Anthemideae | | 14. Eupatorieae |

Publication of these parts will be taking place in reverse order as they are completed.

The Inuleae are treated sensu lato (including former tribal splits of the Plucheeae and the Gnaphalieae). The Heliantheae are treated sensu lato (including the Helenieae) and not split as under following the proposals of Panero (see his accounts in the FGVP volume)

---

[1] By D.J.N. Hind and B. Loeuille

where 20 tribes were recognized, instead of one. The other two genera of the Athroismeae (*Athroisma* and *Blepharispermum*) key out next to the Heliantheae because of the presence of black, carbonized achenes. We were tempted to opt for Jeffrey's proposal to go to the level of recognising 'supersubtribes' because of the conceived problem with the Eupatorieae being nested within the Perityleae-Millerieae-Madieae clade in the so-called 'Heliantheae alliance'! However, this was a step too far: the concept of the Eupatorieae is one that is easy to communicate to students; how many know of supersubtribes?

As originally acknowledged by Pope, the 'foundations [of several generic treatments] were laid by Professor H. Wild', with the significant exception of the Senecioneae, which has been worked on 'from scratch'. Hiram Wild's 'foundations' are also gratefully acknowledged in the remaining treatments where published, or manuscript, accounts were available. There are, needless to say, many generic and tribal revisions that have been published since Wild's accounts appeared in Kirkia. Full advantage has been taken of these, but, in addition, taxonomic problems unearthed in writing several accounts has resulted in several synopses being produced in advance of the Flora accounts. One or two modified concepts will be presented where necessary.

1. Capitula ligulate; corollas all ligulate and ligule with 5 apical teeth. . . . . **4. Lactuceae**
 –  Corollas pseudoligulate or radiate, ray limb with 4 or less apical teeth, and disc corollas actinomorphic, or all florets actinomorphic and capitula discoid, or all corollas actinomorphic and outer radiant, or capitula disciform and outer florets filiform and female and disc florets cylindrical and hermaphrodite or functionally male . . . . . . . . 2
2. Style arms usually long, well exserted from corolla and with conspicuous papillose appendages; capitula homogamous and discoid; florets all hermaphrodite; mature achenes black . . . . . . . . . . . . . . . . . . . . . . . . . . . . . . . . . . . **14. Eupatorieae**
 –  Style arms short or long, but lacking papillose appendages; capitula heterogamous and radiate, radiant, or disciform, or homogamous and discoid; mature achenes of various colours, sometimes black . . . . . . . . . . . . . . . . . . . . . . . . . . . . . . . . . . . . . . 3
3. Phyllaries uniseriate, coherent by overlapping margins, or partially or wholly connate, calyculate or ecalyculate; pappus present. . . . . . . . . . . . . . . . . . . . . . . . . . . . . . 4
 –  Phyllaries imbricate, 2- or more- seriate, free or connate, if uniseriate then free or pappus absent, or capitula unisexual or achenes densely villous and long-hairy from base . . . . 5
4. Phyllaries with evident elongated oil glands; achenes black when mature. . . . . . . . . . .
. . . . . . . . . . . . . . . . . . . . . . . . . . . . . . **13. Heliantheae** (*Tagetes*, cult. *Dyssodia*)
 –  Phyllaries lacking elongated oil glands; achenes brownish when mature. . . . . . . . . . . .
. . . . . . . . . . . . . . . . . . . . . . . . . . . . . . . . . . . . . . . . . . . .**11. Senecioneae**
5. Style arms long, gradually attenuate-acute, short-hairy abaxially; style shaft similarly hair on upper part; capitula homogamous, and florets hermaphrodite **3. Vernonieae**
 –  Style and capitula not showing above combination of character states, or capitula unisexual. . . . . . . . . . . . . . . . . . . . . . . . . . . . . . . . . . . . . . . . . . . . . . . . . 6
6. Capitula with all or only outer floret corollas bilabiate, corollas with 3-toothed outer lip and 2-lobed inner lip . . . . . . . . . . . . . . . . . . . . . . . . . . . . . . . . . . . . **1. Mutisieae**
 –  Capitula with florets lacking any bilabiate corollas . . . . . . . . . . . . . . . . . . . . . . . . . 7
7. Capitula homogamous and discoid; phyllaries lacking scarious unlobed appendages or apices, and lacking scarious markings, or capitula heterogamous and radiant with outer florets enlarged and sterile, or capitula unisexual and plants dioecious; phyllary appendages, if present, spiniform and/or pinnately divided. . . . . . . . . . . . . . . . . . . . 8
 –  Capitula heterogamous and radiate or disciform but not radiant, or capitula homogamous and phyllaries with unlobed scarious often white or coloured appendages or apices, or with scarious usually brownish margins, but not with spiniform and/or pinnately divided appendages; or capitula unisexual and plants monoecious . . . . . . 13

8.    Leaves spiny or bristly-spiny at least towards base . . . . . . . . . . . . . . . . . . . . . . . . . . . 9
–    Leaves spineless or not bristly-spiny . . . . . . . . . . . . . . . . . . . . . . . . . . . . . . . . . . . . 10
9.    Anthers tailed; receptacle densely setose . . . . . . . . . . . . . . . . . . . . . **2. Cardueae**
–    Anthers tail-less; receptacle alveolate with fringed alveolae . . . . . . . . **5. Arctotideae**
10.    Leaves rosettiform; inflorescences of a dense glomerule of small capitula on crown of plant . . . . . . . . . . . . . . . . . . . . . . . . . . . . . . . . . . . . . . . . . . . **6. Platycarpheae**
–    Leaves alternate or opposite; inflorescences of solitary capitula or capitula variously aggregated into lax to relatively dense, variously branched, inflorescences, or if glomerulate then not sessile on crown of plant . . . . . . . . . . . . . . . . . . . . . . 11
11.    Corolla lobes of tubular florets much longer than wide; leaves alternate; capitula of more than one floret, not aggregated into glomerules . . . . . . . . . . . . . . . . . . . . . . . 12
–    Corolla lobes about as long as wide, or if much longer than wide then leaves opposite; capitula sometimes of only one floret, or capitula aggregated into glomerules . . . . . 14
12.    Receptacle densely setose; inner florets hermaphrodite, outer sterile and usually corollas radiant. . . . . . . . . . . . . . . . . . . . . . . . . . . . . . . . . . . . . . . . . . . . . . **2. Cardueae**
–    Receptacle naked or scaly, not setose; all florets hermaphrodite or unisexual, corollas never radiant. . . . . . . . . . . . . . . . . . . . . . . . . . . . . . . . . . . . . . . . **1. Mutisieae**
13.    Style arms of disc florets connate in lower part, connate part thicker than style shaft and abruptly marked off from it by a ring of short hairs at base; capitula always hermaphrodite . . . . . . . . . . . . . . . . . . . . . . . . . . . . . . . . . . . . **5. Arctotideae**
–    Style arms of disc florets long to short, or absent, but connate lower part of style arms if present thicker than style shaft, or capitula unisexual and plants monoecious. . . . 14
14.    Receptacle scaly and leaves opposite, or leaves alternate and mature achenes black and/or capitula glomerulate, or receptacle not scaly and pappus of scales and/or leaves opposite and/or mature achenes black and/or capitula unisexual and/or capitula glomerulate; pappus of scales or often barbellate coarse setae or bristles but not of slender setae, or if of slender setae then setae plumose . . . . . . . . . . . . . . . . . . . . . 15
–    Receptacle not scaly (although sometimes hairy or fimbrillate), or scaly and leaves alternate (although sometimes opposite on upper stem) and achenes variously brown, sometimes dark brown, or pale-coloured when mature, or capitula not unisexual; capitula glomerulate or inflorescences variously branched and capitula lax or dense on branches; if lower leaves opposite then pappus of slender barbellate setae, at least in disc florets, or pappus absent, receptacle naked and capitula not glomerulate . . . . . 16
15.    Anthers tailed . . . . . . . . . . . . . . . . . . . . **8. Athroismeae** (*Athroisma, Blepharispermum*)
–    Anthers not tailed . . . . . . . . . . . . . . . . . . . . . . . . . . . . . . . . . **13. Heliantheae**
16.    Style arms of hermaphrodite or functionally male florets each with a subulate to triangular papillose appendage. . . . . . . . . . . . . . . . . . . . . . . . . . . . . . . . . . . . . 17
–    Style arms of hermaphrodite or functionally male florets acute to rounded, or truncate and fringed with short hairs or papillae, or short-conical at apex with subdistal fringe of hairs, unappendaged; or style undivided . . . . . . . . . . . . . . . . . . . . . . . . . . . . . . 18
17.    Receptacle alveolate; pappus a deeply and unequally laciniate corona or cupule; capitula disciform . . . . . . . . . . . . . . . . . . . . . . . . . . . . . . . . . . **13. Heliantheae**
–    Receptacle usually alveolate or foveolate, usually glabrous, sometimes fimbriate or paleaceous; pappus usually of barbellate capillary setae, sometimes reduced to scales or awns, or pappus absent; capitula radiate, cryptically radiate, discoid (and then capitula unisexual), or disciform (but then pappus never a corona). . . . . . . . . . . . . **9. Astereae**
18.    Phyllaries with scarious usually brown often erose margins, not appendaged; leaves often pinnatipartite; style arms apically truncate and fringed, or style undivided and truncate; pappus absent, or if present then a short lacerate crown or lobed auricle, or leaves pinnatipartite and outer pappus of 5 white petaloid scales . . . . . . . . **10. Anthemideae**
–    Phyllaries green and herbaceous (though often with hyaline margins), or with scarious membranous often white or coloured appendages or apices, or rarely appendages

foliose; leaves never pinnatipartite; pappus of hairs or scales present in at least some florets, or if absent then style arms acute, obtuse, rounded or truncate, sometimes with external sweeping hairs, but not truncate and fringed . . . . . . . . . . . . . . . . . . . . . . 19

19.  Pappus absent; fruits large, curved or angular, or winged achenes, or smooth drupes; capitula radiate . . . . . . . . . . . . . . . . . . . . . . . . . . . . . . . . .**12. Calenduleae**

–  Pappus present or absent; fruits small achenes, fusiform, rarely distinctly beaked, terete, ribbed or angled, very rarely flattened; capitula radiate or disciform . . . . . . . . . . . 20

20.  Capitula radiate; ray florets neuter, or if fertile then style arms short-conical at apex with a subdistal fringe of hairs, never apically papillate. . . . . . . . . . . . . **1. Mutisieae**

–  Capitula radiate or disciform; ray florets if present female; style arm apices acute, obtuse or truncate, and with sweeping stigmatic hairs, sometimes apically papillate. 21

21.  Carpopodium glabrous on upper margin; capitula radiate or disciform; pappus setae usually of capillary barbellate or plumose setae, or of rigid awns or scales, sometimes absent . . . . . . . . . . . . . . . . . . . . . . . . . . . . . . . . . . . . . . . . . . . . . . . 22

–  Carpopodium setuliferous on upper margin; capitula disciform; pappus setae biseriate, outer series of scales, inner series of few, free, rapidly caducous setae. . . . . . . . . . . . . . . . . . . . . . . . . . . . . . . . . . . . . . . . . . . . . . . . . . . .**8. Athroismeae** (*Artemisiopsis*)

22.  Receptacle epaleaceous (if paleaceous achenes dimorphic and stems winged or wingless {*Callilepis, Neojeffreya*}, or achenes monomorphic and stems winged and receptacle covered in numerous bristles {*Geigeria*}, or plants spiny and leaves decussate or alternate on main stems or on brachyblasts {*Rosenia*}); phyllaries papery or sometimes herbaceous; pappus of capillary barbellate or plumose setae, or of rigid awns or scales, connate at base or free, sometimes absent . . . . . . . . . . . . . . . **7. Inuleae sensu lato**

–  Receptacle usually paleaceous, sometimes epaleaceous; phyllaries herbaceous; stems never winged; achenes monomorphic, dark brown or blackish; pappus of short scales or absent. . . . . . . . . . . . . . . . . . . . . . . . . . . . . . . . **8. Athroismeae** (*Anisopappus*)

# Tribe 11. **SENECIONEAE** Cass.[2]

**Senecioneae** Cass. in J. Phys. Chim. Hist. Nat. Arts **88**: 196–198 (1819). —Jeffrey in Kew Bull. **41**(4): 873–943 (1986). —Lisowski, (Asterac. Fl. Afr. Cent. 2) Fragm. Flor. Geobot. **36** Suppl. 1: 255–448 (1991). —Jeffrey in Kew Bull. **47**(1): 49–109 (1992). —Bremer, Asterac. Cladist. Classif.: 479–520 (1994). —Beentje & Jeffrey in F.T.E.A., Compositae **3**: 547–702 (2005). —Nordenstam in Kubitzki, Fam. Gen. Vasc. Pl. **8**: 208–241 [2006](2007). — Nordenstam, Pelser, Kadereit & Watson in Funk *et al.*, Syst. Evol. Biogeogr. Compositae: 503–525 (2009). —Chen *et al.* in Fl. China **20-21**: 371–544 (2011).

Annual or perennial herbs or shrubs, rarely trees, sometimes scandent, sometimes succulent, occasionally with a woody xylopodiaceous rootstock. Leaves usually alternate, rarely opposite or rosulate, simple, sessile or petiolate (rarely prehensile), herbaceous to coriaceous or fleshy, entire, serrate, deeply lobed to pinnate or pinnatifid. Inflorescence commonly compound cymose, corymbose, paniculate, or occasionally scapose or scapiform. Capitula heterogamous and radiate or disciform, or homogamous and discoid, very rarely plants dioecious; involucre hemispherical to obconical or cylindrical, calyculate or ecalyculate, calycular bracts few to several, scarcely discernible at base of involucre or forming an obvious subinvolucre, often accompanied by further bracteoles in upper part of pedicel; phyllaries usually uniseriate, free or sometimes connate; receptacle flat or convex, sometimes alveolate, epaleaceous. Outer florets female, rayed, rays usually yellow, sometimes white or purple, or filiform and rayless, or outer florets absent; disc florets hermaphrodite or functionally male; corollas yellow, white, purple, pink, red, orange or brownish, usually 5-lobed, lobes short or sometimes comparatively long and narrow; filament collar (= anther collar, or

---

2  *Crassocephalum, Kleinia,* and *Mikaniopsis* by B. Loeuille; *Emilia* and *Senecio* by H. Beentje. All other genera by D.J.N. Hind

antheropodium) cylindrical to flattened, sometimes dilated and balusterform;[3] basal anther appendages sagittate, less often rounded or tailed, sometimes caudate; apical anther appendages ovate, lanceolate or oblong, obtuse to acute, endothecial tissue of anthers radially or transversely thickened; style base often swollen; style arms with separate, contiguous or confluent stigmatic lines, truncate, conical or variously appendaged, sometimes apically penicillate and with or without a corona of sweeping hairs beneath apex. Achenes usually distinctly 5– or 8–10-ribbed, usually cylindrical, sometimes flattened or smooth, sometimes obcompressed (and then margins usually ciliate), sometimes glandular-punctate, glabrous or setuliferous, setulae non-myxogenic, rarely myxogenic; carpopodium an annulus, or rarely absent (or apparently so); pappus setae uni- to multi-seriate, barbellate, or very rarely paleaceous (and of one or few laciniate scales, these often barbellate on surface), persistent (sometimes lengthening considerably in fruit) or deciduous, rarely caducous or fragile, white, straw-coloured, pinkish, reddish or purple, rarely pappus absent.

A large tribe of c. 150 genera and c. 3500 spp., of worldwide distribution, including several pantropic weeds. Many spp. are extremely toxic and poisonous to cattle and horses when eaten. There are 123 spp. in the Flora area, with 55 spp. in *Senecio* L., 18 spp. in *Emilia* (Cass.) Cass. and 11 in *Crassocephalum* Moench.

Species of several genera are cultivated in the Flora Zambesiaca area including: *Cineraria saxifraga* DC., *Crassothonna cacalioides* (L.f.) B.Nord. (formerly *Othonna carnosa* Less. var. *carnosa*), *Curio rowleyanus* (H.Jacobsen) P.V.Heath, *Euryops chrysanthemoides* (DC.) B.Nord., *E. pectinatus* (L.) Cass. subsp. *pectinatus*, *E. virgineus* (L.f.) DC., *Gynura aurantiaca* (Blume) DC., *Kleinia chimanimaniensis* van Jaarsv., *K. fulgens* Hook.f., *K. longiflora* DC., *Ligularia tussilaginea* (Burm.f.) Makino, *Roldana petasitis* (Sims) H.Rob. & Brettell, *Pericallis* × *hybrida* (Bosse) B.Nord., *Pseudogynoxys chenopodioides* (Kunth) Cabrera, *Senecio bicolor* (Willd.) Tod. subsp. *cineraria* (DC.) Chater, *Senecio decaryi* Humbert, *Senecio elegans* L., *Senecio macroglossus* DC., *Senecio tamoides* DC., *Senecio viravira* Hieron.

1. Styles of disc florets undivided, sterile . . . . . . . . . . . . . . . . . . . . . . . **105. Othonna**[4]
 – Styles of disc florets divided, fertile or sterile . . . . . . . . . . . . . . . . . . . . . . . . . . . . . . . . 2
2. Pappus paleaceous or sometimes of scale-like elements; achenes relatively long (when compared to pappus elements) and linear to very narrowly cylindrical; involucre suturing on one side at achene maturity . . . . . . . . . . . . . . . . . . . . . . .**109. Emiliella**
 – Pappus of capillary (usually fine, sometimes coarse) setae, or absent; achenes broadly cylindrical, subtriquetrous, slightly flattened or obcompressed; phyllaries all separating at achene maturity . . . . . . . . . . . . . . . . . . . . . . . . . . . . . . . . . . . . . . . . . . . . . . . . . . . . . . 3
3. Filament collar straight, uniform, cylindrical (in some spp. of *Kleinia* filament collar appears cylindrical but then plants stem and leaf succulents) . . . . . . . . **104. Euryops**
 – Filament collar balusterform or subcylindrical. . . . . . . . . . . . . . . . . . . . . . . . . . . . . . . . 4

---

[3] In many accounts of the Senecioneae a major division of genera is provided based on the appearance of the filament collar (= anther collar or antheropodium), and whether it is evident (termed balusterform) or the filament is cylindrical or flattened. Those with a conspicuous filament collar are often termed 'Senecioid', those without 'Tussilaginoid'. However, during the course of writing the Senecioneae account this Flora it became obvious that this is a simplistic statement. All of the genera present in the Flora are considered 'Senecioid' genera, none are 'Tussilaginoid'. *Euryops* (Cass.) Cass. notably has cylindrical filament collars, and is usually placed in the subtribe Othonninae (alongside *Gymnodiscus* Less., *Hertia* Less., *Lopholaena* DC. and *Othonna* L.) all noted for the presence of balusterform filament collars. Loeuille and I have also been conscious of the range of descriptions and illustrations of the filament collars in spp. of *Kleinia* Mill. These vary from being described as 'slightly balusterform', to observations showing others are conspicuously cylindrical, or possibly even flattened, and thus not typically 'Senecioid'. In all other respects, *Kleinia* sits comfortably within the subtribe Senecioninae, along with *Gynura* Cass. and *Solanecio* (Sch.Bip.) Walp.

[4] The succulent terete-leaved taxa have relatively recently been removed from *Othonna* into the genus *Crassothonna* B. Nord., originally with 13 spp., but with a 14th added by Swanepoel & de Cauwer in Phytotaxa **427**(3): 209–215 (2019). It is worth noting that Swanepoel & de Cauwer indicated that the genus occurs in Botswana, yet Nordenstam (Compositae Newsletter **50**: 70 – 77. 2012) listed none of the original 13 spp. as occurring in our Flora area, and I have found material of none.

4.  Capitula calyculate . . . . . . . . . . . . . . . . . . . . . . . . . . . . . . . . . . . . . . . . . . . . . . . . 5
–   Capitula ecalyculate . . . . . . . . . . . . . . . . . . . . . . . . . . . . . . . . . . . . . . . . . . . . . . . . 13
5.  Achenes obcompressed and margins usually conspicuously ciliate or glabrous . . . . . .
. . . . . . . . . . . . . . . . . . . . . . . . . . . . . . . . . . . . . . . . . . . . . **97. Cineraria**
–   Achenes cylindrical or subtriquetrous, rarely somewhat compressed, and then lacking
ciliate margins and phyllaries ± biseriate . . . . . . . . . . . . . . . . . . . . . . . . . . . . . . 6
6.  Plants stem and leaf succulents . . . . . . . . . . . . . . . . . . . . . . . . . . **101. Kleinia**
–   Plants with woody or herbaceous stems, leaves herbaceous, coriaceous rarely
somewhat fleshy . . . . . . . . . . . . . . . . . . . . . . . . . . . . . . . . . . . . . . . . . . . . . . . . 7
7.  Phyllaries ± biseriate, inner series with a midrib with a thin blackish resin vein and
2 lateral blackish resin ducts (resin ducts of inner phyllaries broad and gland-dotted
above) . . . . . . . . . . . . . . . . . . . . . . . . . . . . . . . . **98. Mesogramma**
–   Phyllaries uniseriate and lacking any resin ducts . . . . . . . . . . . . . . . . . . . . . . . . . 8
8.  Capitula disciform . . . . . . . . . . . . . . . . . . . . . . . . . . **110. Mikaniopsis**
–   Capitula radiate or discoid . . . . . . . . . . . . . . . . . . . . . . . . . . . . . . . . . . . . . . . . . 9
9.  Petiole bases thickened and prehensile, auriculate . . . . . . . . . . **111. Austrosynotis**
–   Petiole bases not prehensile, exauriculate . . . . . . . . . . . . . . . . . . . . . . . . . . . . . 10
10. Style arm apices truncate and penicillate . . . . . . . . . . . . . . . . . . . . . . . . . . . . . . . 11
–   Style arms truncate and lacking apical tuft of papillae, or long and tapering . . . . . . . 12
11. Scrambling perennial herbs, or shrubs or small trees; calycular bracts few; involucre
narrowly cylindrical; ovary cell walls with compound drusiform crystals; capitula
discoid; corollas yellow . . . . . . . . . . . . . . . . . . . . . . . . . . . . **100. Solanecio**
–   Annual or perennial herbs; calycular bracts usually numerous; involucre broadly
cylindrical; ovary cell walls with simple crystals; capitula radiate or discoid; corollas white,
yellow, pink red, mauve, purple or blue . . . . . . . . . . . . . . . . **103. Crassocephalum**
12. Style arms tapering. . . . . . . . . . . . . . . . . . . . . . . . . . . . . . . . . **102. Gynura**
–   Style arms truncate . . . . . . . . . . . . . . . . . . . . . . . . . . . . . . . . . . . . . . . **99. Senecio**
13. Style arms with elongate tapering appendages densely long-papillose abaxially, papillae
longer at mid-point and much short towards apices; plants perennial herbs or suffrutices
with a xylopodiaceous rootstock, or shrubs or treelets . . . . . . . . . . **106. Lopholaena**
–   Style arms truncate with short sweeping hairs, sometimes with a penicillate apical
tuft; plants glabrous, annual, semi-aquatic or aquatic herbs (often rooting at nodes) or
terrestrial annual or perennial herbs . . . . . . . . . . . . . . . . . . . . . . . . . . . . . . . . . 14
14. Ray florets white, pink or pale mauve; plants annual, semi-aquatic or aquatic herbs,
often rooting at nodes; receptacle conical; pappus absent and leaves always linear to
narrowly elliptic-oblong . . . . . . . . . . . . . . . . . . . . . . . . . . . . . **107. Stenops**
–   Ray florets when present yellow or orange, otherwise capitula usually discoid; plants
annual or perennial herbs, never rooting at nodes; receptacle flat to slightly convex;
pappus of capillary setae, rarely absent and then leaves obovate, ovate or elliptic . .
. . . . . . . . . . . . . . . . . . . . . . . . . . . . . . . . . . . . . . . . . . . . . **108. Emilia**

# 97. **CINERARIA** L.

**Cineraria** L., Sp. Pl., ed. 2, **2**: 1242 (1763). —Dyer, Gen. S. Fl. Afr. Pl. **1**: 713–714 (1975).
—Jeffrey in Kew Bull. **41**(4): 873–943 (1986). —Beentje in F.T.E.A., Compositae **3**: 363–365
(2005). —Cron *et al.* in Kew Bull. **61**(2): 167–178 (2006a). —Cron *et al.* in Kew Bull. **61**(4):
449–535 (2006b). —Nordenstam in Kubitzki, Fam. Gen. Vasc. Pl. **8**: 229 [2006](2007). —
Cron *et al.* in Bot. J. Linn. Soc. **154**(4): 497–521 (2007); in Taxon **57**(3): 1–20 (2008); in Bot.
J. Linn. Soc. **160**(2): 130–148 (2009).
*Xenocarpus* Cass. in Dict. Sci. Nat. **59**: 108 (1829).

Perennial, rarely short-lived (and appearing annual), herbs or subshrubs or shrublets. Stems slender to robust, usually erect, rarely stoloniferous or tuberous, often woody near base, or with a woody rootstock. Leaves alternate, petiolate, petiole usually auriculate, lamina 5–7-lobed, typically deltoid or reniform, occasionally pinnatifid, palmately-veined, glabrous or more usually variously hairy (pilose to arachnoid or tomentose) and often long stipitate-glandular, margins usually coarsely dentate. Inflorescence of few to many capitula on long or short terminal pedicels, or of long-pedicellate solitary capitula, pedicels bracteolate, bracteoles scale-like, absent, or few to several along pedicel (and merging with calycular bracts at base of involucre). Capitula heterogamous and radiate, homochromous; involucre campanulate, calyculate, calyculular bracts few, scale-like, herbaceous, margins laciniate, glabrous or arachnoid-pubescent; phyllaries uniseriate, glabrous or variously pubescent, margins scarious; receptacle flat, epaleaceous. Ray florets functionally female, ray limb bright yellow, patent, narrowly elliptic to oblanceolate (drying with conspicuous amber-coloured venation), apex 3-toothed, glabrous or sparsely setuliferous at base, setulae of glandular twin-hairs; style arm apices truncate. Disc florets hermaphrodite, corollas bright yellow (drying with amber-coloured venation becoming apparent, each vein terminating at apex of each corolla lobe), tubular, base flared (and enclosing basal stylar node), limb dilated above, short-5-lobed, lobe apices papillose; apical anther appendages obtuse, anther cylinder partially, or ± conspicuously, exserted from corolla throat, endothecial tissue of anthers radially thickened, basal anther appendages minutely sagittate; filament collars balusterform; style arm apices obtuse to truncate, apex penicillate, peripheral sweeping hairs forming a distinct corona, stigmatic surfaces discrete. Achenes obovate, obcompressed, brown (to black), glabrous or setuliferous, sparsely or ± densely so in upper part of achene body, setulae of myxogenic twin-hairs (i.e. mucilaginous when wetted), achene body distinctly margined or winged, margins glabrous or conspicuously ciliate, cilia of twin-hairs, dense, apices acute; carpopodium distinct, glabrous; pappus setae capillary, finely barbellate, white, caducous or deciduous.

A genus of c. 35 spp. in Africa and Madagascar. Six spp. are native to the Flora area, and *C. saxifraga*, a native of the Eastern Cape, South Africa, is only known from cultivation (in Harare). *Cineraria saxifraga* is included within the key, simply to enable identification of material. Putative hybrids, between *C. pulchra* and *C. deltoidea*, were listed by Cron *et al.* (2006b), and included here, but the hybrids are rare, have no formal standing, and would be difficult to key out. The key to the spp. and descriptions are based on, and modified from, those in Cron *et al.* (2006b).

1. Achenes glabrous . . . . . . . . . . . . . . . . . . . . . . . . . . . . . . . . . . . . . . . . . . **1.** *deltoidea*
– Achenes setuliferous or ciliate on margins and/or setuliferous on faces . . . . . . . . . . 2
2. Leaves white, grey or greyish-green from arachnoid or woolly indumentum . . . . . . . 3
– Leaves green, glabrous or hairy (but not arachnoid) . . . . . . . . . . . . . . . . . . . . . . . 7
3. Achenes broad-winged (either both ray and disc achenes or ray achenes only). . . . .
. . . . . . . . . . . . . . . . . . . . . . . . . . . . . . . . . . . . . . . . . . . . . **2.** *vallis-pacis*
– Achenes margined or with narrow wings. . . . . . . . . . . . . . . . . . . . . . . . . . . . . . . 4
4. Ray florets 3–8 (never more than 8). . . . . . . . . . . . . . . . . . . . . . . . . . . . . . . . . 5
– Ray florets 8–13 (occasionally more than 13). . . . . . . . . . . . . . . . . . . . . . . . . . . 6
5 Leaves deltoid-reniform, distinctly lobed, sinuses deeply rounded between lobes. . . . .
. . . . . . . . . . . . . . . . . . . . . . . . . . . . . . . . . . . . . . . . . . . . . . . **3.** *mazoensis*
– Leaves deltoid (occasionally deltoid-reniform), sometimes distinctly lobed and then lacking rounded sinuses between lobes. . . . . . . . . . . . . . . . . . . . . . . . . . . **1.** *deltoidea*
6. Capitula large; phyllaries 12–17, ray florets 12–16; disc florets c. 70 . . . . . **5.** *magnicephala*
– Capitula medium-sized; phyllaries ± 13; ray florets 8–13; disc florets 25–60 . . . . . . .
. . . . . . . . . . . . . . . . . . . . . . . . . . . . . . . . . . . . . . . . . . . . . . . . . **4.** *pulchra*
7. Petiole base exauriculate; leaves ± succulent . . . . . . . . . . . . . . . . . . . . . . . . . . . .
. . . . . . . . . . . . . . . . . . . . . . . . . . . . . . *saxifraga* (a cultivated plant in Flora area)
– Petiole base usually conspicuously auriculate (or auricles small and lanceolate); leaves herbaceous or papyraceous . . . . . . . . . . . . . . . . . . . . . . . . . . . . . . . . . . . . . . . . 8
8. Uppermost leaves lyrate-pinnatifid/pinnatisect, middle to lower leaves similar or deltoid to deltoid-reniform to reniform, usually with lateral pinnae on petiole; auricles small and lanceolate; plants of coastal grassland or seasonal wetlands; 20–70 m. . . . . . . . **6.** *pinnata*

–    Uppermost and lower leaves deltoid to deltoid-reniform or reniform, lamina not lyrate-pinnatifid or pinnatisect, occasionally petiole with 1 or 2 lateral pinnae; auricles conspicuous; plants of forest margins, montane grassland, old lava flows; 500–4300 m  . . . . . . . . . . . . . . . . . . . . . . . . . . . . . . . . . . . . . . . . . . . . . . . **1.** *deltoidea*

1. **Cineraria deltoidea** Sond. in Linnaea **23**(7): 68 (1850). —Harvey in Harvey & Sonder, Fl. Cap. **3**: 312 (1865). —Hilliard, Compositae Natal.: 379 (1977). —Maquet in Fl. Rwanda **3**: 670 (1985). —Jeffrey in Kew Bull. **41**(4): 930 (1986). —Lisowski, (Asterac. Fl. Afr. Cent. 2) Fragm. Flor. Geobot. **36** Suppl. 1: 434 (1991). Type: [South Africa: Natal.] 'Port Natal. *Gueinzius* No. 343.' (S holotype, MEL, P00117589, W0006820, W1889-0287114 – s.n.).[5]

 *Cineraria grandiflora* Vatke in Linnaea **39**(6): 503 (1875), nom. illeg. non Spreng. ex DC. (1838: 314). —Hedberg, Symb. Bot. Uppsala **15**: 222 & 349 (1957). Type: [Ethiopia:] 'l. in convallium marginibus ad Dschan Mèda 8600' a. m. 15. sept. 1863. [*Schimper*] n. 1517.' (B† holotype, BM000924590, E00414016, K000306889, S05-9703, S05-9704, US00604241).

 *Cineraria abyssinica* Sch.Bip. ex A.Rich. f. *longiradiata* sensu Oliver in Trans. Linn. Soc. London, Bot., **2**: 340 (1887), based on *Schimper* 1517 (citing '7000–10,000 ft.; barren, 13,200 ft.'), non *Cineraria abyssinica* Sch.Bip. ex A.Rich.

 *Cineraria kilimandscharica* Engl. in Abh. Königl. Akad. Wiss. Berlin **1891**(2): 439 (1892). Types: [Tanzania:] 'KILIMANDSCHARO, um 1300–2300 m (*Johnston*); an der oberen Waldgrenze um 3000 m (*Dr. Hans Meyer* n. 84, 239).' *Johnston* 4 (K000306884 – 'No 4/ 7000 ft.', mounted with *Johnston* 120 and *Johnston* 129, syntype); *Johnston* 128 (K000306883 – 'No. 128. corolla yellow 10,000 ft.' syntype); *Johnston* 129 (K000306882 – '129. found in stream valley 13,200 ft.' syntype); *Meyer* 84 (?B† syntype); *Meyer* 239 (?B† syntype).[6,7]

 *Cineraria prittwitzii* O.Hoffm ex Engl. in Götzen, Afrika Ost nach West: 375 (list) & 383 (descr.) (1895). Type: 'In der Ebene am Kirunga, bei 2000 m. – [*Götzen*] No. 29.' (B† holotype).

 *Cineraria bracteosa* O.Hoffm. ex Engl. in Götzen, Afrika Ost nach West: 377 & 383 (1895). Types: [D.R. Congo:] 'Kirunga, im Hochwalde bei 2500 m. [*Götzen*] No. 64; am Kraterrande auf Lava, um 3300 m. – [*Götzen*] No. 106.' *Götzen* 64 (B† syntype); *Götzen* 106 (B† syntype).

 *Cineraria buchananii* S.Moore in J. Linn. Soc., Bot. **35**(245): 352 (1902). Type: [Malawi:] 'Hab. Nyassaland; *J. Buchanan*, 1895, no. 10.' (BM000797528 holotype, GRA0003138-0, MO391506, MO391507, NY00167507, PRE, SAM).

 *Cineraria gracilis* O.Hoffm. in Bot. Jahrb. Syst. **38**(2): 206 (1906). Type: [Ethiopia, Sidamo Province, 'Gallahochland', Djam- Djam Mt range:] 'Gallahochland (Dr. Ellenbeck in Exped. Baron C. v. ERLANGER. – Ohne Nummer und Fundort' *Ellenbeck* s.n. (B† holotype).

 *Senecio kirschsteineanus* Muschl. in Wiss. Ergebn. Deut. Zentr.-Afr. Exped. (1907-1908), Bot. **2**: 405 (1911). Type: [Rwanda:] 'NO – Kiwu: Lichtungen mit Hypericum-Gebüsh im Bambusmischwald südöstlich Karissimbi, westlich vom Karago-See (blühend und fruchtend Ende November 1907 – [*Mildbraed*] n. 1647).' (B† holotype, BR5111572 – a capsule containing 2 leaves and the multi-headed apex of an inflorescence).

 *Senecio schubotzianus* Muschl. in Wiss. Ergebn. Deut. Zentr.-Afr. Exped. (1907-1908), Bot. **2**: 405 (1911). Type: [D.R. Congo:] 'Vulkan-Gebiet: Ninagongo, von der Strauchformation bis fast

---

[5]  Jeffrey (1986: 930) cited the holotype as in MEL; however, Cron *et al.* (2006b: 474) stated that the ('reinstated') holotype was in S, providing a useful explanation of the history of the MEL 'falsification' as the holotype.

[6]  Jeffrey & Beentje (2005: 565) oddly cited 'Tanzania, Kilimanjaro, *Johnston* 129 or 4 (K!, lecto, chosen by Jeffrey)'; none of the three *Johnston* collections on the sheet have been determined as the lectotype, although it can be presumed that *Johnston* 4 was selected by Jeffrey in the event that the *Meyer* collections were destroyed during World War 2.

[7]  Oddly, Cron *et al.* (2006) made no mention of the *Meyer* syntypes, nor any mention as to why they cited three *Johnston* collections from K; their citation of *Johnston* '120' is incorrect; the material is clearly of *Johnston* 128, showing a significant difference in the writing of the '0' and '8', and no mention was made of the prior lectotypification by Jeffrey.

zur Vegetationsgrenze, 3200 m ü. m. Blüten goldgelb (blühend 6. Oktober 1907 – [*Mildbraed*] n. 1416).' (B† holotype, BR5112494 – a capsule containing 2 leaves, 2 flowering capitula, plus a loose capitulum in bud).

    *Cineraria laxiflora* R.E.Fr. in Acta Horti Berg. **9**: 146 (1928). Type: [Kenya:] 'Mt. Aberdare: Ostabhang des Sattima in Gebüschen von Cliffortia aequatorialis, ca. 3000 m ü. m. (blühend und fruchten 17.III.1922. – *Rob. E. und Th. C. E. Fries* n. 2651).' (UPS holotype, S18-13508).

    *Cineraria densiflora* R.E.Fr. in Acta Horti Berg. **9**: 147 (1928). Type: [Kenya:] 'Mt Kenia: Nordseite bei dem Kongoni-Fluss im unteren Teil des montanen Regenwaldes auf adgebranntem Boden (blühen und fruchtend 13.III.1922. – *Rob. E. und Th. C. E. Fries* n. 1555).' (UPS holotype, S-G-1368).

    *Cineraria monticola* Hutch. in Bull. Misc. Inf., Kew **1931**(5): 251 (1931). Type: [South Africa: Limpopo Province.] 'Northern Transvaal: Zoutspansberg; ascent to Wylie's Poort from Louis Trichardt, herb 2-3 ft., ray-flowers deep yellow, June, *Hutchinson and Gillett* 3201.' (K000306898 – the main label clearly indicated as 'type' in Hutchinson's hand, holotype; K000306899 – with only a numbered ticket label attached to the sheet).

    *Cineraria bequeartii* De Wild., Pl. Bequaert. **5**: 441 (1932). Type: [D.R. Congo:] 'Entre Tongo et Mokule 25 septembre 1914 (*J. Bequaert*, n. 5863. – Plaine de lave; fleurs jaune).' (BR6423025 holotype).

Perennial herb, erect or scandent, often climbing, to 1.5 m (scrambler), often 0.6 m tall/long. Stems slender, herbaceous, woody and branching towards base, glabrous or sparsely hairy or very hairy or arachnoid-pubescent, glabrescent. Leaves petiolate, petiole (5)11–41(68) mm long, sparsely or densely hairy or cobwebby, auricles small or large, auriform, coarsely dentate, leaf lamina 10–60(80) × 11–65(73) mm, deltoid to deltoid-reniform, distinctly to shallowly lobed, upper leaves often with lateral pinnae below main lamina, occasionally lyratiform, bright green, glabrous or hairy or arachnoid-pubescent, glabrescent above, usually hairy below especially on veins, or densely arachnoid-pubescent, apex acute to obtuse, margins coarsely dentate, base truncate, subcordate or cordate. Inflorescences with few to many capitula in lax corymb or more compact corymbose panicle, capitula pedicellate, pedicels (2)6–44 mm long, glabrous or hairy and glabrescent, or cobwebby, bracteolate, bracteoles scale-like, c. 2 mm long. Capitula heterogamous and radiate; involucre ecalyculate or calyculate, calycular bracts few, to c. 2 mm long; phyllaries 8–13(14), 3.5–8(13) mm long, glabrous, occasionally arachnoid-pubescent between calycular bracts, margins scarious, apices sometimes purple. Ray florets 5–8(13) (rarely 16), 4.8–12(16) mm long, ray limbs (3.5)4–10.5(14) mm long, 4(6)-veined. Disc florets 14–32(39), corollas 3.5–5.5 mm long. Achenes obovate, obcompressed, margined (to narrow-winged), dark brown (to black), often with paler margins or wings (these often appearing reddish-brown in some material), (1.8)2–3(3.5) mm long, glabrous or ciliate on margins and/or body to varying degrees; pappus setae c. 5 mm long.

**Zambia**. E: Nyika, 29.xii.1962, *Fanshawe* F7927 (K). **Zimbabwe**. C: 71 km from Rusape on Rusape-Juliasdale-Nyanga road, 1925 m, 17.v.1998, *Cron & Balkwill* 497 (B, CM, E, J, K, MO, PRE, RSA, S). E: Nyanga, Mtenderere Source, [2134 m,] 4.ix.1954, *Wild* 4591 (K, MO). **Malawi**. N: Nyika Plateau, North Nyasa Dist., 2350 m, 17.viii.1946, *Brass* 17299 (K, MO, US). S: Blantyre Dist., Ndirande Mt, 1370–1530 m, 28.vi.1970, *Brummitt* 11715 (K). **Mozambique**. Z: Zambesia, Guruè, Pico, Namuli, 23.ix.1944, *Mendonça* 2247 (LISC). MS: Manica e Sofala Dist., southern tip of Chimanimani Mts, ± 2000′[610 m], 31.v.1969, *Müller* 1248 (LISC, K, PRE).

Also in Ethiopia, Sudan, D.R. Congo, Rwanda, Angola, Uganda, Kenya, Tanzania, South Africa. Common in forest margins straggling or climbing over vegetation, also on bushy margins to rivers and streams, roadsides and forest clearings, especially in moist places, in lava crevices, on disturbed ground, on basalt derived soils in E Africa; 500–4300 m; flowering sporadically throughout the year.

Conservation Status: LC (Least Concern) following Cron *et al.* (2006b: 474), and clearly well-collected, but often not that widespread in much of the Flora area.

One of the most widespread of the spp. ranging from just above sea level in South Africa to over 3000 m in E Africa and Ethiopia. Variable in terms of habit, leaf shape and size, pubescence, and number of capitula but as yet no infraspecific taxa have been recognized.

Although Jeffrey & Beentje (2005: 505) had originally included the name in the synonymy of *C. deltoidea*, Cron *et al.* (2006b: 509) considered that *Cineraria foliosa* O.Hoffm. was a separate, narrow endemic sp. of Tanzania. No affinity between *C. deltoidea* and *C. foliosa* was suggested by Cron *et al.* (2006b), the two easily separated on the basis of the achene characters early in their key, and the 5- to 7-lobed leaves in *C. foliosa*. The presence of two Angolan collections, *Borges* 189 (LISC, LUA) and *Pritchard* 361 (BM, LISC), under the herbarium name of '*Cineraria paracanescens* Torre', were discussed by Cron *et al.* (2006b: 474). They were considered variants of *Cineraria deltoidea*, but requiring 'further investigation before it is formally named'; the species, however, was not recorded from Angola. In the event that these do prove to be exauriculate material of *C. deltoidea*, I have included Angola in the distribution of the species.

Although described as possessing calyculate involucres by Cron *et al.* (2006b) it is clear that some determined material is ecalyculate.

2. **Cineraria vallis-pacis** Dinter ex Merxm. in Mitt. Bot. Staatssamml. München **3**(3): 605 (1960). —Merxmüller in Prodr. Fl. Südwestafr. **139**: 42 (1967). Type: [Namibia:] 'Südwestafrika: Distr. Rehoboth: Friedental, Nordbastardland, [1.i.1935], *Dinter* 7989' (M0105348 holotype, B100088286, B100088287, BM000924576, BOL150167, BOL150168, K000306891, HBG504171, M, PRE0211032-1, PRE0211032-2, S-G-1372).[8]

　　*Cineraria vallispacis* Range in Feddes Repert. Spec. Nov. Regni Veg. **43**: 268 (1938), nom. nud.

Subshrub, usually 0.3–0.7 m tall, sometimes to 2 m (although noted for 'falling over' in some material); rootstock woody. Stems woody towards base, 3–7 mm diam., branching, arachnoid-pubescent at first, glabrescent, striate. Leaves petiolate, petiole 4–15 mm long excluding pinnate section, 8–70 mm long including pinnate section, arachnoid-pubescent, glabrescent, auricles large and conspicuous, margins dentate, leaf lamina deltoid-reniform to ovate, 13–52 × 15–55(75) mm, total leaf length 21–85 × 16–100 mm, shallowly or distinctly 5–7 lobed with 1, 2 or 3 pairs of subopposite lateral pinnae (in our material usually lacking any pinnae), uppermost leaves often pinnatifid (to lyrate-pinnatifid) (but not in our material), sometimes decurrent, young leaves densely tomentose, glabrescent to arachnoid-pubescent above and below, apex obtuse to rounded, margins coarsely or minutely dentate, base truncate to subcordate, to cordate in lower leaves. Inflorescence of many capitula (14–140 per branch) in subcorymbose panicles; pedicels 2–12 mm long, arachnoid-pubescent, glabrescent, bracteate, basal bract subtending pedicel c. 22 mm long. Capitula heterogamous and radiate; involucre calyculate, calycular bracts few, c. 2 mm long, lanceolate, margins laciniate, midrib conspicuously amber coloured (in herbarium material); phyllaries 8–10(12), 3.5–4.5 mm long, often glabrous, arachnoid-pubescent at base amongst calycular bracts, or occasionally densely arachnoid-pubescent, margins scarious. Ray florets 5–8, 5–7 mm long, ray limb 3–5 mm, 4-veined. Disc florets c. 25–30, corollas 3.5–4.2 mm long. Achenes obovate, obcompressed, ray achenes broad-winged (wings c. 0.5 mm wide), disc achenes broad-winged or narrow-winged, dark brown with paler brown wings, 2–2.8 mm long, usually densely ciliate on wing, glabrous on inner face, sparsely hairy on central rib of outer face; pappus setae c. 4.5 mm long.

**Botswana**. SW: Ghanzi & Kgalagadi Dist., Okwa valley, 1 km NE of Namibian border, 22°23'S, 20°00'E, 2.vii.1978, *Skarpe* S-290 (K, PRE, UPS).

Also known from Namibia and South Africa. Sometimes found growing luxuriantly on the banks of rivers in Namibia, commonly on well-drained red-brown sandy soil (Kalahari sands) in the Northern Cape, and on gentle, north-facing, often stony slopes, usually abundant at the base of slopes, also in shady areas, sprawling beneath bushes and trees, and in deep sandy soil; 1140–2000 m across its range; flowering December to April, but flowering sporadically throughout the year.

Conservation Status: A plant of the Griqualand West Centre of Endemism in the Northern Cape (Cron *et al.* 2009: 139), although not an endemic; LC (Least Concern) following Cron *et al.* (2006b: 499), but clearly rarely collected in the Flora area.

---

8　The holotype is the sheet *Dinter* 7989/I determined as '*Senecio vallis-pacis* Dinter' as specified by Merxmüller; *Dinter* 7989/II, determined as '*Cineraria vallis-pacis* Dinter' is taken as an isotype [although it could be taken as a paratype], although not mentioned by Cron *et al.* (2006b), and does not appear in the M virtual herbarium – yet.

3. **Cineraria mazoensis** S.Moore in J. Bot. **46**(542): 43 (1908). Type: [Zimbabwe:] 'Hab. Mazoe, Iron Mask Hill, 1525–1585 m [5000 to 5200 ft., April 1906], *F. Eyles* 345.' (BM000924587 holotype, BOL139086).

Perennial, sometimes appearing short-lived, herb, to c. 1 m tall. Stems single or several, herbaceous, slightly woody and sometimes branching near base, arachnoid-pubescent, sometimes glabrescent. Leaves deltoid-reniform to reniform, distinctly (3)5–7-lobed with deep, rounded sinuses, occasionally with lateral pinnae, 10–66 × 13–72 mm, green and arachnoid-pubescent above, glabrescent, white or grey and thickly tomentose below, apex obtuse to acute, margins dentate, base truncate to subcordate to cordate, uppermost leaf forming bract-like below pedicels, base acute, petiole 10–43(56) mm long, younger leaves tomentose, mature leaves sparsely cobwebby, auricles conspicuous or very small (rarely absent), ovate to auriform, sometimes procurrent on petiole (especially in uppermost leaves). Inflorescences with few (4–17 per stem branch) to many capitula (20–30 per stem branch) arranged in lax corymbose panicles, capitula pedicellate, pedicels (2)10–47(67) mm long, glabrous or arachnoid-pubescent, especially towards calycular bracts, ebracteolate. Capitula heterogamous and radiate; involucre calyculate, calycular bracts few, linear-lanceolate, c. 2 mm long; phyllaries 10–13, 4–5 mm long, arachnoid-pubescent or glabrous, margins scarious. Ray florets 7 or 8, 6–8.5 mm long, tube sparsely pubescent, ray limb 3.5–5 mm long, 4-veined (rarely more), veins conspicuous when dried. Disc florets 25–40, corollas 3.5–5(6) mm long, glabrous. Achenes obovate, obcompressed, margined to narrow-winged, brown, 2.2–2.8 mm long when mature, faces sparsely to densely setuliferous, some almost glabrous; pappus setae c. 4 mm and clearly shorter than corolla tube (but recorded as long as or slightly longer than disc corolla by Cron *et al.* 2006a).

Endemic, but scattered throughout the Flora area and mainly known from Zimbabwe, but found in Malawi and Zambia; 1100–1905 m; flowering April to June.

Conservation Status: The species is best regarded as DD (Data Deficient) following Cron *et al.* (2006a: 171; 2009: 132), although both varieties are at risk from restricted distribution, small populations and threats from human activities.

Phyllaries and pedicels aranchoid-pubescent, glabrescent, abaxial surfaces of leaves do not stick together . . . . . . . . . . . . . . . . . . . . . . . . . . . . . . . . . . . . . . . . . . a) var. *mazoensis*
Phyllaries and pedicels glabrous, abaxial surfaces of leaves stick like Velcro® when fresh . . . . . . . . . . . . . . . . . . . . . . . . . . . . . . . . . . . . . . . . . . . . . . . . . .b) var. *graniticola*

a) Var. **mazoensis**

Perennial herb, to c. 0.4 m tall. Leaves (3)5–7-lobed, densely tomentose beneath, trichomes on dorsal surface agranular with narrow, slightly tapering basal cells and long multi-cellular apical appendage, trichomes on ventral surface with 2–4 narrow basal cells (cylindrical) and a long apical appendage. Pedicels densely or sparsely arachnoid-pubescent, especially towards calycular bracts. Phyllaries 12–13, arachnoid-pubescent, sometimes glabrescent, but remaining arachnoid-pubescent amongst calycular bracts. Ray florets usually 8 (rarely 7). Disc floret corollas 3.5–3.8 mm long.

**Zambia**. S: Mazabuka, 20.v.1961, *Fanshawe* F6591 (BR, K). **Zimbabwe**. N: Mt Darwin, Mvuradona Mts, 16.iv.1964, *Wild* 6528 (K). C: Romorehota, northern summit of Wedza Mt, 1753 m, 16.v.1998, *Cron & Balkwill* 486 (J, K, MO, PRE). E: Mutare, Spinney Hill, Christmas Pass, 3800' [1158 m], 22.vi.1946, *Chase* 223 (BM, K). **Malawi**. C: Mchinji Hills near Fort Manning, 6000 ft. [1829 m], 7.viii.1936, *Burtt* 6200 (BR, K).

Endemic to the Flora area, this the more widespread variety. On slopes of hills, on river gorge walls, among boulders, on pyroxenite hills; up to 1905 m in the Mchinji Mts, Malawi.

Conservation Status: DD (Data Deficient), but some collections have described this variety as 'locally common'.

b) Var. **graniticola** Cron in Cron *et al.* in Kew Bull. **61**(2): 171 (2006a). Type: 'Zimbabwe: S of Lake Mutirikwi, 21 May 1998, 1064 m, *Cron & Balkwill* 532 (holotype J; isotypes B, E, K!, PRE, S, SRGH).' (J holotype, B, E, K001418105, PRE, S, SRGH).

Perennial herb to c. 1 m tall. Leaves 5–7-lobed, sparsely tomentose to cobwebby below; trichomes on dorsal surface usually with granular, broader tapering basal cells with long apical appendage, trichomes on ventral surface with 4–8 basal cells (narrowly tapering or cylindrical) with long apical, corkscrew-like, appendage creating a Velcro® effect when similar surfaces touch. Pedicels glabrous. Phyllaries glabrous, 10–13. Ray florets usually 7, occasionally 8. Disc floret corollas 4–5 mm long.

**Zimbabwe**. W: Matobo Dist. Farm Quaringa, 4900′ [1494 m], iv.1955, *Miller* 2776 (K). E: Mutare, Dora Farm, 26.vi.1948, *Chase* 802 (BM, K). S: Western side of Lake Mutirikwi, 9.viii.1988, *Carter & Coates Palgrave* 2224 (K).

Apparently endemic to Zimbabwe. Cron *et al.* (2006a: 176) suggested that this variety may also occur in Malawi, where floristically similar inselbergs occur. However, Cron *et al.* (2006b: 509) added nothing to this suggestion, implying more fieldwork was necessary. Restricted to granite inselbergs – often recorded as growing under rocks, or in rock clefts on inselbergs; 1100–1494 m; flowering between late March and June.

Conservation Status: Best recorded as DD (Data Deficient), Cron *et al.* (2006a: 171) noting only that 'var. *graniticola* is at greater risk ... due to its more restricted distribution and even smaller populations.' Label data suggests that amongst collections it is either rare or locally frequent.

4. **Cineraria pulchra** Cron in Cron *et al.* in Kew Bull. **61**(2): 168 (2006a). Type: 'Zimbabwe, Vumba, summit of Castle Beacon, 19 May 1998, 1900 m, *Cron & Balkwill* 510c (holotypus J; isotypi K, MO, PRE, SRGH).' (J holotype, E00815192, K000037056, MO2246672, PRE, SRGH). FIGURE 6.4.**1B**.

Perennial suffrutex or spreading shrub, to 1.2 m tall, often spreading or rambling. Stems woody and branching near base, rooting along decumbent stems, tomentose to cobwebby, glabrescent. Leaves petiolate, petiole 20–72 mm long (including portion supporting lateral pinnae), tomentose, sometimes glabrescent, auricles very conspicuous, auriform and procurrent, leaf lamina deltoid-reniform to reniform, deeply 5–7-lobed with rounded sinuses between lobes, with 1 or 2 pairs of lateral pinnae, lamina procurrent along petiole, 26–55 × 26–63 mm (excluding lateral pinnae at base), green, arachnoid-pubescent, glabrescent above, tomentose white or grey below with very prominent venation, apex acute to obtuse, margins dentate, base truncate to subcordate (to cordate), often merging with lateral pinnae. Inflorescences of many capitula (12–60 per inflorescence branch), in a compact corymbose cyme, capitula pedicellate, pedicels 3–14 mm long, densely arachnoid-pubescent, glabrescent, distinctly bracteolate near capitula, bracteoles lanceolate to linear, 3–6 mm long. Capitula heterogamous and radiate; involucre calyculate, calycular bracts c. 2 mm long, narrow-lanceolate, margins laciniate and interspersed with arachnoid pubescence; phyllaries 8–12, 4–5 mm long, densely or sparsely arachnoid-pubescent, occasionally glabrescent, margins scarious. Ray florets 8(13), 5–9(11) mm long, ray limb 4–6(8.5) mm long, 4(5)-veined. Disc florets 26–43, corollas 4(5) mm long. Achenes obovate, obcompressed with distinct median rib on inner surface, margined, sometimes appearing narrow-winged when immature, dark brown when mature, c. 2 mm long, ciliate and sparsely to moderately setuliferous on faces, rarely glabrous; pappus setae 3.5–4 mm long.

**Zimbabwe**. E: Nyanga, Nyangani, 2483 m, 17.v.1998, *Cron & Balkwill* 499 (E, J, K, MO, PRE, SRGH). **Mozambique**. MS: Chimanimani Mts, 4.vi.1948, *Munch* 93 [Government Herb. No. 21038] (K, LISC, SRGH).

Endemic to the Flora area, from Nyangani, Vumba and the Chimanimani Mts in the eastern highlands of Zimbabwe and from Mt Gorongosa and the Chimanimani Mts in Mozambique. Growing between rocks on mountain summits or on south-facing, south-eastern and eastern slopes, usually in the mist belt near the summits, in ericoid scrub above the forest on Mt Gorongoza in Mozambique, on quartzite and granite. Most localities in the Chimanimani Mts in Zimbabwe have corresponding localities inside the Mozambique border with one outlier collection; 1640–2540 m; flowering May to July, but flowering sporadically at other times.

Conservation Status: An endemic of the Chimanimani-Nyanga Centre of endemism; LC (Least Concern) following Cron *et al.* (2006a: 168).

The following putative *Cineraria pulchra* × *C. deltoidea* hybrids were listed by Cron *et al.* (2006b: 511), although of limited duplicate distribution: **Zimbabwe**. E: Cashel-Chimanimani Road, *Levyns* 9935 (BOL); Cashel-Chimanimani road, 20.v.1998, *Cron & Balkwill* 519 (J, K); ibid., *Cron & Balkwill* 520 (J, PRE); ibid., 20.v.1998, *Cron & Balkwill* 525 (J, MO); between Cashel and Chimanimani (Melsetter), 10.vii.1953, *Schelpe* 4019 (BM).

5. **Cineraria magnicephala** Cron in Cron *et al.* in Kew Bull. **61**(4): 511 (2006b). Type: 'Malawi: Northern Province, Mzimba Dist., E of Champira, rocky summit of Lwanjati Hill, 1830 m [6000'], 14 July 1975, *Pawek* 9875 (holotypus: WAG!, isotypus K!).' (WAG0249909 holotype, K000324553, MO, SRGH, UC). FIGURE 6.4.**1A**.

Perennial suffrutex, to c. 0.5 m tall. Stems woody, branched, cobwebby, glabrescent. Leaves petiolate, petiole (9)18–34 mm long, tomentose, slightly glabrescent, winged, auricles auriform, coarsely dentate, leaf lamina deltoid to deltoid-reniform, (11)20–44 × (14)25–40 mm, (3)5–7-lobed, some with a pair of lateral pinnae, arachnoid-pubescent, glabrescent (green) above, thickly white tomentose below, apex acute to obtuse; margin coarsely dentate, revolute, base subcordate to cordate. Inflorescences of solitary or few to many (c. 28 per inflorescence branch) capitula arranged in a lax corymbose cyme, capitula pedicellate, pedicels 8–20 mm long (70 mm when solitary), arachnoid-pubescent, sparsely bracteolate, bracteoles 10–15 mm long, narrowly-lanceolate, arachnoid-pubescent. Capitula heterogamous, radiate; involucre calyculate, calycular bracts c. 4 mm long, narrowly-lanceolate, sparsely to densely arachnoid-pubescent; phyllaries 12, 13 or 14 (17), 5–6 mm long, densely arachnoid-pubescent, glabrescent, margins scarious. Ray florets 12 or 13(16), 9–10 mm long, ray limbs 6–7 mm long, 4- or 5-veined. Disc florets c. 70, corolla 4.2–5.2 mm long. Achenes obovate, somewhat concave on inner surface, obcompressed with strong median rib on inner and outer surfaces, margined, blackish-brown, margins paler, (2.0)2.5–2.8 mm long, ciliate and sparsely setuliferous on faces towards apex; pappus setae 3.5–4.5 mm long.

**Malawi**. N: 4 miles SW of Chikangawa, 4.vii.1977, *Phillips* 2587 (K).

A Malawi endemic. Rocky summit of hill, amongst rocks; 1900–2030 m; flowering in July.

Conservation Status: An endemic to the Lwanjati Hill (Mzimba Dist.) Centre of endemism, and presently only known from two sites; DD (Data Deficient) following Cron *et al.* (2006b: 512; 2009: 132). Fieldwork assessing the population sizes is clearly needed to provide a meaningful statement of the Conservation Status, especially since this species appears to be so rare amongst collections.

6. **Cineraria pinnata** O.Hoffm. ex Schinz in Mém. Herb. Boissier **10**: 73 (1900). Type: [Mozambique. Maputo:] 'Delagoa Bay: [Taillis á Rikatla. 9.1890.] *Junod* 91.' (Z000003239 holotype, BR5105922).

Annual or possibly short-lived perennial herb, erect to 40 cm tall. Stems herbaceous, but slightly woody towards base, slender, 2–3.5 mm in diameter at base, unbranched or with few branches, glabrous. Leaves petiolate, petiole 7–35(50) mm long, glabrous, with arachnoid-pubescent axils; auricles small, lanceolate, sometimes dentate, caducous, lamina lyrate-pinnatifid (uppermost) or deltoid to deltoid-reniform, usually with 1–4 pairs of lateral leaflets below lamina, terminal leaflet largest, lamina of terminal leaflet 7–26 × 7–21 mm; lamina (including first pair of leaflets) 10–32(45) × 8–30 mm, total leaf length up to 90 mm long with petiole accounting for ½ to ⅓ of length, glabrous, usually with few short stipitate-glandular hairs in sinuses of lobes, occasionally sparsely hairy on veins below, texture usually papyraceous, apex acute, margins coarsely dentate, base cuneate to truncate to subcordate. Inflorescences a lax corymbose panicle, capitula usually few (2–12 per branch), rarely as many as 40 per branch; pedicels 5–32 mm long, glabrous, bracteoles few, minute descending (rarely densely bracteate). Capitula heterogamous, radiate; involucre calyculate, calycular bracts few, c. 2 mm, lanceolate, glabrous; phyllaries (7)8–12, 4 mm long, glabrous, margins scarious. Ray florets 5–8, 4–6 mm long, ray limb 3–4.2 mm long, 4-veined. Disc florets c. 24, corolla 3–3.5 mm long. Achenes narrowly obovate, compressed, margined, brown, 2.2–2.5 mm long, sparsely to moderately ciliate, body usually sparsely setuliferous, occasionally glabrous; pappus setae approximately equal to length of disc floret corolla.

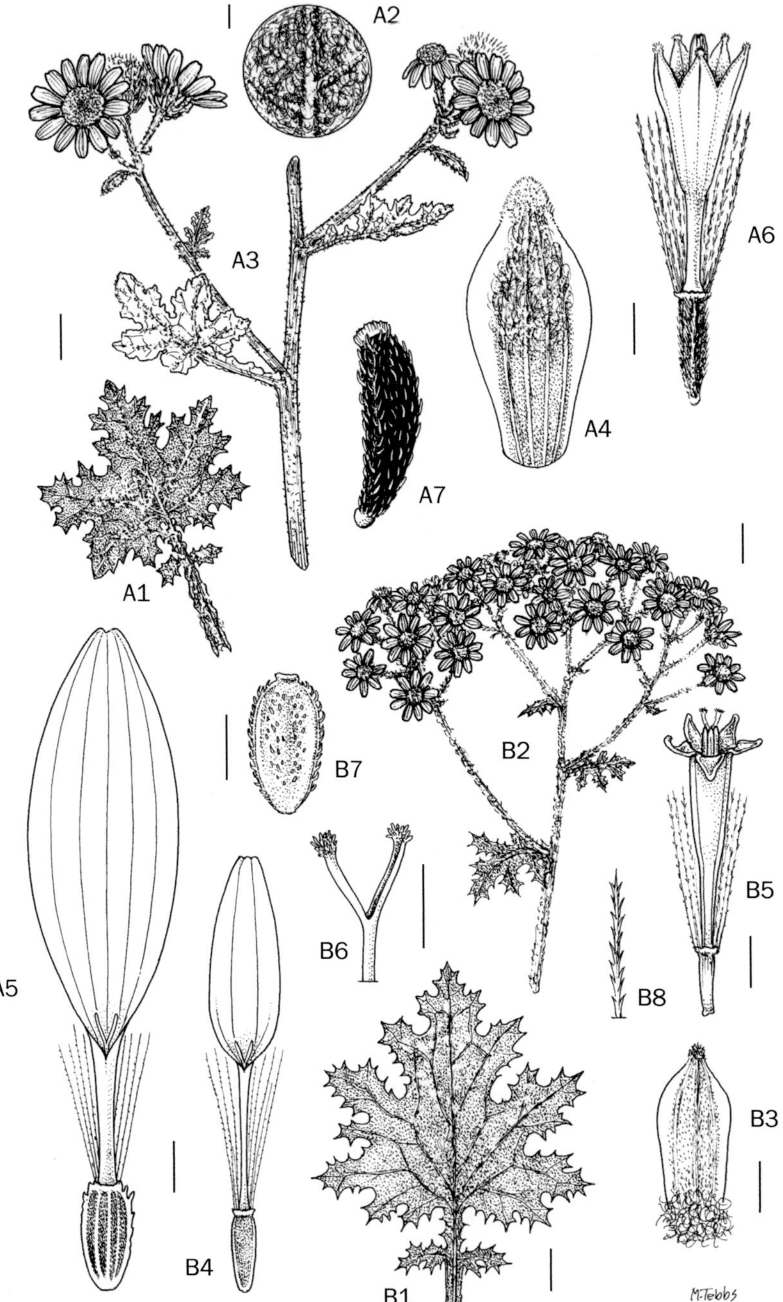

Fig. 6. 4. **1.** A. CINERARIA MAGNICEPHALA. A1, leaf; A2, details of lower leaf surface; A3, portion of inflorescence with two flowering branches; A4, inner phyllary; A5, ray floret; A6, disc floret; A7, achene. B. CINERARIA PULCHRA. B1, leaf; B2, inflorescence; B3, phyllary; B4, ray floret; B5, disc floret; B6, disc floret style arms; B7, achene; B8, details of apex of pappus seta. A1, 3–6 from *Pawek* 9875; A2, 7 from *Phillips* 2587; B6 from *Chase* 6127; B7–8 from *Wild* 2867. Scale bars: A1–2, A4–7, B3–8 = 1 mm; A3, B1–2 = 10 mm. Drawn by Margaret Tebbs.

**Mozambique**. GI: Gaza, between Chiconela and Gumbe, c. 14 km from Chiconela, 26.v.1965, *Pereira, Marques & Balsinhas* 478 (WAG). M: Sul do Save, Marracuene, 1.viii.1952, *Barbosa & Myre* 215 (BR, NU, PRE).

Also from KwaZulu Natal, South Africa. Coastal grassland or bush, seasonal wetlands, in sandy soil, also under trees; 20–70 m; flowering between July and November, but sporadically throughout the year. Cultivated in Zimbabwe (Harare, 20.i.1976, *Biegel* 5199 (K)).

A native of the Eastern Cape, South Africa, predominantly in the Albany and Uitenhage Dist., also in the Zuureberg. Cultivated in St. Helena and Zimbabwe; 400–1000 m in habitat, but clearly cultivated at much higher alitudes in Zimbabwe (c. 1490 m); flowering between (August) October and January.

Conservation Status: An endemic of the Maputaland Centre of Endemism, and considered by Cron *et al.* (2009: 136) as a 'near endemic'; LC (Least Concern), based on *Cron et al.* (2006b: 525; 2009: 132), globally and in the Flora area, although considered as 'NT-D2' (Near Threatened) for South Africa.

## 98. **MESOGRAMMA** DC.

**Mesogramma** DC., Prodr. **6**: 304 (1838). —Nordenstam & Pelser in Compositae Newslett. **42**: 74–88 (2005). —Nordenstam in Kubitzki, Fam. Gen. Vasc. Pl. **8**: 229 (2006). —Cron *et al.* in Kew Bull. **61**(4): 527–528 (2006). —Nordenstam & Cron in Compositae Newslett. **45**: 1–7 (2007).

Glabrous annual or short-lived perennial herbs. Leaves alternate, petiolate, ovate-lanceolate, markedly dentate and lobate. Inflorescence of few- to many-headed lax corymbs, rarely of solitary capitula. Capitula radiate, heterogamous; involucre calyculate; phyllaries ± biseriate, inner with 2 distinct dark resin ducts. Ray limbs yellow. Disc floret corollas black-veined; basal anther appendages sagittate; style arms apices truncate; stigmatic areas separate. Achenes triquetrous or compressed, black, setuliferous, setulae of twin-hairs, few-seriate along angles; carpopodium annuliform; pappus setae white, fragile, deciduous although possibly caducous.

A genus with a single sp. from southern Africa, N to Angola and Botswana.

**Mesogramma apiifolium** DC., Prodr. **6**: 304 (1838). —DC. in Delessert, Ic. Select. Pl. **4**: 25, t. 58 (1838). Type: [South Africa:] '... in Africae Capensis regione Gariepinâ vix suprà maris altitudineum legit cl. *Drege* [2823]! (v. s.)' (G-DC-G00470114 – '2823. Ufern des Garip. R. I.' lectotype, E00239783, HAL0112285 – s.n., s.loc., HBG505264 – s.n., s.loc., K000378001, K000378002 – both specimens s.n., s. loc., although the former has the locality written onto the sheet in a third hand, P00119575 – '16/9 30. Am Ufern der Garip. 300' (III, R).', s.n., P00119576, s.n., s. loc., P00119578, s.n. s. loc., S07-3038 – collection number and locality details not on sheet, but label is de Candolle's distribution label, S07-3041, s.loc., s.n., TUB005755 – s.n.), lectotypified by Nordenstam & Pelser (2005: 79).[9] FIGURE 6.4.2.

    *Cineraria microglossa* DC., Prodr. **6**: 305 (1838). —Harvey in Harvey & Sonder, Fl. Cap. **3**(1): 313 (1865). Type: [South Africa: Northern Cape:] 'in Africae Capensis regione Geriepinâ legit cl. *Drege* [5926, 19 Sept. 1830]! ... (v.s.)' (G-DC-G00470123 – '5926. Ufer des Garip. R. II.' holotype, K – '*Drege* 5926' has been written onto the capsule, P00117663 – '19/9 30. Am Ufern des Garip. – 300' (III, R).', with '5926' on a larger label, P00117664).

    *Senecio apiifolius* (DC.) [Benth. & Hook.f. ex] O.Hoffm., Nat. Pflanzenfam. **4**, 5(74): 298 (1892).

    *Senecio apiifolius* (DC.) [Benth. & Hook.f. ex] Mendonça, Cont. Conhec. Fl. Angola **1**: 119 (1943), isonym, nom. & comb. superfl.

---

9  Following the details of the disposition of the Delessert herbarium (vide TL II: 614) it is quite likely that, as indicated, 'A.P. de Candolle's types of the Icones Selectae are at G in the general herbarium, ...'

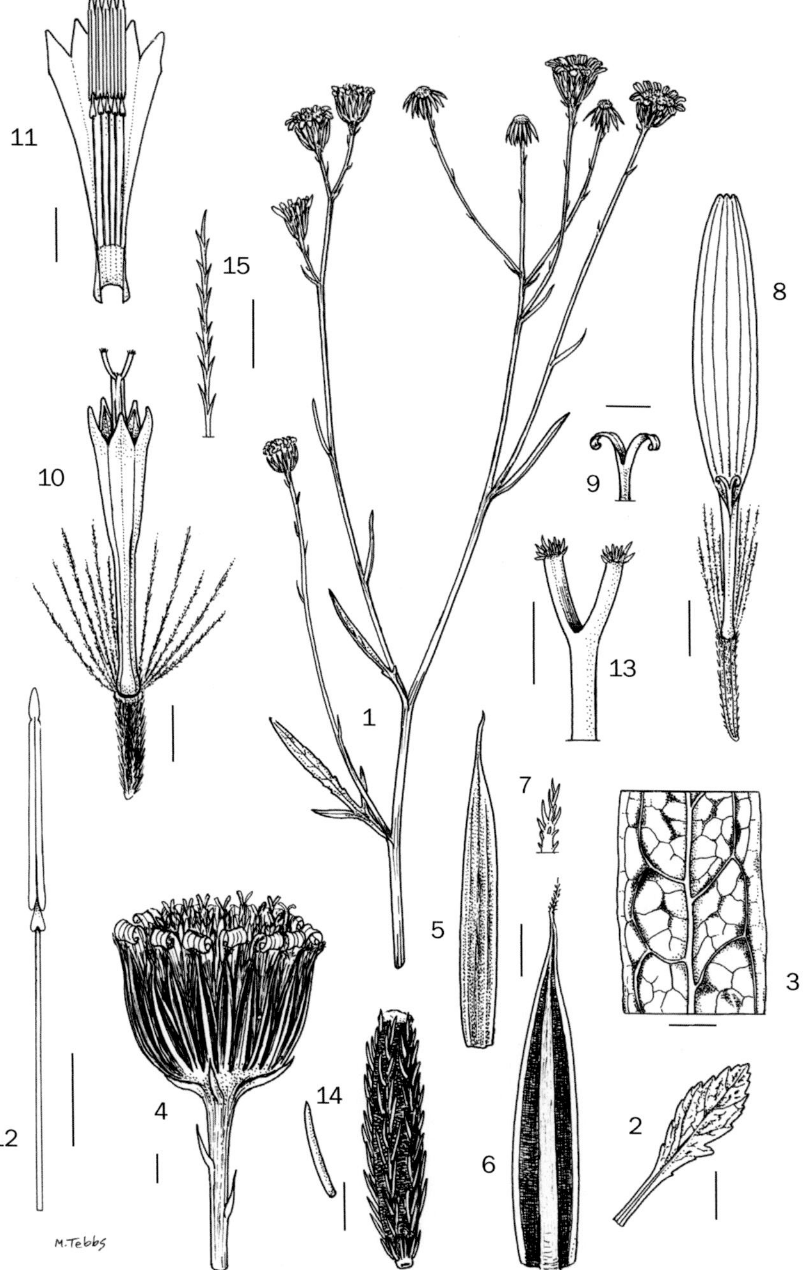

Fig. 6.4.2. **MESOGRAMMA APIIFOLIUM**. 1, habit of flowering branch; 2, leaf; 3, detail of leaf surface; 4, capitulum; 5, calycular bract; 6, phyllary; 7, detail of phyllary apex; 8, ray floret; 9, detail of ray floret style arms; 10, disc floret; 11, disc floret corolla opened out to show attachment points of anther filaments; 12, stamen; 13, details of disc floret style arms; 14, achene and detail of setula; 15, detail of apex of pappus seta. 1, 3, 4 from *Smith* 1784; 2 from *Smith* 1795; 5–7, 14 from *Mott* 306; 8–13 from *Lambrecht* 309; Scale bars:1–2 = 10 mm; 3–6, 8, 10–12 = 1 mm; 9, 13, 14 = 0.5 mm. Drawn by Margaret Tebbs.

*Senecio peculiaris* Dinter, Feddes Repert. Spec. Nov. Regni Veg. **30**(1–8): 94 (1932). Type: [Namibia:] 'Groß-Namaland: Garius bei Warmbad, im Rivier an dauern feuchten Stellen des "Wasserfalles", 30. Nov. 1922, *Dinter* 4252.' (B† holotype, HBG505315).

Annual herb (or short-lived perennial), to 60 cm tall, glabrous, erect, branching from base. Leaves cauline, alternate, petiolate, petiole to 3 cm long, upper leaves sessile, leaf-blade ovate-lanceolate, 3–7 ×1–3 cm, grossly dentate or pinnatilobate with 5–6 pairs of lobes, young stem leaves dentate, apex mucronate; upper leaves gradually smaller. Inflorescence laxly corymbose, capitula 1 or 2(12), pedicels 2–8 cm long, bracteolate, bracteoles 1 or 2, subulate, c. 1 mm long. Capitula radiate; involucre 4–8 mm diam., calyculate, calycular bracts 2–5, subulate, c. 1 mm long; phyllaries (8)10–21, ± biseriate, lanceolate, 4–6 × 0.8–1 mm, green, margins scarious, midrib with a thin blackish resin vein and two lateral blackish resin ducts, lateral resin stripes of inner phyllaries broad and gland-dotted; receptacle slightly convex, glabrous, minutely foveolate. Ray florets 8–13, female, fertile, ray limb oblong, 4–5 × 1.5 mm, 4-veined, yellow, tube cylindrical, c. 2 mm long; style bilobed. Disc florets numerous (c. 30–70), hermaphrodite; corollas yellow, c. 4 mm long, tubular, gradually widening above, 5-lobed, distinctly 5-veined with black veins from corolla tips down to base of tube; corolla lobes triangular-ovate, 0.5–0.7 mm long, thin lateral veins in addition to dark mid-vein; basal anthers appendages sagittate or minutely caudate, apical anther appendages narrowly ovate, apices obtuse; filament collar distinctly balusterform; style with dark resin vein branching into style branches, style base with distinct node; style arms linear-oblong, stigmatic areas separate, apices truncate with rather long sweeping-hairs. Achenes subtriquetrous or somewhat compressed, constricted apically and basally, c. 1 mm long, black, setuliferous, setulae of twin-hairs in few lines on ribs/angles, white, apices connate, obtuse, myxogenic and mucilaginous when wetted, achene body apically and basally with a crown of longer hairs, apices rounded; carpopodium distinct; pappus setae numerous (c. 20), uniseriate, basally connate, c. 3 mm long, coarsely barbellate, fragile, deciduous, although possibly caducous, white.

**Botswana**. N: Xigera Island, 19°20'S, 22°42.5'E, 25.ix.1976, *Smith* 1795 (K, SRGH). SE: Road from Francistown to Gaberones, ix.1967, *Lambrecht* 309 (K, SRGH).

Also known from Angola, Namibia and South Africa (Cape, Orange Free State, Transvaal).

Growing in a wide range of damp soils (sandy, loamy, clay, silt), frequently on river margins, shallow pans or in damp places in dry riverbeds, disturbed areas near dams, roadsides and in cultivated or fallow fields; 1020–1200m; flowering from March to November, although probably flowering throughout the year.

Comments on one or two specimens from South Africa have associated the plant with suspected tainting of milk, as well as livestock poisoning.

Conservation Status. Although poorly represented, or collected, in the Flora area it is a relatively widespread species: LC (Least Concern).


## 99. **SENECIO** L.

**Senecio** L., Sp. Pl. **2**: 866 (1753). —Humbert, Fl. Madagasc. Composées **189**: 677–804 (1963). —Hilliard & Burtt in Notes Roy. Bot. Gard. Edinburgh **32**(3): 374–384 (1973); in Notes Roy. Bot. Gard. Edinburgh **34**(1): 86–99 (1975). —Dyer, Gen. S. Afr. Fl. Pl. **1**: 714 (1975). —Hilliard, Compositae Natal: 387–502 (1977). —Lisowski, (Asterac. Fl. Afr. Cent. 2) Fragm. Flor. Geobot. **36** Suppl. 1: 256–328 (1991). —Jeffrey in Kew Bull. **41**(4): 873–943 (1986); in Kew Bull. **47**(1): 49–109 (1992). —Jeffrey & Beentje in F.T.E.A., Compositae 3: 617–669 (2005).

Annual, biennial, triennial or perennial herbs, subshrubs, shrubs, rarely climbers or trees. Stems simple, poorly branched or well-branched, glabrous or variously pubescent. Leaves alternate, very rarely opposite, sometimes radical and rosulate, sessile or petiolate, glabrous or variously pubescent, margins entire or serrate. Inflorescence terminal or axillary, scapose or corymbose. Capitula homogamous and discoid, or heterogamous and disciform or radiate; involucre calyculate, cylindrical or campanulate, sometimes ± hemispherical; phyllaries uniseriate; receptacle flat or + convex, glabrous, alveolate. Ray florets when present female, uniseriate, fertile, sometimes lacking limb, limb when present small to conspicuous, white, yellow, orange, red, or purple. Disc florets hermaphrodite, many, corollas tubular,

(4)5-lobed; anther bases sagittate; style arms truncate or triangular at apex, penicillate. Achenes cylindrical, ribbed, glabrous or setuliferous; pappus setae multiseriate, capillary, barbellate, fragile and usually caducous, usually white.

About 1000 (possibly as many as 1250) spp. worldwide; it should be noted that some segregates are not recognised by the present author (e.g. *Jacobaea* Mill.). *Senecio* contains some notable pernicious, noxious and toxic weeds; some spp. are sometimes lethal (highly hepatoxic) to grazing livestock.

Two spp., *S. pachyrhizus* and *S. strictifolius*, may flower before leaves are present; the first has no ray florets (and has phyllaries 15–22 mm long), the second one does (and has phyllaries 4.5–5.5 mm long).

1. Disc and ray (if present) corollas all same colour . . . . . . . . . . . . . . . . . . . . . . . . . 2
– Ray corollas mauve/violet, disc corollas yellow . . . . . . . . . . . . . . . . . . . . . . . **12.** *eenii*
2. Corollas mauve, reddish, pink or purple . . . . . . . . . . . . . . . . . . . . . . . . . . . . . 3
– Corollas yellow, orange, cream or white . . . . . . . . . . . . . . . . . . . . . . . . . . . . . 10
3. Rays present . . . . . . . . . . . . . . . . . . . . . . . . . . . . . . . . . . . . . . . . . . . . . . . . . 4
– Ray absent, only disc florets present. . . . . . . . . . . . . . . . . . . . . . . . . . . . . . . . . 7
4. Plants from below 200 m altitude; Mozambique . . . . . . . . . . . . . . . . . . . . . . . . . 5
– Plants from above 1000 m altitude; Mozambique or Botswana . . . . . . . . . . . . . . . . 6
5. Rays ± 13, 1–2 mm long; seashore . . . . . . . . . . . . . . . . . . . . . . . . . . . . . **2.** *arenarius*
– Rays 8–20, ± 8 mm long; damp or swampy grassland . . . . . . . . . . . . . . . **42.** *speciosus*
6. Leaves 1–4 cm long; phyllaries 5–5.5 mm long; Botswana, on Kalahari sand at 1250 m . . . . . . . . . . . . . . . . . . . . . . . . . . . . . . . . . . . . . . . . . . . . . **12.** *eenii*
– Leaves usually more than 5 cm long; phyllaries 7.8–9 mm long; Mozambique, only known from Mt Namuli at ± 1950 m. . . . . . . . . . . . . . . . . **42.** *speciosus* (see note there)
7. Stem and leaves glabrous; below 100 m altitude . . . . . . . . . . . . . . . . . . . **31.** *ngoyanus*
– Stem and leaves glabrous to glandular-pubescent; above 1000 m altitude . . . . . . . . . 8
8. Achenes glabrous (rarely slightly pilose) . . . . . . . . . . . . . . . . . . . . . . . . . . . . . . 9
– Achenes hairy . . . . . . . . . . . . . . . . . . . . . . . . . . . . . . . . . . . . . . . . . **13.** *erubescens*
9. Stem and leaves glandular-hairy; leaves usually pinnately lobed, more than 9 cm wide . . . . . . . . . . . . . . . . . . . . . . . . . . . . . . . . . . . . . . . . . . . . . . . . . . . . **39.** *purpureus*
– Stem and leaves glabrous; leaves narrowly elliptic with dentate margin and less than 3 cm wide . . . . . . . . . . . . . . . . . . . . . . . . . . . . . . . . . . . . . . . . . . . **30.** *ngandae*
10. Leaves lobed, lyrate, pinnatisect/pinnatifid to bipinnatisect . . . . . . . . . . . . . . . . . 11
– Leaves more or less entire (margins may be dentate or serrate); hastate and triangular-leaved taxa also grouped here. . . . . . . . . . . . . . . . . . . . . . . . . . . . . . . . . . . . . 41
11. Leaves in two parts, upper almost reniform with dentate margin, lower auriculate and surrounding stem; ray florets present; forest margin climber. . . . . . . . . . . . . . . . . . . . . . . . . . . . . . . . . . . . . . . . . . . . . . . . . . . . . . . . . . . **3.** *auriculatissimus*
– Leaves not divided like this (although *S. sp. A* has a blade in two parts) . . . . . . . . . . . 12
12. Leaves with two pairs of minute lobes at very base only . . . . . . . . . . . . **54.** *sp.* aff. *serra*
– Leaves with main lamina lobed . . . . . . . . . . . . . . . . . . . . . . . . . . . . . . . . . . . . 13
13. Leaves divided halfway or almost to midrib, pinnatipartite to pinnatisect. . . . . . . . . 14
– Leaves shallowly lobed to lyrate . . . . . . . . . . . . . . . . . . . . . . . . . . . . . . . . . . . 28
14. Capitula discoid (or ray florets < 1 mm long in *S. cryphiactis*) . . . . . . . . . . . . . . . . . 15
– Capitula radiate . . . . . . . . . . . . . . . . . . . . . . . . . . . . . . . . . . . . . . . . . . . . . . 19
15. Phyllaries 18–22, 3–4 mm long; (ray florets present but minute) disc floret corollas 1.4–3.3 mm long; widespread. . . . . . . . . . . . . . . . . . . . . . . . . . . . . . . **9.** *cryphiactis*
– Phyllaries 8–19, 4–9 mm long; disc floret corollas 4–10 mm long; *S. hochstetteri* widespread, *S. lisowskii* and *S. teixeirae* Zambia only, *S. vulgaris* Zimbabwe only. . . . . . . . . . . . . . . . 16

16. Leaves glandular-pubescent; basal leaves present and persistent . . . . . . . . . . . . . . . 17
 — Leaves pubescent, puberulous or glabrous, but without glands; basal leaves absent or short-lived (in *S. vulgaris*) . . . . . . . . . . . . . . . . . . . . . . . . . . . . . . . 18
17. Basal leaves 7–38 × 1–10 cm; disc florets 5–10 mm; widespread . . . . . . . . **20.** *hochstetteri*
 — Basal leaves 7–12 × 3–5.4 cm; disc florets 4.8–6 mm; Zambia . . . . . . . . . . **49.** *teixeirae*
18. Leaves puberulous; phyllaries ± 15, 7–8 mm long; disc florets 4–4.5 mm; Zambia . . . . . . . . . . . . . . . . . . . . . . . . . . . . . . . . . . . . . . . . . . **26.** *lisowskii*
 — Leaves with scattered long white hairs; phyllaries ± 19, 6.5–7.5 mm long; disc florets 5 mm; occasional weed of cultivation . . . . . . . . . . . . . . . . . . . . . . . . . **53.** *vulgaris*
19. Seashore sp., glandular-hairy on stem and leaves; ray florets 13, limb 1–2 mm long . . . . . . . . . . . . . . . . . . . . . . . . . . . . . . . . . . . . . . . . . . . . . . **2.** *arenarius*
 — More inland taxa without such indument; if near sea, ray florets 8–15, limb more than 3 mm long . . . . . . . . . . . . . . . . . . . . . . . . . . . . . . . . . . . . . . . . . 20
20. Ray florets minute, less than 1 mm long . . . . . . . . . . . . . . . . . . . . . . . . . . 21
 — Ray florets more than 3 mm long . . . . . . . . . . . . . . . . . . . . . . . . . . . . . . 22
21. Leaves arachnoid to glabrous; phyllaries 18–22, 3–4 mm long; disc florets narrowly infundibuliform, 1.4–3.3 mm long; pappus setae 2–4.3 mm . . . . . . . . . . **9.** *cryphiactis*
 — Leaves glandular-pubescent; phyllaries 10–12, 7–8 mm long; disc florets narrowly cylindrical, 5.3 mm long; pappus setae 6 mm . . . . . . . . . . . . . . . . . . . . . **16.** *glutinosus*
22. Lowland spp. found below 150 m . . . . . . . . . . . . . . . . . . . . . . . . . . . . . . 23
 — Inland spp. found above 1000 m . . . . . . . . . . . . . . . . . . . . . . . . . . . . . . 25
23. Plants from sea level; leaves grey-velutinous beneath; ray florets 8, limb 3–4 mm long . . . . . . . . . . . . . . . . . . . . . . . . . . . . . . . . . . . . . . **36.** *polyanthemoides*
 — Plants from altitudes 0–150 m; leaves glabrous or nearly so; ray florets 12–15 or (*sp. A*) unclear, limb 3–6 mm long . . . . . . . . . . . . . . . . . . . . . . . . . . . . . . . 24
24. Phyllaries 19–21, 4–7 mm long; ray florets limb 4.5–6 mm long; achenes hairy . . . . . . . . . . . . . . . . . . . . . . . . . . . . . . . . . . . . . . . **27.** *madagascariensis*
 — Phyllaries 11–14, 3.8–4.3 mm long; ray florets limb 3.2 mm long; achenes glabrous . . . . . . . . . . . . . . . . . . . . . . . . . . . . . . . . . . . . . . . . . . . . . **55.** *sp. A*
25. Stem and leaves glandular-hairy . . . . . . . . . . . . . . . . . . . . . . . . . . . . . . 26
 — Stem and leaves short-villous or puberulous, but not glandular . . . . . . . . . . **1.** *aetfatensis*
26. Phyllaries 13–14; achenes 3–3.2 mm long; Botswana . . . . . . . . . . . . . . **6.** *consanguineus*
 — Phyllaries 12–18; achenes 2 mm long; Zimbabwe . . . . . . . . . . . . . . . . . . . . 27
27. Phyllaries 7.5 mm long; ray florets ?8; ± 1250 m altitude . . . . . . . . . . . . . **18.** *hastatus*
 — Phyllaries 4 mm long; ray florets 10; ± 1950 m altitude . . . . . . . . . . . . . **1.** *aetfatensis*
28. Capitula radiate . . . . . . . . . . . . . . . . . . . . . . . . . . . . . . . . . . . . . . 29
 — Capitula discoid . . . . . . . . . . . . . . . . . . . . . . . . . . . . . . . . . . . . . . 33
29. Leaves peltate . . . . . . . . . . . . . . . . . . . . . . . . . . . . . . . . . . . . . . . 30
 — Leaves petiolate . . . . . . . . . . . . . . . . . . . . . . . . . . . . . . . . . . . . . . 31
30. Leaf 3.5–9 cm across; phyllaries 6.5–9 mm long; pappus setae 4.5–6 mm . . **29.** *milanjianus*
 — Leaf 1.5–3 cm across; phyllaries 4–5.5 mm long; pappus setae 3–3.5 mm . . **35.** *peltophorus*
31. Ray limbs 5–11 mm long; phyllaries bearded at apex; disc florets 9–35 (very occasionally lobed) . . . . . . . . . . . . . . . . . . . . . . . . . . . . . . **41.** *ruwenzoriensis*
 — Ray limbs less than 5 mm long; phyllaries glabrous; disc florets 30–110[10,11] . . . . . . . . 32
32. Plants glabrous or sparsely hairy . . . . . . . . . . . . . . . . . . . . **27.** *madagascariensis*
 — Stems minutely glandular, sometimes arachnoid . . . . . . . . . . . . . . **36.** *polyanthemoides*

———————————————————

[10] In *S. madagascariensis*, unknown for *S. polyanthemoides*.

[11] *S. hastatus* may also key here, as it has shallowly lobed upper leaves (but deeply lobed leaves lower down on the plant); ray floret lamina is 4.5–5 mm long; phyllaries are glandular-pubescent; disc florets 37–40 in number; Zimbabwe.

33.   Leaves peltate, circular to triangular-hastate . . . . . . . . . . . . . . . . . . . . . .**33.** *oxyriifolius*
  –   Leaves not peltate . . . . . . . . . . . . . . . . . . . . . . . . . . . . . . . . . . . . . . . . . . . . . . . . 34
34.   Erect herbs with leaves longer than wide . . . . . . . . . . . . . . . . . . . . . . . . . . . . . . . . 35
  –   Climbers with ± deltate or ovate-deltate leaves (or lanceolate in *S. brachypodus*
      var. *brachypodus*) . . . . . . . . . . . . . . . . . . . . . . . . . . . . . . . . . . . . . . . . . . . . . . . . 37
35.   Stem, leaves and phyllaries glandular; perennial herbs with fleshy rootstock . . . . . . . .
      . . . . . . . . . . . . . . . . . . . . . . . . . . . . . . . . . . . . . . . . . . . . . .**20.** *hochstetteri*
  –   Stem and leaves pilose or puberulous, but eglandular; phyllaries glabrous; ± annual
      herbs with thin root system. . . . . . . . . . . . . . . . . . . . . . . . . . . . . . . . . . . . . . . . 36
36.   Achenes 4–5 mm long; Zambia, annual at stream-sides . . . . . . . . . . . . . . **26.** *lisowskii*
  –   Achenes 2–2.5 mm long; Zimbabwe, a weed of nurseries. . . . . . . . . . . . . . .**53.** *vulgaris*
37.   Leaves lanceolate, only slightly lobed . . . . . . . . . . . . . . . . . .**5.** *brachypodus* var. *brachypodus*
  –   Leaves deltate or ovate-deltate . . . . . . . . . . . . . . . . . . . . . . . . . . . . . . . . . . . . . . 38
38.   Capitula 1–6; phyllaries 10–12 mm long; achenes 2.8–6 mm long . . . . .**19.** *helminthioides*
  –   Capitula many; phyllaries < 8.5 mm long; achenes < 4 mm long . . . . . . . . . . . . . 39
39.   Stem finely pubescent when young; phyllaries 8–13, 3–5 mm long; achenes 2.5–3 mm
      long. . . . . . . . . . . . . . . . . . . . . . . . . . . . . . . . . . . . . . . . . . . . . . . .**10.** *deltoideus*
  –   Stem glabrous; phyllaries 5–8, 3–8.5 mm long; achenes 1.2–4 mm long. . . . . . . . . 40
40.   Phyllaries 5, 3–4 mm long; achenes 1.3–1.5 mm long . . . . . . . . . . . . . . . . . . . . . . . .
      . . . . . . . . . . . . . . . . . . . . . . . . . . . . . . . . .**5.** *brachypodus* var. *pleistocephalus*
  –   Phyllaries 8, 5–8.5 mm long; achenes 3–4 mm long . . . . . . . . . . . . . . **46.** *syringifolius*
41.   Leaves circular, triangular/deltate or narrow and linear (and then less than 2 mm
      wide) . . . . . . . . . . . . . . . . . . . . . . . . . . . . . . . . . . . . . . . . . . . . . . . . . . . . . . . . . 42
  –   Leaves ovate, eliptic, obovate or lanceolate, if narrow more than 4 mm wide (only
      *S. striatifolius* sometimes 2–8 mm wide) . . . . . . . . . . . . . . . . . . . . . . . . . . . . . . . . 53
42.   Leaves circular or nearly so, and peltate. . . . . . . . . . . . . . . . . . . . . . . . . . . . . . . . . 43
  –   Leaves of different shape, not peltate. . . . . . . . . . . . . . . . . . . . . . . . . . . . . . . . . . . 45
43.   Capitula radiate; leaf margins with few large teeth . . . . . . . . . . . . . . . . . . . . . . . . . 44
  –   Capitula discoid; leaf margins with many small teeth/serrations . . . . . . .**33.** *oxyriifolius*
44.   Leaves 35–90 mm across; phyllaries 6.5–9 mm long; pappus setae 4.5–6 mm. . . . . . .
      . . . . . . . . . . . . . . . . . . . . . . . . . . . . . . . . . . . . . . . . . . . . . . **29.** *milanjianus*
  –   Leaves 15–30 mm across; phyllaries 4–5.5 mm long; pappus setae 3–3.5 mm . . . . . . .
      . . . . . . . . . . . . . . . . . . . . . . . . . . . . . . . . . . . . . . . . . . . . . . . . .**35.** *peltophorus*
45.   Leaves triangular/deltate . . . . . . . . . . . . . . . . . . . . . . . . . . . . . . . . . . . . . . . . . . . 46
  –   Leaves a different shape . . . . . . . . . . . . . . . . . . . . . . . . . . . . . . . . . . . . . . . . . . . . 50
46.   Erect herb . . . . . . . . . . . . . . . . . . . . . . . . . . . . . . . . . . . . . . . . .**33.** *oxyriifolius*
  –   Climbing herbs or lianas. . . . . . . . . . . . . . . . . . . . . . . . . . . . . . . . . . . . . . . . . . . . 47
47.   Capitula discoid; achenes hairy, or glabrous in *S. syringifolius* . . . . . . . . . . . . . . . . . 48
  –   Capitula radiate; achenes glabrous. . . . . . . . . . . . . . . . . . . . . . . . . . . . . . . **48.** *tamoides*
48.   Capitula 1–6; phyllaries 10–11 mm long; achenes 5–6 mm long. . . . . . . .**19.** *helminthioides*
  –   Capitula many; phyllaries less than 8.5 mm long; achenes less than 4 mm long . . . . . 49
49.   Stem finely pubescent when young; phyllaries 8–13, 3–5 mm long; achenes 2.5–
      3 mm long. . . . . . . . . . . . . . . . . . . . . . . . . . . . . . . . . . . . . . . . . . . . . .**10.** *deltoideus*
  –   Stem glabrous; phyllaries 8, 5–8.5 mm long; achenes 3–4 mm long . . . . . **46.** *syringifolius*
50.   Leaves ericoid and linear, less than 2 mm wide . . . . . . . . . . . . . . . . . . . . . . . . . . . 51
  –   Leaves wider, not linear (except *S. striatifolius*, 400 × 4–8 mm, and *S. barbertonicus* with
      succulent terete leaves 50–90 × 5 mm) . . . . . . . . . . . . . . . . . . . . . . . . . . . . . . . . . 52
51.   Capitula radiate; phyllaries 7–10; pappus setae 2.2–4.5 mm; Zambia, Zimbabwe and
      Malawi . . . . . . . . . . . . . . . . . . . . . . . . . . . . . . . . . . . . . . . . . . . . . .**15.** *gazensis*
  –   Capitula discoid; phyllaries 20–22; pappus setae 5.5–7 mm; Zimbabwe . . . **50.** *torticaulis*
52.   Capitula discoid . . . . . . . . . . . . . . . . . . . . . . . . . . . . . . . . . . . . . . . . . . . . . . . . . . 53
  –   Capitula radiate (ray limbs may be as small as 1.5 mm in *S. maranguensis* from Malawi,
      or even 0.5 mm in *S. cryphiactis*) . . . . . . . . . . . . . . . . . . . . . . . . . . . . . . . . . . . . . 69

53. Leaves succulent and terete, 5 mm in diameter; in sand forest and on rock outcrops . . . . . . . . . . . . . . . . . . . . . . . . . . . . . . . . . . . . . . . . . . . . . . . . . . . . . . . . **4.** *barbertonicus*
– Leaves not succulent (or sometimes succulent in *S. crassorhizus*) flat, over 10 mm wide (except in *S. kayomborum* and *striatifolius*); grassland or wooded grassland or upland forest, with only *S. brachypodus* sometimes in sand forest and *S. diphyllus* sometimes in rock crevices . . . . . . . . . . . . . . . . . . . . . . . . . . . . . . . . . . . . . . . . . . . . . . . . . . . 54
54. Large leaves present near base of stem, and larger than leaves higher up stem (if these are present); elliptic to obovate with attenuate base (linear in *S. striatifolius*; sometimes orbicular in *S. diphyllus*) . . . . . . . . . . . . . . . . . . . . . . . . . . . . . . . . . . . . . . . . . . . . 55
– Leaves only on stem, no leaves near ground; largest leaves in mid-stem . . . . . . . . . . 61
55. Leaves linear, 8–40 × 0.2–0.5(0.8) cm . . . . . . . . . . . . . . . . . . . . . . . .**43.** *striatifolius*
– Leaves not linear, more than 0.6 cm wide . . . . . . . . . . . . . . . . . . . . . . . . . . . . . . . . 56
56. Capitula solitary (rarely in 2); phyllaries 15–22 mm long. . . . . . . . . . . . . **34.** *pachyrhizus*
– Capitula in larger groups; phyllaries 4–9 mm long . . . . . . . . . . . . . . . . . . . . . . . . . 57
57. Perennial herbs of grassland and miombo. . . . . . . . . . . . . . . . . . . . . . . . . . . . . . . . 58
– Annual herb, weed of nurseries . . . . . . . . . . . . . . . . . . . . . . . . . . . **53.** *vulgaris*
58. Stem and leaves glandular-pubescent; corollas white to yellow . . . . . . . . . . **37.** *polyodon*
– Stem and leaves eglandular (rarely few glands present); corollas yellow, sometimes orange. . . . . . . . . . . . . . . . . . . . . . . . . . . . . . . . . . . . . . . . . . . . . . . . . . . . . . . . . . 59
59. Phyllaries 16–22; achenes glabrous . . . . . . . . . . . . . . . . . . . . . . . . . **17.** *gramineticola*
– Phyllaries 7–14; achenes hairy . . . . . . . . . . . . . . . . . . . . . . . . . . . . . . . . . . . . . . . 60
60. Basal leaves 1–2; phyllaries 7–8; pappus setae 4–4.5 mm . . . . . . . . . . . . . **11.** *diphyllus*
– Basal leaves usually more; phyllaries 8–14; pappus setae 5–7.5 mm . . . . . **52.** *urundensis*
61. Phyllaries 5–7 . . . . . . . . . . . . . . . . . . . . . . . . . . . . . . . . . . . . . . . . **5.** *brachypodus*
– Phyllaries 8 or more . . . . . . . . . . . . . . . . . . . . . . . . . . . . . . . . . . . . . . . . . . . . . . 62
62. Phyllaries 16–20 . . . . . . . . . . . . . . . . . . . . . . . . . . . . . . . . . . . . . . . . . . . . . . . . . 63
– Phyllaries 8–14 . . . . . . . . . . . . . . . . . . . . . . . . . . . . . . . . . . . . . . . . . . . . . . . . . . 64
63. Disc florets 4.5–7 mm; achenes glabrous (basal leaves usually present). . . . . . . . . . . . . . . . . . . . . . . . . . . . . . . . . . . . . . . . . . . . . . . . . . . . . . . . . . . . . . . . . . . . **17.** *gramineticola*
– Disc florets 8.5–10.2 mm; achenes pilose between ribs[12]. . . . . . . . . . . . . .**23.** *kayomborum*
64. Climbing plants; leaves triangular, rhomboid or hastate . . . . . . . . . . . . . . . . . . . . . 65
– Erect herbs; leaves elliptic, obovate or lanceolate. . . . . . . . . . . . . . . . . . . . . . . . . . 66
65. Achenes 'ciliate'; forest margin or miombo; 1050–1200 m . . . . . . . . . . . . .**10.** *deltoideus*
– Achenes glabrous; evergreen forest; 1550–2350 m . . . . . . . . . . . . . . . . **46.** *syringifolius*
66. Stem and leaves glandular-pubescent; widespread. . . . . . . . . . . . . . . . . .**20.** *hochstetteri*
– Stem and leaves hairy or glabrous, but without glands . . . . . . . . . . . . . . . . . . . . . . . 67
67. Leaves with several veins parallel to midrib; achenes hairy; Zambia. . . . . . . . . . . . . 68
– Leaves with all veins at angles to midrib; achenes glabrous; Malawi, Mt Mulanje[13] . . . . . . . . . . . . . . . . . . . . . . . . . . . . . . . . . . . . . . . . . . . . . . . . . **28.** *maranguensis*
68. Plant unbranched outside inflorescence; phyllaries 7–10 mm long; corollas greenish white; Zambia W . . . . . . . . . . . . . . . . . . . . . . . . . . . . . . . . . . . . . . . . **8.** *crassorhizus*
– Plant often branched from near base; phyllaries 6 mm long; corollas pale yellow; Zambia N . . . . . . . . . . . . . . . . . . . . . . . . . . . . . . . . . . . . . . . . . . .**22.** *katangensis*
69. Large leaves present near base of stem, and larger than leaves higher up stem (if present), these lower leaves elliptic to obovate with attenuate base (or linear in *S. striatifolius*) . . . 70
– Leaves only on stem, no leaves near ground (except in *S. fanshawei* with small, virtually linear basal leaves). . . . . . . . . . . . . . . . . . . . . . . . . . . . . . . . . . . . . . . . . . . . . . . . . 75

---

[12] Also *S. cryphiactis* with minuscule ray florets, only cauline leaves, 18–22 phyllaries

[13] This species does have minute ray florets, but these are easily overlooked.

70.  Leaves linear, 8–40 × 0.2–0.5(0.8) cm; Zambia . . . . . . . . . . . . . . . . . . . .**43.** *striatifolius*
 –   Leaves wider than linear (though 16–50 × 0.5–2.5 cm in *S. proprior*); more widespread
     spp. . . . . . . . . . . . . . . . . . . . . . . . . . . . . . . . . . . . . . . . . . . . . . . . . . . . . . . . 71
71.  Plant with woolly root-crown at base of stem; achenes hairy (rarely glabrous in *S.
     proprior*)[14] . . . . . . . . . . . . . . . . . . . . . . . . . . . . . . . . . . . . . . . . . . . . . . . . . . . 72
 –   No woolly root-crown present, except possibly in *S. swynnertonii*; achenes glabrous (rarely
     slightly hairy in *S. inornatus*). . . . . . . . . . . . . . . . . . . . . . . . . . . . . . . . . . . . . . 73
72.  Phyllaries 18–23, 8–12 mm long; ray florets 10–12, limb 7–18 mm long . . . .**7.** *coronatus*
 –   Phyllaries 11–17(18), 5.3–7.5 mm long; ray florets absent or 7–9, limb 5–11 mm long
     . . . . . . . . . . . . . . . . . . . . . . . . . . . . . . . . . . . . . . . . . . . . . . . . . . . .**38.** *proprior*
73.  Phyllaries 10–16, 4–7.5 mm long, glabrous (except for papillate apex); ray florets 5–9
     . . . . . . . . . . . . . . . . . . . . . . . . . . . . . . . . . . . . . . . . . . . . . . . . . . . .**21.** *inornatus*
 –   Phyllaries 10–26, 7.5–14 mm long, with some hairs at least near base; ray florets 8–14
     . . . . . . . . . . . . . . . . . . . . . . . . . . . . . . . . . . . . . . . . . . . . . . . . . . . . . . . . . . . . 74
74.  Phyllaries 22–26 . . . . . . . . . . . . . . . . . . . . . . . . . . . . . . . . . . . . . .**45.** *swynnertonii*
 –   Phyllaries 10–22 . . . . . . . . . . . . . . . . . . . . . . . . . . . . . . . . . . . . . . . .**47.** *tabulicola*
75.  Leaves felted beneath; annual of clay soils and abandoned cultivation at low (0–50 m)
     altitude in Mozambique . . . . . . . . . . . . . . . . . . . . . . . . . . . **36.** *polyanthemoides*
 –   Leaves glabrous or hairy, but not very densely so (sometimes pubescent in *S. maranguensis*,
     *S. latecorymbosus*); herbs or scandent shrubs usually from higher altitudes . . . . . . . . . . . 76
76.  Scandent shrub of forest margins at 0–1100 m altitude; ?Botswana, ?Mozambique . .
     . . . . . . . . . . . . . . . . . . . . . . . . . . . . . . . . . . . . . . . . . . . . . . . . . . .**5.** *brachypodus*
 –   Erect herbs or, when scandent (in *S. maranguensis*) from above 1800 m. . . . . . . . . . . . 77
77.  Achenes/ovary glabrous. . . . . . . . . . . . . . . . . . . . . . . . . . . . . . . . . . . . . . . . . . . 78
 –   Achenes/ovary hairy . . . . . . . . . . . . . . . . . . . . . . . . . . . . . . . . . . . . . . . . . . . . . 84
78.  Leaves with two narrow 1–6 mm long lobes at very leaf-base . . . . . . . **54.** *S.* sp. aff. *serra*
 –   Leaves without lobes at base. . . . . . . . . . . . . . . . . . . . . . . . . . . . . . . . . . . . . . . . 79
79.  Stem base with woolly root crown; leaf base cordate. . . . . . . . . . . . . . . . . **25.** *latifolius*
 –   Stem base without woolly crown; leaf base cuneate, rounded or (sometimes in
     *S. maranguensis*) cordate . . . . . . . . . . . . . . . . . . . . . . . . . . . . . . . . . . . . . . . . . . 80
80.  Leaves 0.2–1 cm wide; ray florets 6–13 . . . . . . . . . . . . . . . . . . . . . . . . . . . . . . . . 81
 –   Leaves nearly always more than 1 cm wide; ray florets 5–9. . . . . . . . . . . . . . . . . . . . 82
81.  Leaves 0.3–1 cm wide; ray florets 8–13; phyllaries 4.5–5.5 mm long. . . . . . .**44.** *strictifolius*
 –   Leaves 0.2–0.4 cm wide; ray florets 6–7; phyllaries 8.5 mm long . . . . . . . **14.** *fanshawei*
82.  Phyllaries 7–8, 2.5–5 mm long; achenes 1.5–2 mm long. . . . . . . . . . . . .**28.** *maranguensis*
 –   Phyllaries 10–16, 4–9 mm long; achenes 1.5–4 mm long . . . . . . . . . . . . . . . . . . . . . 83
83.  Leaves succulent; disc florets 9–35 . . . . . . . . . . . . . . . . . . . . . . . . . . **41.** *ruwenzoriensis*
 –   Leaves herbaceous; disc florets 50–60 (occasional specimens lacking usual basal
     leaves) . . . . . . . . . . . . . . . . . . . . . . . . . . . . . . . . . . . . . . . . . . . . . **21.** *inornatus*
84.  Herbs of forest margins; achenes 5–5.5 mm long, pappus setae 8–9 mm; phyllaries
     8–11 mm; Malawi, Zambia . . . . . . . . . . . . . . . . . . . . . . . . . . . . . . **24.** *latecorymbosus*
 –   Herbs of grassland, heath, miombo, dambos or cultivation; achenes less than 5 mm long,
     pappus setae less than 8 mm (except in *S. nyangani*, 8–9 mm), phyllaries less than 7.5 mm
     long (except in *S. ruwenzoriensis*, 6–9 mm and *S. coronatus*, 8–12 mm) . . . . . . . . . . . . . 85
85.  Phyllaries 8–16; ray florets 3–9. . . . . . . . . . . . . . . . . . . . . . . . . . . . . . . . . . . . . . . 86
 –   Phyllaries 18–23; ray florets 7–15. . . . . . . . . . . . . . . . . . . . . . . . . . . . . . . . . . . . . . 88
86.  Phyllaries 2.8–4 mm long; ray florets 3–4 . . . . . . . . . . . . . . . . . . . . . . .**51.** *triactinus*
 –   Phyllaries 5–9 mm long; ray florets (5)7–9 . . . . . . . . . . . . . . . . . . . . . . . . . . . . .87[15]

---

[14] If *S. swynnertonii* has a woolly root crown [specimens are incomplete], it still has glabrous achenes!

[15] And the rare radiate form of *S. urundensis*, with phyllaries 4.5–7 mm long; ray florets 4.

87. Leaves 1–5.3 cm wide; ray limbs 5–11 mm long; pappus setae 4–7 mm; widespread (occasional sparsely hairy achenes)............................. **41.** *ruwenzoriensis*
  –  Leaves 0.4–0.7 cm wide; ray limbs 12–13.5 mm long; pappus setae 8–9 mm; only known from Inyangani area ...............................**32.** *nyangani*
88. Stem arachnoid-lanate; phyllaries 8–12 mm long, ray limbs 7–18 mm long, achenes 4–5 mm long.......................................**7.** *coronatus*
  –  Stem glabrous or sparsely puberulous; phyllaries less than 7 mm long, ray limbs less than 6 mm long, achenes less than 2.5 mm long ........................ 89
89. Phyllaries 3–4 mm long; ray limbs 0.5–0.8 mm long; pappus setae 2–4.3 mm...... ............................................................ **9** *cryphiactis*
  –  Phyllaries 4–7 mm long; ray limbs 4.5–6 mm long; pappus setae 4.5–7 mm ...... 90
90. Leaves 0.3–3 cm wide; plants from below 150 m altitude ......... **27.** *madagascariensis*
  –  Leaves 0.2–0.6 cm wide; plants from altitudes over 1050 m ............. **40.** *randii*

1. **Senecio aetfatensis** B.Nord. in Novon **9**(2): 245 (1999). —Mapaura & Timberlake, Checkl. Zimb. Vasc. Pl.: 28 (2004). Type: Zimbabwe, Chimanimani National Park, along footpath from entrance office to Mountain Hut, 11.ii.1997, *Nordenstam* 9292 (S-G-11009 holotype, MO, SRGH).

Perennial herb 0.6–1.5 m tall. Stems well-branched, short-villous to glandular-pubescent. Leaves sessile, elliptic to slightly obovate in outline, 1.5–8 × 0.5–3.5 cm, (bi)pinnatisect, base narrowed and pseudopetiolate, half-clasping and auriculate, auricles dissected or dentate, lamina margins 3–7-lobed, lobes 3–5-lobed or coarsely dentate, lobules oblong, apices acute; short-hirsute or glandular-hairy to almost glabrous, discolorous, green above, paler beneath. Inflorescences terminal, in lax small groups in a larger terminal corymbose panicle or solitary and terminal on small leafy branches; peduncles of individual capitula 1–5 cm long, slender. Capitula radiate, heterogamous; involucre campanulate, calyculate, calycular bracts 5–8, 1–2 mm long, narrow-triangular, flat, apices acute; phyllaries 13–21, linear-lanceolate, 4–6 × 0.5–1.5 mm, green with narrow membranous margins, 1- or 2-veined, resiniferous, apices acute to acuminate, tips puberulous or papillate, otherwise glabrous; receptacle flat, glabrous, alveolate. Ray florets (5)8–12, corollas yellow, tube 2–3 mm long, limb elliptic-oblong, 4–8 × 2.2–3.5 mm, spreading or rolled; disc florets 34–60, corollas yellow, tube narrowly infundibuliform, 3.5–4.5 mm long, lobes 0.5–0.7 mm long, minutely papillate towards apex. Achenes elliptic-oblong, terete, 2–2.5 × 0.6 mm, 8-ribbed, setuliferous on and between ribs, setulae mucilaginous when wetted; pappus setae many, white, 3–4.5 mm long, caducous.

**Zimbabwe**. E: Chimanimani Mts, Mt Peza, 1980 m, 15.x.1950, *Wild* 3621 (BR, K, S, SRGH). **Mozambique**. I have not seen herbarium specimens, but the images by B. Wursten on the Flora of Zimbabwe website show this species from a ridge north of Mt Peza in the Chimanimani Mts on the Mozambique side of the border.

Growing in rock crevices; 1400–2550 m; flowering October –February.

Endemic to the Flora area; apparently still only known from four collections.

Conservation Status: While only four collections are known, these both occur in well-protected National Parks, and therefore the assessment will be Least Concern (Conservation Dependent).

*Wild* 3621 differs from the type in having a glandular-pubescent indumentum rather than a short-villous one. Two further specimens, from the Nyangani [Inyangani] summit ridge, *Whellan & Davies* 986 (K, SRGH) and *Wild* 4898 (K, SRGH), are again very similar but differ from the other two specimens in having the capitula solitary on small leafy branches.

A specimen from **Zambia**: Barotseland, Bulozi Plain, 23.xii.1959, *Gilges* 756 (K, SRGH) is close to this taxon, but differs in several characters: the stems are glabrous, the leaves have fewer and much narrower lobes (these less lobed themselves), the phyllaries are longer (7 mm), disc florets are about 30, and the achenes are 5–7 mm long (but also puberulous between the ribs) with pappus setae 6–7 mm long. As the inflorescences look fairly similar I tentatively include this specimen here. There is no label info on habitat or altitude, but the Bulozi Plain (= Barotse floodplain) lies at about 1000 m altitude.

2. **Senecio arenarius** Thunb., Prodr. Pl. Cap. **2**: 158 (1800), non M.Bieb (1816), nec Salzm. ex DC. (1838), nec Lange ex Nyman (1879). —Merxmüller, Prodr. Südwestafr. **139**. Asteraceae: 163 (1967). —Setshogo, Prelim. Checkl. Pl. Botsw.: 38 (2005). Type: Not stated in protologue.

*Senecio elegans* Thunb., Fl. Cap., ed. 2: 685 (1823), nom. illeg., non L.(1753). Type: South Africa, 'crescit in Leuwekopp et in Groenekloof. Floret Julio et sequentibus mensibus.' (UPS holotype).

*Senecio myrrhifolius* Thunb., Fl. Cap., ed. 2: 685 (1823). Type: South Africa, 'Crescit in arenosis Groene kloof et Swartland. Floret Septembri, Octobri.', *Thunberg* s.n. (UPS holotype, LD, S).

*Senecio cakilefolius* DC., Prodr. **6**: 408 (1838). —Merxmuller, Prodr. Südwestafr. **139** Asteraceae: 163 (1967). —Setshogo, Prelim. Checkl. Pl. Botswana: 38 (2005). Type: South Africa, Cape, Zilverfontein [Silverfontain], 2.ix.1830, *Drège* s.n. [2816] (G-DC holotype, HAL, P00119762, P00119763, P00119764).

Annual herb to 40 cm high, multistemmed and much branched with erect or spreading branches; branches thinly glandular-hairy. Leaves oblong and lobed to almost pinnatifid, 15–70 × 4–6 mm, base clasping stem and auriculate, margins sometimes revolute, apex acute, lobes deltate and callus-tipped, glandular-hairy. Capitula with both ray and disc florets (radiate and heterogamous), in corymbiform panicles. Involucre cylindrical to almost conical; calycular bracts few, minute; phyllaries 10–11, linear, 4.8–5.5 × 0.3–0.5 mm, with pale margins and often with darker apex, glandular-hairy. Ray florets ± 13, corollas mauve or magenta (rarely white; our single Mozambique specimen has "heads pale yellow"), tube ± 3 mm long, ray limb 1–2 mm long; disc florets 14–15, corollas yellow, narrowly infundibuliform to almost tubular, 3.2–4 mm long, lobes 0.2 mm long. Achenes 1.5–2 mm long, hairy between ribs; pappus setae 5–6 mm long, white.

**Mozambique**. M: Inhaca Island, Lighthouse, 30.ix.1958, *Mogg* 28427 (K, PRE).

Also known from Namibia and South Africa; no specimens seen from Botswana, though mentioned in Setshogo (2005). Seashore vegetation on sand; near sea level.

Conservation Status: I have seen only a single Mozambican specimen. Due to a wide distribution in a common habitat, this has been assessed as Least Concern in the Red List of South African Plants (2011).

3. **Senecio auriculatissimus** Britton in Trans. Linn. Soc. London, ser. 2, Bot. **4**(1): 21 (1894) —Mapaura & Timberlake, Checkl. Zimb. Vasc. Pl.: 28 (2004). Type: Malawi, Mt Mlanje [Milanji], 1891, *Whyte* 83/s.n. anno 1891 (BM000609372, K001094203 syntypes).[16]

*Senecio whyteanus* Britton in Trans. Linn. Soc. London, ser. 2, Bot. **4**(1): 21 (1894). Type: Malawi, Mt Mlanje [Milanji], x.1891, *Whyte* 109 (K holotype).

Erect herb, 0.6–1.5 m, or climbing to 1–5 m, multi-branched; stem glabrous, purple-maroon. Leaves slightly succulent, two-parted with upper part very broad and with apical notch, 1.5–7 × 2.4–8.5 cm, base with a deep narrow notch, apex shallowly notched with small dentate tip, margins crenate; lower part (divided from upper part by petiole-resembling purple stalk 1–3 cm long) sessile on stem, ovate with deep basal auricles, this part 1–4.8 × 0.9–4 cm; upper part with several main veins from base; blade with purplish tinge on lower surface or only round margins; glabrous or with sparse scattered hairs underneath. Capitula with both ray and disc florets, in corymbiform panicles, sometimes the whole upper part of the plant seeming to form one large corymb. Involucre cylindrical to slightly conical, 4.5–7 mm long; calycular bracts few and minute; phyllaries 14–16, 4.5–7 × 0.6–1.2 mm, acute, glabrous but for a tiny tuft at apex. Ray florets 8–13, corollas yellow, tube 3.5 mm long, limb 6–9 × 1.5–3 mm; disc florets 80–100, corollas yellow, 4–5.3 mm long, of which narrowly infundibuliform tube 3.6–4.8 mm. Achenes cylindrical, 1.7–2.6 mm, shortly hairy; pappus setae white, 4–6 mm long.

**Zimbabwe**. E: Nyanga (Inyanga), between Gleneagles and Aberfoyle, 29.x.1976, *Müller* 2658 (K, SRGH). **Malawi**. S: Mt Zomba, Queen 's View, 9.viii.1967, *Salubeni* 790 (K, MAL,

---

[16] I take Britton's remark 'also from Makua country (*J. T. Last*, 1887) and Shiré Highlands (*Buchanan*, 1881).' to mean that these two specimens are paratypes.

SRGH). **Mozambique**. Z: Namuli, Makua country, *Last* s.n. (K). MS: Mt Gorongosa Gogogo summit area, x.1971, *Tinley* 2200 (K, SRGH).

Not known elsewhere. Forest edges, occasionally in giant heath or thicket; 1500–2700 m.

Conservation Status: This endemic taxon occurs in a fairly common habitat and has a wide altitude range; with more than 25 specimens known, this must be Least Concern.

4. **Senecio barbertonicus** Klatt, in Bull. Herb. Boissier **4**: 840 (1896). —Macnae & Kalk, Nat. Hist. Inhaca Isl., revis. ed.: 155 (1969). —Hilliard, Compositae Natal: 497 (1977). —Mapaura & Timberlake, Checkl. Zimb. Vasc. Pl.: 28 (2004). —Bandeira *et al.*, Wild Flowers South. Mozamb.: 213 (2007). Type: South Africa, Transvaal: Barberton, Highland Creek, 20.viii.1890, *Galpin* 1000 (G holotype, GRA, K, NH, PRE, Z000003861).

*Kleinia barbertonica* (Klatt) Burtt Davy in Bull. Misc. Inform. Kew 1935(10): 569 (1935).

Succulent shrub 0.5–3 m, prostrate or scandent, well-branched; old stems rough with leaf scars, branchlets closely leafy; all parts glabrous. Leaves spreading to porrect, sessile, terete and slightly curved, 5–9 × 0.5 cm, base narrowed, apex acute and mucronate. Capitula discoid, in small corymbose clusters often grouped in divaricate panicles at branch tips; involucre narrowly campanulate; calycular bracts few, small; phyllaries 5–8, 11–15 mm long with 3–5 pale resinous veins. Ray florets absent; disc florets ?10–12, corollas bright yellow-orange, to 8 mm long, corolla lobes with median yellow resinous line. Achenes cylindrical, c. 3.5 mm long, ribbed, hispid between ribs; pappus setae 8.5–14 mm long.

**Zimbabwe**. W: Dist. Insiza, near Dhlo Dhlo Ruins, Fort Rixon, vi.1965, *Walter* 16 (K, SRGH). C: Marondera (Marandellas) Dist., 6.iv.1950, Wild 3325 (SRGH, fide M. Hyde). E: Mutare (Umtali) Dist., 30.vii.1952, *Chase* 4619 (K, SRGH). S: West side of the Lake Kyle after crossing the dam wall, 54.5 km from Masvingo, 9.ix.1988, *Carter & Coates Palgrave* 2225 (K). **Malawi**. C: Dedza Dist., Kapirimutu Hill, c. 22 km from Dedza on Lilongwe-Dedza road, 36XLV2423, 1400 m, 4.viii.1978, *Iwarsson & Ryding* 998 (K). **Mozambique**. M: Maputo (Lourenço Marques), Marracuene, Maxaquene, 27.viii.1958, *Macuácua* 67 (K, LMJ).

Also known from South Africa (Transvaal, Natal) and Eswatini.

Sandy areas amongst rock outcrops, clambering over trees and shrubs in sand forest; flowering from August to September.

Conservation Status: Due to a wide distribution in a common habitat, this should be Least Concern.

5. **Senecio brachypodus** DC., Prodr. **6**: 403 (1838). – -Hilliard, Compositae Natal: 490 (1977). —Da Silva *et al.*, Prelim. Checkl. Vasc. Pl. Mozamb.: 34 (2004). —Setshogo, Prelim. Checkl. Pl. Botsw.: 38 (2005). Types: South Africa, Cape, ceded territories, *Ecklon* s.n. [415] (G-DC00486281 lectotype), lectotypified by Beentje in Kew Bull. **74**-67: 3 (2019); on the Zwarte Omsanwoubo River, *Drège* s.n. [3768 p.p.] (G-DC syntype).

*Senecio natalensis* Sch.Bip., in Flora **27**(40): 700 (1844). Type: South Africa, near Natal Bay [Natalbai], vi.1839, vi.1839, *Krauss* 279 (BM, G, K, P, TUB syntypes).

Heads radiate . . . . . . . . . . . . . . . . . . . . . . . . . . . . . . . . . . . . . . . . . . . . . . . . .a) var. *brachypodus*
Heads discoid . . . . . . . . . . . . . . . . . . . . . . . . . . . . . . . . . . . . . . . . . . . . .b) var. *pleistocephalus*

### a) Var. **brachypodus**

Succulent scandent shrub; stem to 1 cm diam.; glabrous in all parts. Leaves petiolate, petiole to 2 cm long, lamina elliptic, 12 × 5 cm, base tapering, margins ± entire, repand, crenate-serrate or denticulate, sometimes more coarsely toothed. Inflorescences of congested corymbs or corymbose-paniculate. Capitula heterogamous, radiate; involucre narrow-campanulate, calycular bracts few, short; phyllaries 5–8, 3–4 mm long, keeled, veins resinous. Ray florets 2–5, limbs bright yellow. Disc florets c. 30 corollas bright yellow, lobes with weak resinous median line. Achenes c. 3 mm long, cylindrical, ribbed, glabrous or setuliferous between ribs, pappus setae 4–6 mm long.

**Botswana**. No specimens seen, but likely to occur. **Mozambique**. No specimens seen, but likely to occur.

Also known from South Africa (Cape Province, Natal) and Eswatini. Sea level to c. 1100 m; forest margins on the coast, gallery forest more inland.

Conservation Status: Due to a wide distribution in a common habitat, this has been assessed as Least Concern in the Red List of South African Plants (2020).

b) Var. **pleistocephalus** (S.Moore) Beentje in Kew Bull. **74**-67: 6 (2019). Type: Eswatini [Swaziland], near Miles, i.1905, *Burtt-Davy* 2906 (BM holotype, PRE).

> *Senecio pleistocephalus* S.Moore in J. Bot. **43**: 170 (1905). —Merxmüller, Prodr. Südwestafr. **139** Asteraceae: 167 (1967). —Hilliard, Compositae Natal: 491 (1977). —Da Silva *et al.*, Prelim. Checkl. Vasc. Pl. Mozamb.: 34 (2004). —Setshogo, Prelim. Checkl. Pl. Botsw.: 38 (2005). —Bandeira *et al.*, Wild Flowers South. Mozamb.: 214 (2007).
>
> *Senecio callimocephalus* S.Moore in J. Bot. **56**: 231 (1918). —Turton & Blomberg-Ermatinger, Some Flower. Pl. Southeast, Botsw.: 115 (1988). —Setshogo, Prelim. Checkl. Pl. Botsw.: 38 (2005). Type: Zimbabwe, 'Victoria', ix.1909, *Monro* 1956 (BM 924653 holotype).

Shrubby scrambler with long branches, 2–5 m high; stems fleshy, more than 5 mm across, whitish green, glabrous. Leaves semi-succulent, elliptic to obovate, 1.5–7 × 0.7–3.7 cm, base attenuate, margin almost entire to slightly lobed with several up to 7 mm long narrow lobes near leaf base, apex acute to obtuse, glabrous. Inflorescences of dense corymbs or corymbose-paniculate, whole to 18 × 16 cm. Capitula homogamous, dark yellow with strong unpleasant smell; calycular bracts few, to 1.5 mm long; phyllaries 5–7 (in ours), 3–4 mm long, with thin membranous margins, ?glabrous; disc florets 8–13, corollas presumably yellow, narrowly infundibuliform, 4.4–6 mm long. Achenes 1.2–1.5 mm long, glabrous; pappus setae 4–6 mm long.

**Botswana**. No specimens seen, but likely to occur. Setshogo (2005) lists *S. callimocephalus*, but does not cite specimens. **Zimbabwe**. S: Masvingo (Fort Victoria), '1909–1912', *Monro* 1956 (BM). **Mozambique**. M: Maputo, Baía de Maputo (Delagoa Bay), *Monteiro* 25 (K).

Also known from South Africa (Natal, Transvaal) and Eswatini. Sea level to c. 1100 m; thicket margins, woodland and disturbed areas, sometimes covering shrubs and small trees.

Conservation Status: Due to a wide distribution in a common habitat, this should be Least Concern.

6. **Senecio consanguineus** DC., Prodr. **6**: 381 (1838). —Merxmüller, Prodr. Südwestafr. **139** Asteraceae: 165 (1967). —Mapaura & Timberlake, Checkl. Zimb. Vasc. Pl.: 28 (2004). —Setshogo, Prelim. Checkl. Pl. Botsw.: 38 (2005). Type: South Africa, Gariep [Garip] river bank, 16.ix.1830, *Drège* s.n. [5894] (G-DC holotype, P).

Annual (?) herb 20–45 cm high; stem branched from near base, with fairly dense stalked glandular hairs and multicellular simple hairs. Leaves elliptic to obovate in outline, pinnatisect, 2–4 × 0.3–1.6 cm, base attenuate (in lower leaves) to auriculate (in median and upper leaves), with 5–12 linear lobes on each side, margins callose-dentate, occasionally lobes themselves pinnatisect, apex callose-acute; sticky, with fairly dense stalked glandular hairs and multicellular simple hairs. Capitula few together in lax corymbose panicles ending a branch, sometimes whole plant looking like a single large leafy inflorescence, slightly aromatic; peduncles of individual heads slender, with indument like stem, 1–3 cm long; involucre cylindrical; calycular bracts few, linear, 1–1.5 mm long; phyllaries 13–14, green with darker tips, 5.5–7.2 mm long, fairly densely covered in white multicellular hairs. Ray florets 8–11, corollas yellow, tube 3.3–4 mm long, lamina 3.5–4 × 1.7–2 mm; disc florets 25–30, corollas yellow, narrowly infundibuliform, 4.8–5.5 mm long, of which lobes 0.5–0.7 mm. Achenes dark, 3–3.2 mm long, ribbed, densely white-pilose; pappus setae white, 5.5–6 mm long.

**Botswana**. SW: 4 km N of Dondong borehole, 29.iv.1976, *Skarpe* 66 (GAB, K, SRGH); SE: between Sekhuma Pan and Ghansi, ix.1967, *Lambrecht* 369 (K, SRGH). **Zimbabwe** – I have not seen any specimens, though this taxon is listed in the Zimbabwe checklist.

Also widespread in Namibia and South Africa. Wooded grassland or heavily grazed scrub on Kalahari sand; 1150–1200 m

Conservation Status: Due to a wide distribution in a common habitat, this should be Least Concern.

7. **Senecio coronatus** (Thunb.) Harv. in Harvey & Sonder, Fl. Cap. **3**: 369 (1865). — Hilliard, Compositae Natal: 462 (1977). —Jeffrey in Kew Bull. **41**(4): 896 (1986). — Lisowski, (Asterac. Fl. Afr. Cent. 2) Fragm. Flor. Geobot. **36** Suppl. 1: 268 (1991). —Mapaura & Timberlake, Checkl. Zimb. Vasc. Pl.: 28 (2004). —Jeffrey & Beentje in F.T.E.A., Compositae **3**: 647 (2005). —Phiri, Checkl. Zamb. Vasc. Pl.: 33 (2005). — Setshogo, Prelim. Checkl. Pl. Botsw.: 38 (2005). —Bandeira *et al.*, Wild Flowers South. Mozamb.: 215 (2007). Type: none indicated in the protologue. South Africa, near Gamtoos [Camtours] rivier, *Thunberg* s.n. (UPS lectotype), lectotypified by Hilliard (1977: 462).

*Cineraria coronata* Thunb., Prodr. Pl. Cap. **2**: 154 (1823).

*Senecio lasiorhizus* DC., Prodr. **6**: 387 (1838). —Oliver & Hiern, F.T.A. **3**: 415 (1877). Types: South Africa, 'Cape, Albany District', 22.x.1829, *Drège* s.n. [2092 'Albany'] (G-DC, P syntypes); *Ecklon* s.n. [2092] (G-DC, ?HAL – numbered 549 syntypes); Uitenhagen, *Ecklon* s.n. [1346] (G-DC syntype); Stormberg, Fishrivier et Zeeurebergen, *Drège* s.n. [2092 'Stormberg'] (G-DC syntype); ibid., *Drège* s.n. [2092 'Zw. Vischrivier & Zuurebergen'] (G-DC syntype).

*Senecio lasiorhizus* [var.] β *lasiorhizoides* Sch.Bip. in Flora **27**(40): 696 (1844). Type: South Africa, 'Natal, Tafelberge', viii.1839, *Krauss* 438 (P00119865 holotype, BM924784, G00018445, G00018446, K000377859, M0105174, TCD0003021, TUB005681).

*Senecio lasiorhizoides* (Sch.Bip.) Harv. in Harvey & Sonder, Fl. Cap. **3**: 370 (1865).

*Senecio coronatus* var. β *minor* Harv. in Harvey & Sonder, Fl. Cap. **3**: 370 (1865). Types: South Africa, 'Cafferland', 1860, *Cooper* 150 (TCD syntype); Camperdowm (*sic*) and Bulu, *Gerrard & McKen* 1053 (TCD syntype).

*Othonna plantaginea* Hiern, Cat. Afr. Pl. **1**(3): 604 (1898). Type: Angola, Huila, above Mumpulla, x.1859, *Welwitsch* 3684 (BM000924624 holotype, K000311514, LISU218544).

*Senecio lachnorhizus* O.Hoffm., in Bot. Jahrb. Syst. **32**(1): 150 (1903). Type: Angola, Huilla, *Antunes* 148 (B† holotype).

Perennial herb 20–60 cm high. Rootstock of several long (>10 cm) fleshy tubers (usually missing from herbarium material), crown of rootstock densely long-woolly and covered with remnants of old (usually burnt) leaf-bases; stems 1–2, erect, 20–60 cm high, arachnoid-lanate. Basal leaves (missing in much herbarium material) to 29 cm long, extremely long-petiolate, petiole 2–16 cm long, wingless at base but gradually winged from about half way to attenuate leaf base, lamina 4–16 × 2–7 cm, margins crenate-dentate to subentire, apices acute to subobtuse with acuminate tip, or obtuse; lower stem leaves sessile, oblanceolate, 5–16 × 0.4–3 cm, gradually tapered into a basally slightly expanded and sheathing petioloid base, margins subentire or indistinctly crenate-dentate distally, apex shortly acuminate, green, initially densely villous but glabrescent and eventually almost glabrous above, green and initially densely villous beneath and becoming sparsely so with age, midrib prominent beneath; middle stem leaves smaller and becoming progressively smaller upward, sessile, oblong-lanceolate, midrib prominent beneath, base semi-amplexicaul, apex shortly acuminate. Inflorescence of lax, few-headed corymbs, rarely of solitary terminal capitulum, peduncle few-bracted, lower bracts reduced and leaf-like, upper bracts subtending pedicels narrow-lanceolate, to 20 mm long, peduncles of capitula woolly, ebracteolate or few-bracteolate these tending to calycular bracts beneath involucre, scale-like and narrow-lanceolate. Capitula heterogamous, radiate, ± erect, often with a strong honey smell; involucre broadly hemispherical, 9–11 × 10–15 mm, base woolly; calycular bracts c. 10, narrowly lanceolate, 6–7 mm long; phyllaries 18–23, lanceolate, 8–12 × 0.8–1.3 mm, laxly villous, shortly bearded at apex. Ray florets (8)10–12(15), ray limb yellow or lemon-yellow (apparently fading), 7–18 × 3–3.5 mm, 10-veined, corolla tube 5–6 mm long, strongly pubescent. Disc florets (20)44–67, corollas deep yellow, 6–9 mm long, tube slightly expanded above middle, pubescent at and below middle, lobes 1–1.5 mm long, triangular. Achenes 4–5 mm long, shortly hairy between ribs; pappus setae 5–8 mm long, barbellate, white.

**Botswana**. N: near Francistown, viii.1911, *Rogers* 6036 (K). ?SE: 'Makalaka', 6.ix.1871, *Baines* s.n. (K). **Zambia**. N: Abercorn Dist., old Kasama Road, Mbala (Abercorn), 1500 m, 21.vii.1960, *Richards* 12882 (K). W: Chingola, 9.xii.1972, *Fanshawe* 11697 (K, NDO). **Zimbabwe**. N: Tatagura Valley 3 miles W Mazowe, 9.xi.1962, *Angus* 3400 (FHO, K). W: Matobo Dist., Quaringa Farm, 1463 m, xii.1957, *Miller* 4848 (K, SRGH). C: Makoni Dist., Diana Farm, 1650 m, 15.x.1961, *Chase* 7546 (K, SRGH). E: Nyanga (Inyanga), Mt. Inyangani, 25.x.1946, *Wild* 1503 (K, SRGH). **Malawi**. S: Mt. Mlanje, Chambe-Tuchila path, c. 2286 m, 23.i.1967, *Hilliard & Burtt* 4604 (K). **Mozambique**. fide Bandeira *et al.* (2007) but no specimens seen.

Also known from D.R. Congo, Tanzania, Angola, Eswatini, South Africa. Grassland (coming up where recently burnt), open miombo, dambos and seasonally boggy sites; may be locally common; 1200–1900(2250) m.

Conservation Status: Due to a wide distribution in a common habitat, this should be Least Concern.

8. **Senecio crassorhizus** De Wild. in Bull. Jard. Bot. État Bruxelles **5**: 54 (1915). —Lisowski, (Asterac. Fl. Afr. Cent. 2) Fragm. Flor. Geobot. **36** Suppl. 1: 299 (1991). Types: D.R. Congo, Biano Plateau, near Katentania, xi.1912, *Homblé* 759 (BR syntype); ibid., *Homblé* 845 (BR syntype); Biano Plateau, xii.1912, *Homblé* 899 (BR syntype).

Perennial erect herb 40–80 cm, unbranched but for inflorescence; roots of several tubers to 3 × 1.4 cm; stem sometimes magenta at base, becoming green higher up and 'winged on four corners' (fide Merrett, visible in photo!, no longer visible in dried material); glabrous. Leaves all cauline, porrect, fleshy to succulent, glaucous green, lanceolate to narrowly obovate, 1.5–9 × 0.5–1 cm, base sessile and cuneate, margin entire, apex acute, with three light yellow-green parallel veins; glabrous. Capitula discoid, 7–12 in long-stalked terminal panicles; calycular bracts ± 6, linear, to 10 mm; involucre campanulate, phyllaries 10–14, 7.2–10 × 0.6–1.1 mm, with glandular tips. Florets 15–20, corollas yellow-cream or greenish-white, 15–18, 5.7–7 mm long, of which lobes 1.5 mm; anthers brown; style branches greenish white. Achenes 1.6–2 mm long, puberulous; pappus setae white, 6–6.3 mm long.

**Zambia**. N: SW of Dobeka Bridge, 19.xi.1937, *Milne-Redhead* 3315 (K); Mutinondo, Mayense camp, 14.xii.2008, *Merrett* 275 (K); ibid., 12.xii.2012, *Merrett* 1064 (K).

Also known from SE D.R. Congo (Biano & Manika plateaus). *Cryptosepalum* woodland on sand, miombo on sandy loam, or on granite rock, or border zone rock/woodland; 1350–1500 m.

Conservation Status: Due to a limited distribution and not many specimens known, with a limited altitude range in a habitat which is under threat in part of its range; this taxon might well be Vulnerable.

Our material agrees quite well with the Congolese material, though our specimens have slightly shorter phyllaries and shorter achenes: 1.6–2 mm in ours, 4 mm according to Lisowski. The Congolese sites are not too far from ours, and occur in the same habitat type at the same altitude.

9. **Senecio cryphiactis** O.Hoffm. in Warburg Kunene-Sambesi-Exped.: 423 (1903). —Merxmüller, Prodr. Südwestafr **139** Asteraceae: 166 (1967). —Gibbs Russell in Kirkia **10**(2): 502 (1977). —Mapaura & Timberlake, Checkl. Zimb. Vasc. Pl.: 28 (2004). —Setshogo, Prelim. Checkl. Pl. Botsw.: 38 (2005). Type: Angola, between Ediva and Humbe on the Caculovar, 6.ix.1899, *Baum* 68 (?B holotype, E, W, Z000003868).

Annual herb 20–60 cm high, much branched, with erect or decumbent branches, branches glabrous or nearly so; stem sometimes reddish at base. Leaves oblong, pinnatisect with narrow oblong dentate lobes or entire with dentate margins, 4–15 × 0.3–4 cm, base clasping and auriculate, margins entire (except for lobes) to callose-dentate, apex obtuse, with some arachnoid hairs but glabrescent (and looks like might be glaucous on lower surface). Capitula radiate, in small paniculate to corymbose groups ending branches; peduncles of individual heads with scattered pale hairs; involucre ± squat-cylindrical, 3–4 × 4 mm; calycular bracts 6–12, 1–1.5 mm long, glandular-ciliate; phyllaries 18–22, 3–4 × 0.3–0.5 mm, with pale margins, acute, glabrous but for a tuft of hairs near apex. Ray florets minute, hardly overtopping phyllaries, possibly 10–16, ray 0.5–0.8 × 0.3 mm. Disc florets 55–80, corollas yellow, 1.4–3.3 mm long, very narrowly infundibuliform, of which lobes 0.2–0.4 mm. Achenes 1.5–2.5 mm long, ribbed, puberulous; pappus setae white, 2–4.3 mm long.

**Botswana**. fide Setshogo (2005) but no specimens seen. **Zambia**. B: Machili, 21.ix.1969, *Mutimushi* 3689 (K, NDO). C: Mwomboshi River 12 km N of Chipembi, 26.viii.1972, *Kornas* 2013 (K). S: Mapanza N, 20.ix.1953, *Robinson* 36. **Zimbabwe**. N: Chimanda Reserve, 10.ix.1958, *Phipps* 1335 (K, SRGH). W: Bulawayo, x.1957, *Miller* 4589 (K, SRGH). C: Chinamora Reserve, Domboshawa, 2.xi.1960, *Rutherford-Smith* 342 (K, SRGH). E: Mutare [Umtali], Vumba Mts, 1.ix.1925, *Peter* 30587 (B, K). S: Dopsoa Dam near Devuli HQ, 22.vii.1958, *Chase* 6951 (K, SRGH). **Mozambique**. T: Zambesi River, Msusa, 25.vii.1950, *Chase* 2803 (K, SRGH).

Also known from Angola and Namibia. Damp sand or mud on river banks or dam edges; 200–1400 m.

Conservation Status: Due to a wide distribution in a common habitat, this should be Least Concern.


10. **Senecio deltoideus** Less., Syn. Gen. Compositae: 392 (1832). —Oliver & Hiern, F.T.A. 3: 420 (1877). —Hilliard, Compositae Natal: 492 (1977). —Jeffrey in Kew Bull. **41**(4): 886 (1986). —Da Silva *et al.*, Prelim. Checkl. Vasc. Pl. Mozamb.: 34 (2004). —Mapaura & Timberlake, Checkl. Zimb. Vasc. Pl.: 28 (2004). —Jeffrey & Beentje in F.T.E.A., Compositae 3: 625 (2005). Type: South Africa, Cape of Good Hope, *Sonnerat* s.n. (P holotype).[17],[18]

    *Eupatorium auriculatum* Lam., Encycl. 2: 411 (1788), non *Senecio auriculatus* Burm.f. (1768). Type as for *S. deltoideus*.

    *Cacalia scandens* Ait., Hort. Kew. 3: 157 (1789), non *Senecio scandens* Buch.-Ham. ex D.Don (1825). Type: South Africa, Cape of Good Hope, *Masson* s.n. [introd. 1774] (BM or ?K holotype).

    *Mikania auriculata* (Lam.) Willd., Sp. Pl., ed. 4, **3**: 1745 [1803](1804).

    *Cacalia fimbrillifera* Cass. in Dict. Sci. Nat., ed. 2, **48**: 460 (1827), nom. illeg. (based on *Eupatorium auriculatum* Lam.).

    *Senecio mikaniae* DC., Prodr. **6**: 405 (1838). Type: South Africa, Zwellendam Dist., 25.vii.1831, *Drège* s.n. [5622] (G-DC holotype, P00122010, P00122011).

    *Senecio mikaniiformis* DC., Prodr. **6**: 405 (1838). Type: South Africa, Basche, *Drège* s.n. (G-DC holotype, HAL, HBG, LECB, P00122004, P00122005, P00122006, P00122007).

    *Senecio mikaniiformis* [var.] β *octosquamosus* Sch.Bip. in Flora 27(40): 700 (1844). Type: South Africa, Natal Bay [Natalbai], vi.1839, *Krauss* 318 (P00122019, G, M syntypes).

    *Senecio fimbrillifer* (Cass.) B.L.Rob. in Proc. Amer. Acad. Arts **47**: 215 (1911), nom. et comb. illeg. (based on *Cacalia fimbrillifera* Cass.).

    *Senecio sarmentosus* O.Hoffm. in Bot. Jahrb. Syst. **20**: 236 (1894), non (DC.) Sch.Bip. (1867). —Brenan in T.T.C.L.: 159 (1949). Types: Tanzania, Usambara, Mlalo, v.1892, *Holst* 621 (B† syntype); ibid., v.1892, *Holst* 439 (B† syntype); Lutindi, *Holst* 3256ᵃ (B†, HBG, M, P00229986, P00119987, Z000003977 syntypes); Kwa Mshusa, Handeï, viii.1893, *Holst* 8914ᵇ (B†, K syntypes).

    *Senecio durbanensis* Gand. in Bull. Soc. Bot. France **65**: 46 (1918). Types: South Africa, near Durban, *Schlechter* 2884 (LY syntype); *Wood* 6430 (LY syntype).

    *Senecio tanzaniensis* Cufod., (Enum. Pl. Aethiop.) Bull. Jard. Bot. Natl. Belg. **37**(3, suppl.): 1159 (1967), nom. nov pro *S. sarmentosus*.

---

[17] At P I have seen a sheet marked *Senecio deltoideus* Less. DC.! Cap., pre-1837, which is possibly this type; there are no other *Sonnerat* specimens of this taxon that I could find.

[18] Although accepted by Charles Jeffrey this name when proposed by Lessing included the synonyms *Cacalia scandens* Thunb., *Cacalia fimbrillifera* Cass., *Eupatorium auriculatum* Lam. and *Eupatorium scandens* Link, but with the comment '(syn. excl. de quo cf. supra.)'.

Scandent perennial herb or liana; stems 2–7 m long, green, ribbed, finely pubescent, often glabrescent. Leaves ovate-triangular or triangular, 4.3–12 (including petiole) × 2–6.5 cm, base truncate to cordate, margins coarsely apiculate-dentate or bidentate to slightly lobed with shallow lobes, apex obtuse to subacute or acuminate, sometimes with 2 (or very rarely 4) much smaller lateral lobes, finely pubescent, mostly on veins, especially beneath, to ± glabrous, narrowly petiolate with petiole auriculate at base. Capitula discoid, many in rather divaricately-branched terminal and axillary corymbs forming large, lax, pendent thyrsoid panicles; peduncles of individual capitula finely pubescent, glabrescent; involucre cylindrical, 4–5.5 × 2.5–3.5 mm; calycular bracts 3–5, lanceolate, 1.5–2 mm long, floccose or ciliate on margins or glabrous; phyllaries 8–13, green, 3.5–5 × 0.3–0.8 mm, sparsely finely pubescent or glabrous. Florets 10−20, corollas pale creamy yellow to yellow, corolla 3–5 mm long, tube glabrous, expanded above middle, lobes 1–1.2 mm long. Achenes 2.5–3 mm long, ribbed, ciliate; pappus setae white, 3–5.5 mm long.

**Zimbabwe**. N: Mazowe (Mazoe), on banks of river Mazoe, vii.1906, *Eyles* 379 (SRGH, fide M. Hyde).?W: 'Victoria', viii.1927, *Daphne King in Eyles* 5042 (K, NBI). C: Enterprise, 28.vii.1946, *Wild* 1181 (SRGH, fide M. Hyde). E: Umtali, 20.vii.1969, *Cannell* 77 (K, SRGH). S: Zimbabwe ruins, 1.vii.1930, *Hutchinson & Gillett* 3318 (K, PRE). **Malawi**. S: Mpatamanga [Muata Manga] Gorge, 5.ix.1859, *Kirk* s.n. (K). **Mozambique**. MS: Mt Maruma, 13.ix.1907, *Swynnerton* 1873 (K). M: Zitundo, Gala, 19.vii.1967, *Gomes e Sousa & Balsinhas* 4910 (K, LISC).

Also known from Kenya, Tanzania, South Africa (SW Cape, Cape Province, Natal, Transvaal), and Eswatini. Moist or dry forest or forest margins and occasionally in miombo, often on sand; 1050–1600 m.

Conservation Status: Due to a wide distribution in a common habitat, this should be Least Concern.

No ray florets have been seen for this taxon in our area, as is the case in East Africa, but in South Africa some specimens have 1(4) ray florets. Hilliard, Compositae Natal: 492 (1977): "not easily confused with any other species, the zigzag branches and deltate, closely toothed leaves giving a characteristic facies".


11. **Senecio diphyllus** De Wild. & Muschl. in Bull. Soc. Roy. Bot. Belgique **49**: 227 (1913), as '*diphyllus*'.[19] —Phiri, Checkl. Zamb. Vasc. Pl.: 33 (2005). Type: D.R. Congo, Lubumbashi [Élisabethville], xi.1911, *Hock* s.n. (BR8878960 holotype).

Herb 40–70 cm when flowering; rhizome tuberous, a series of disc shaped tubers 1.5–4 cm long and up to 2 cm across (said to be produced in succession laterally); stem solitary, erect, branched in upper part, pubescent; leaves only at base, or 1–2 reduced leaves (to 14 × 4 mm) higher up on what are really flowering scapes [*Milne-Redhead* 3690 has up to 3–4 linear leaves 10–47 × 1 mm per flowering stem]. Non-flowering plants consist of a single fleshy leaf, flat on ground. Leaves often missing in herbarium material; sub-basal leaf one or rarely two, fleshy, pale green with purple margin, tinged purple below, obovate to broadly elliptic-orbicular, 11–21 × 4.5–10.5 cm of which apparent petiole (also purplish) is about half, margin sinuate-dentate; pubescent on both surfaces; a second much smaller leaf often present at about 5 cm above basal one, narrowly obovate, 4.4–7.5 × 1.4–3.2 cm, strongly dentate or feebly lobed. Capitula in lax branched terminal corymbose panicles to 40–50 cm; peduncles of individual capitula 2–10 mm long; capitula homogamous; calycular bracts 2–4, lanceolate, 2–3 mm long; phyllaries 7–8, yellow-green, linear, 4.5–6 × 0.3–0.8 mm. Florets 9−20, all bisexual, tubular; corollas bright or pale dull yellow, once described as cream and once as orange-yellow, 4–4.8 mm long, enlarged at base; lobes ovate, 0.7–1 mm long, papillose; style with short branches. Achenes oblong to slightly obovoid, 1–2.5 mm long, pubescent; pappus of white bristles 4–4.5 mm long.

**Zambia**. N: Mbala, Kasulu Farm, 4.i.1969, *Sanane* 385 (K). W: Solwezi, 6.i.1969, *Mutimushi* 2987 (K, NDO). E: Nyika Plateau, 2.i.1959, *Robinson* 2995 (K, SRGH). **Malawi**. N: Mzimba Dist., Champira Forest, 26.xii.1975, *Pawek* 10540 (K, MAL, MO, SRGH, UC).

Also known from D.R. Congo. Grassland, where it may be frequent, or miombo/woodland; sometimes in rock crevices; 1050–2250 m.

---

[19] Muschler 's determination slip in BR gives yet a further spelling variation as '*Senecio dipyllus* De Wild. & Muschler sp. nov. '.

Conservation Status: While the distribution area is rather limited, the habitat is a common one and the altitude range is considerable; probably Least Concern.

*Richards* 18801 from Lumi River marsh (Zambia N) keys neatly here, except that it has two cauline leaves some distance above the basal ones; it has about 16 disc florets per head, with glabrous achenes 3.2–4 mm long. It has a rhizome that looks very like *S. diphyllus* proper, though I place it here as 'aff. *diphyllus*'.

12. **Senecio eenii** (S.Moore) Merxm. in Mitt. Bot. Staatssamml. München **3**: 608 (1960). — Merxmüller, Prodr. Südwestafr. **139** Asteraceae: 164 (1967). —Setshogo, Prelim. Checkl. Pl. Botsw.: 38 (2005). Type: Namibia, Damaraland, 1879, *Een* s.n. (BM924646 holotype).

    *Cineraria eenii* S.Moore in J. Bot. **37**: 402 (1899).

    *Senecio rautaneni* S.Moore in Bull. Herb. Boiss., ser. 2, **4**(10): 1022 (1904). Type: Namibia [Deutsch-Südwest-Afrika]: 'Hereroland, im Bette des Tsoachaublusses, [Zwackopje]', *Rautanen* s.n. (BM924621 holotype).

    *Senecio lentior* S.Moore, Bull. Herb. Boiss., ser. 2, **4**(10): 1022 (1904). Type: Namibia [Deutsch-Südwest-Afrika]: 'Hereroland, Salem', *Dinter* 120 (BM924630 holotype).

Herb to 15 cm high; stem sparsely branched, glandular-pilose and sticky. Leaves sessile, membranaceous, dark green turning purple when old, oblanceolate to slightly obovate, 1–12 × 0.5–2.7 cm, base attenuate and auriculate, margin irregularly incised-lobed (rarely pinnatifid), apex acute or obtuse, minutely glandular but otherwise glabrous. Capitula few to many; inflorescence axes with few linear bracts and sticky-glandular-hairy; phyllaries 12–14, 5–5.5 × 0.6–0.8 mm, sparsely and minutely glandular. Ray florets 8, 6–10 mm long, corollas mauve or pale violet (though protologue says 'probably yellow'); disc florets 22–32, corollas yellow. Achenes 1.2–2 mm, more or less glabrous in ours (pubescent elsewhere?); pappus setae white, 4.5–5 mm.

**Botswana**. SW: 6 km N Okwa Valley along Mamuno road, 8.vii.1978, *Skarpe* 291 (GAB, K). Also known from Namibia and South Africa. On loose Kalahari sand with *Terminalia sericea*; c. 1250 m.

Conservation Status: Due to a wide distribution in a common habitat, this should be Least Concern.

This is the only species in the Flora Zambesiaca area where ray florets and disc florets have contrasting colours.

13. **Senecio erubescens** Ait., Hort. Kew **3**: 190 (1789). —Hilliard, Comp. Natal: 421 (1977). —Jeffrey in Kew Bull. **41**(4): 903 (1986). —Lisowski, (Asterac. Fl. Afr. Cent. 2) Fragm. Flor. Geobot. **36** Suppl. 1: 318 (1991). —Da Silva *et al.*, Prelim. Checkl. Vasc. Pl. Mozamb.: 34 (2004) —Mapaura & Timberlake, Checkl. Zimb. Vasc. Pl.: 28 (2004). —Jeffrey & Beentje in F.T.E.A., Compositae **3**: 667 (2005). —Burrows & Willis, Pl. Nyika Plateau: 105 (2005). —Phiri, Checkl. Zamb. Vasc. Pl.: 33 (2005). —Gereau *et al.*, Lake Nyasa Climat. Reg. Florist. Checkl.: 31 (2012). Type: South Africa, Cape of Good Hope, introduced to cult. Kew in 1774, *Masson* s.n. (BM holotype).

    *Senecio ianthinus* Mattf., Bot. Jahrb. Syst. **59** (133): 39 (1924). Type: Tanzania, Kyimbila, Ukinga-Mwekalila, 8.i.1914, *Stolz* 2418 (B† holotype, K).

Erect or decumbent perennial herb 30–60 cm high, rosette-forming, subscapigerous, from a stout woody rootstock to 1.5 cm across; with 1-several stems from crown, these simple below, forking into inflorescence, with twisted glandular hairs. Whole plant often aromatic with a sweet smell, persisting in dried specimens. Leaves mostly crowded near base, more distant upwards and reducing into bracts; basal leaves sessile, narrowly obovate or oblanceolate-elliptic in outline, 4–30 × 0.9–3.5 cm (to 7 cm wide elsewhere), attenuate into a long petioloid base, margins often reddish, sinuate-serrate to sinuate-dentate, bidentate-lobate or pinnately lobed with rounded, apiculate-denticulate lobes, often lyrate-pinnatifid towards base, apex rounded to obtuse and apiculate, with twisted glandular hairs above and on margins and especially or at least on midrib beneath; stem leaves few, sessile, oblong or oblanceolate-oblong, 4–8.5 × 0.6–1.5 cm (rarely to 13 × 5 cm in a few Malawi specimens), semi-amplexicaul, sinuate-denticulate. Capitula discoid, 3–many in a rather lax terminal corymbose panicles; inflorescence axes with twisted glandular hairs; involucre cylindrical, 8–13 × 4–6 mm, glandular; calycular bracts 2–4, narrow, lanceolate, 1.5–4 mm long, glandular; phyllaries 12–21, dark green with purplish tips, (5)7.5–11 × 0.5-1 mm long. Florets (28)40–50, corollas purple, mauve or rarely white (once described as yellow), narrowly infundibuliform, (4.2)5.5–10.2 mm long of which lobes 0.6–1.2 mm long, tips of lobes glandular. Achenes cylindric, 3–3.7 mm long, ribbed, shortly hairy in grooves; pappus setae 6–10 mm long.

Phyllaries with broad-based white hairs among glandular ones . . . . . . . . . . . . a) var. *erubescens*
Phyllaries without broad-based white hairs; only glandular hairs present . . . . . . . . . .b) var. *A*

## a) Var. **erubescens**

**Zambia**. N: Nkale dambo, Kwaimbe, 18.i.1971, *Sanane* 1489 (K, UZL). **Zimbabwe**. C: Harare, Makabusi river, 1.x.1945, *Wild* 140 (K, SRGH). E: Inyangani summit ridge, 25.i.1979, *Müller* 3571 (K, SRGH). S: Belingwe, Mt Buhwa summit, 29.iv.1973, *Pope* 997 (K, SRGH). **Malawi**. N: Katete W of Mzuzu, 7.xi.1970, *Pawek* 3937 (K, MAL). S: Mulanje, Lichenya, 14.viii.2005, *Brummitt et al.* 299 (FRIM, K). **Mozambique**. Z: Namuli Mountain, 16.xi.2007, *Timberlake et al.* 5188 (K, LMA).

Also known from Angola, D.R. Congo, Tanzania, South Africa, Eswatini. Grassland, often in rocky or boggy sites such as dambos; 1100–2550 m.

Conservation Status: Due to a wide distribution in a common habitat, this should be Least Concern.

A specimen from Zimbabwe, E: Nyanga (Inyanga), 6 m N of Troutbeck, 20.xi.1956, *Robinson* 1873 (K, NRGH, SRGH) is similar but smaller in several respects, the most important being the phyllaries 5 mm long and the fewer (28) disc florets that are 4.2 mm long. The corolla colour was described as pale pinkish brown. I am treating this specimen here as a slightly aberrant *S. erubescens*.

## b) Var. **A**

**Zambia**. W: Solwezi, 12.x.1953, *Fanshawe* 379 (K, NDO). **Zimbabwe**. C: Harare, Ruwa river, 20.xii.1947, *Wild* 2263 (K, SRGH). S: Makohali Experimental Station, 17.xi.1976, *Senderayi* 60 (K, SRGH). **Malawi**. N: Nyika Platerau, Chelinda, 4.i.1977, *Pawek* 12258 (K, MAL, MO, SRGH, UC).

Swampy grassland; 1350–2400 m.

Conservation Status: While one location is protected in Nyika National Park, other sites are probably under threat from agricultural conversion of its habitat; the taxon is most likely Vulnerable.

Hilliard (1977: 421) says the name *S. erubescens* has become a dumping ground for many grassland senecios with discoid purple heads in SE Africa. This dumping ground includes (apart from several purely South African entities) what should be named *S. variabilis*, *S. ngoyanus*, and *S. polyodon*; Hilliard calls these names "rallying points in the jungle… but by no means all specimens can be placed." In this troublesome group *S. ngoyanus* is distinguished by the presence of rays; I take *S. polyodon* to have yellow corollas; which leaves the intractable *S. erubescens* and *S. variabilis*. Hilliard distinguishes these by the presence/absence of broad-based white hairs giving a scaly look to phyllaries and inflorescence axes; she felt *S. variabilis* had such hairs, and *S. erubescens* did not; also, the specimens with such hairs have a sweetish odour, often persisting in dry specimens. Unfortunately for this theory the type of *S. erubescens* "does" have such hairs (as does the type of *S. variabilis*), and that name has priority over *S. variabilis*. This whole group needs working out, and as all the types are South African, this will need to include another six or eight taxa.

For practical reasons the current flora treatment needs names for almost fifty specimens that previously were named *S. variabilis* and *S. erubescens*. The first group, with purple corollas and broad white-based hairs, I will call *S. erubescens*, as they resemble the type of that name; the second group, with purple corollas but without the white broad-based hairs, needs a working name – but none of the Candolle varieties of *S. erubescens* work, as the types all seem to have the broad-based white hairs! I hesitate to bring yet another name into this already complicated group, so I will call these specimens '*S. erubescens* var. *A*', while the main group will remain as '*S. erubescens* var. *erubescens*'. The solution is unsatisfactory, but the most practical I can think of. The species is quite variable, with the density of hairs and glands in a range of almost none to dense and long; similarly, leaf margins range from serrate to pinnately lobed. Variation is continuous, with no obvious taxa distinguishable.

I do not use, for our flora area, the name *Senecio variabilis* Sch.Bip. in Flora **27**(40): 700 (1844). —Hilliard, Compositae Natal: 425 (1977). Type: South Africa, Natal, Durban Bay, *Krauss* 431 (BM797691, FR30480, G18405, K377788, K377789, P122212, TUB5653, TUB5654 syntypes).

The record for Botswana was based on a *Burchell* specimen from the source of the Kuruman River – which is in South Africa.

*Bidgood, Merrett & Vollesen* 9164 from Zambia, Mutinondo looks very much like this taxon – but has yellow corollas.

14. **Senecio fanshawei** Beentje in Kew Bull. **74**-67: 1 (2019). Type: Zambia, Mufulira, 18.i.1973, *Fanshawe* 11748 (K holotype, NDO, SRGH). FIGURE 6.4.3.

Erect herb to 60 cm, unbranched except for inflorescence; slightly thickened root base with remants of old leaf bases (which probably are quite a different shape compared to cauline leaves); stem glabrous. Basal leaves smaller than cauline ones, all leaves almost appressed to stem, semi-succulent (fide Bingham), linear, (3.5)6–12 × 0.1–0.5 cm, base sessile and cuneate and possibly running down stem in small wings, margins with tiny callous protuberances, apex attenuate, glabrous. Capitula radiate, 2–3 together on peduncles 4.5–6 cm long; phyllaries 11–14, 8.5 × 2.2 mm, with broad pale entire margins and darker hairy and papillate apex. Ray florets 6–7, corollas yellow, 8–9.5 mm long; disc florets ± 60, 7.5–8 mm long; basal node glabrous and slightly tuberculate. Achenes 2 mm long, faintly 10-ribbed, glabrous; pappus setae white, 7 mm long.

**Zambia**. W: Mufulira, 18.i.1973, *Fanshawe* 11748 (K, NDO). Note: M. Hyde (pers. comm. with low-res. photos!) found another three specimens of this taxon. B: Luansongwa, Samakai village, 160 m, 2.i.1991, *Bingham* 7186 (SRGH). N: Saisi valley Mbala, 13.i.1968, *Richards* 22919 (SRGH). C: Kalengwa mine, i.1991, *Bingham* 7200 (SRGH).

Moist dambo and miombo woodland; 1100–1500 m.

Conservation Status: With only a few specimens known, occurring in a habitat that is declining as it is being used increasingly for agriculture, this is likely a species in danger; however, as it might have been confused with more common taxa, I prefer to keep this as Data Deficient (DD), until specific searches for the taxon (as well as research into the status of the habitat in this area) have provided more data.

15. **Senecio gazensis** S.Moore in J. Linn. Soc. Bot. **40**: 121 (1911). —Mapaura & Timberlake, Checkl. Zimb. Vasc. Pl.: 28 (2004). Type: Zimbabwe, Melsetter, 10.x.1908, *Swynnerton* 6143 (BM holotype).

   *Senecio depauperatus* Mattf. in Bot. Jahrb. Syst. **59** (Beibl. 133): 36 (1924). —Jeffrey in Kew Bull. **41**(4): 895 (1986). —Jeffrey & Beentje in F.T.E.A., Compositae 3: 636 (2005). Type: Tanzania, Station Kyimbila, near Madehani, 3.xii.1913, *Stolz* 2336 (B† holotype, BR, EA, K, P).

Shrubby perennial herb 30–75 cm high, probably a pyrophyte, from a creeping woody rootstock; stems one to several, erect, usually densely branched about halfway up and with closely parallel densely leafy branches, each topped by a small group of flowering heads, sparsely arachnoid and glabrescent or glabrous. Leaves dense, small and ericoid, linear to narrowly obovate, 0.8–3.8 × 0.05–0.2 cm, base attenuate, margins entire (in northern populations) and revolute to dentate with a few 0.3–0.5 mm callose

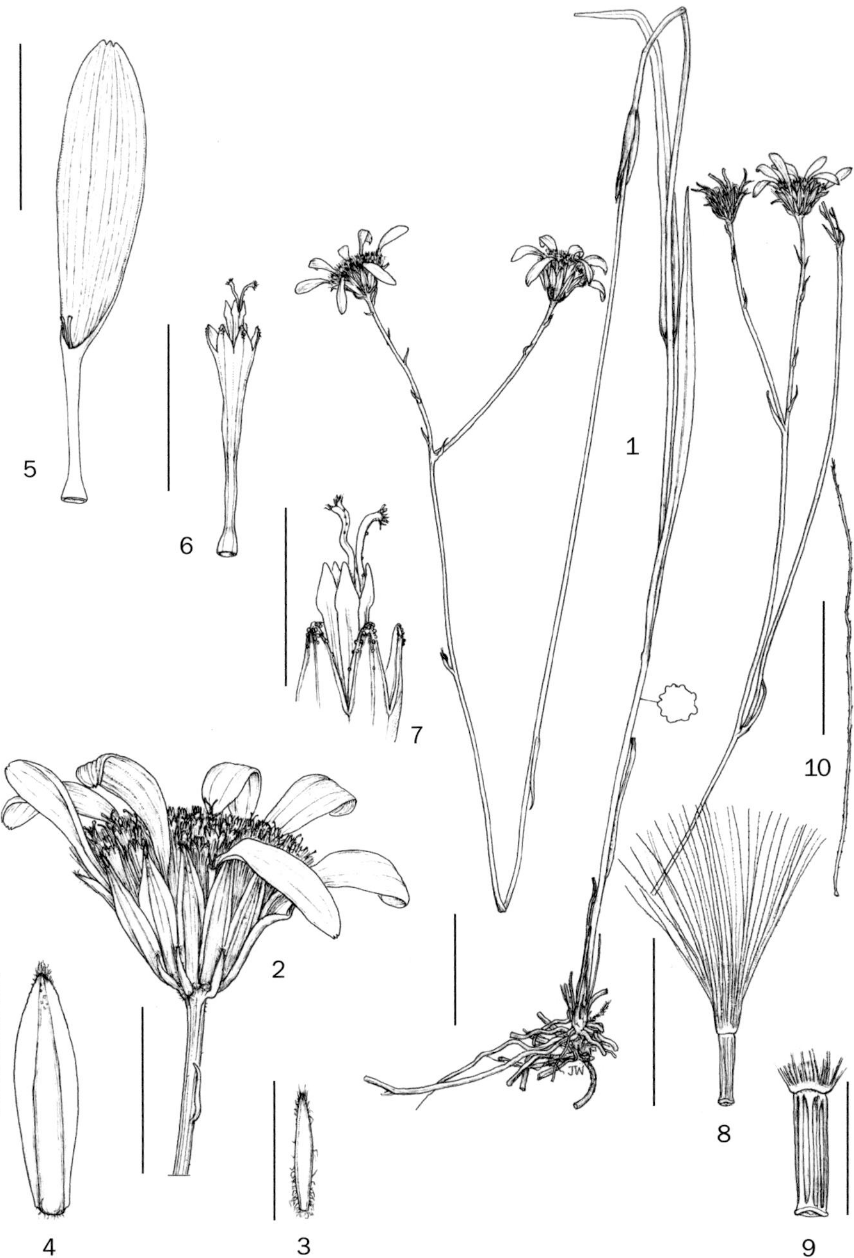

Fig. 6.4.3. SENECIO FANSHAWEI. 1, habit; 2, capitulum; 3, calycular bract; 4, phyllary; 5, ray floret corolla; 6, disc floret corolla; 7, detail of disc floret; 8, achene with pappus setae; 9, achene; 10, pappus seta. All from *Fanshawe* 11748. Scale bars: 7, 9, 10 = 2 mm; 3–6, 8 = 5 mm; 2 = 1 cm; 1 = 3 cm. Drawn by Juliet Williamson. Reproduced from Kew Bulletin (2019) with permission of the artist.

teeth, apex acute and callose; glabrous except for a few hairs at very base. Capitula radiate, few to many in lax to subumbelliform terminal corymbose panicles; axes arachnoid to glabrous; peduncles of individual capitula 5–35 mm long; involucre campanulate, 3.5–5 × 3–10 mm, glabrous or slightly pubescent at base; calycular bracts 2–6, 1–4 mm long; phyllaries 7–10, elliptic, 3–5 × 0.4–0.7 mm, with narrow to wide pale membranous margins, glabrous except for hairy apex; ray florets 4–7, tube 2–2.5 mm, lamina 2.5–6.5 × 1.5–2.5 mm; disc florets 9–20, corollas yellow to light brown, narrowly infundibuliform, 3.8–5 mm long of which lobes 0.5 mm. Achenes oblong, 1.3–2.3 mm long, slightly ribbed to almost smooth, glabrous; pappus setae white, 2.2–4.5 mm long.

**Zambia**. N: Mafinga Mts, NW sloppes in Isoka Dist., 18.xii.1964, *Robinson* 6299 (K, SRGH). **Zimbabwe**. E: Mutare, Banti F.R., 10.xi.1967, *Mavi* 570 (K, SRGH). **Malawi**. N: Misuku Hills, Matipa forest road, 27.xii.1977, *Pawek* 13426 (K, MAL, MO, SRGH, UC).

Also in Tanzania. Open rocky grassland, on granite and schist; 1240–2250 m. **Mozambique**. Manica: Chimanimani N.R., path to Mt Binga, 1241 m, 1.i.2019, *Ballings & Wursten* 3135 (BR, LMA).

Conservation Status: I have seen nine specimens from Zimbabwe, and one each from the other countries; with a fairly large distribution and a moderate altitude range in a common habitat, this is most likely Least Concern.

Hilliard (1977: 421) already remarked that *S. depauperatus* is closely allied (she added *S. baurii* Oliv. from South Africa), and indeed Mattfeld compared his new species to *S. gazensis*. The differences are so small, and of degree rather than real ones, that I have put *S. depauperatus* into synonymy.


16. **Senecio glutinosus** Thunb., Prodr. Pl. Cap. **2**: 158 (1800). Type: South Africa, Cape, 1774, *Thunberg* s.n. (S-G-5602 holotype, SBT).

Annual herb 30–90 cm; stems branched from near base, with short glandular hairs. Leaves all cauline, narrowly obovate in outline, 2–8 × 0.8–2.5 cm (reportedly to 20 cm in lower leaves), pinnatisect, base attenuate and petiole-like in lower leaves, in more upper leaves auriculate and clasping, margins sinuate-lobed, major lobes lobed in turn, margins dentate, apex acute, glandular-pubescent. Capitula radiate, in fairly large sub-corymbose terminal inflorescences, axes glandular-hairy; calycular bracts 1–2, minute; phyllaries 10–12, 7–8 × 0.5–0.6 mm, acute, with some glandular hairs. Ray florets 4+, corollas yellow, rays 0.7–0.8 mm long, spreading or revolute; disc florets 18–26, 5.3 mm long. Achenes 3.2–3.5 mm long, hairy; pappus setae white, 6 mm long.

**Zimbabwe**. W: Bulawayo, Redlcaf, 15.vii.1945, *Hopkins in GH* 13463 (K).

Only habitat indication is 'weed near chicken run'; altitude 1350 m. Also in Namibia and South Africa.

Conservation Status: With a rather large distribution area and a preference for ruderal habitat, this taxon is probably Least Concern.

The specimen is not very good, especially the leaves being rather moth-eaten, but I feel fairly certain in naming it this species. It is the first record for our area.


17. **Senecio gramineticola** C.Jeffrey in Kew Bull. **41**(4): 895 (1986). —Jeffrey & Beentje in F.T.E.A., Compositae **3**: 640 (2005). Type: Tanzania, Ufipa Dist., Nsanga Mt, *Richards* 12130 (K holotype).

Perennial herb, 40–150 cm high, with radiating tuberous roots; stems erect, in our area apparently unbranched apart from inflorescence, with white arachnoid tomentum, glabrescent (sometimes with very few glands), reddish-brown. Basal leaves sessile, elliptic, (11)31–60 × 1.4–5(7) cm including 10–22 cm long petioloid base, base attenuate, margins closely to coarsely dentate or bidentate, apex obtuse; stem leaves sessile, elliptic to lanceolate, 4.5–32 × 0.5–4(10.5) cm, attenuate into an exauriculate or scarcely auriculate petioloid base in lower leaves, expanded and semi-amplexicaul in upper leaves, margins strongly dentate or bidentate, obtuse, shortly acuminate; all leaves arachnoid-pubescent and ± glabrescent. Capitula discoid, erect, 7–many in rather congested terminal corymbose panicles; involucre cylindrical, 7–9 × 5.5–6.5 mm, arachnoid at least at base; calycular bracts 8–10, lanceolate, 3–8 mm long, ± arachnoid; phyllaries 16–22, 6–8 × 0.4–1 mm, arachnoid or thinly so, long-tapered and bearded at apex. Florets 30–60, corollas yellow to orange or brownish, 4.3–7 mm long, tube gradually expanded above middle, glabrous, lobes 0.7–1 mm long. Achenes 2.7–4 mm long, slightly angular, glabrous; pappus setae white, 4.5 –7 mm long.

**Zambia**. N: road Mbala–Lunzua Falls and Kambole, 5000 ft, 24.i.1955, *Richards* 4245 (K). **Malawi**. N: Nyika Plateau, Dembo road, 2300 m, 17.iv.1975, *Pawek* 9294 (K, MAL, MO, SRGH, UC).

Also known from Tanzania. Grassland (in Tanzania only where not burnt); 1500–2550 m.

Conservation Status: While the distribution is rather limited, the habitat is a common one and the altitude range fairly wide. Least Concern.

18. **Senecio hastatus** L., Sp. Pl. **2**: 868 (1753). —Hilliard & Burtt in Notes Roy. Bot. Gard. Edinburgh **32**(3): 378 (1973). —Hilliard, Compositae Natal: 439 (1977). —Mapaura & Timberlake, Checkl. Zimb. Vasc. Pl.: 28 (2004). Type: 'Africa', herb. *van Royen* 900.57-332 (L, lectotype), lectotypified by Hilliard & Burtt (1973: 378).

    *Senecio hastulatus* L., Sp. Pl. ed. 2, **2**: 1218 (1763), probably a spelling error for the same taxon. —Candolle, Prodr. **6**: 383 (1838). —Harvey in Harvey & Sonder, Fl.Cap. **3**: 367 (1865).

Herb 30–60 cm, root crown with 1–several leaf rosettes, flowering stems lateral to each rosette, branching from low down; stem glandular-pubescent. Basal leaves sometimes purplish underneath, oblong in outline, 20 × 4 cm, base attenuate but clasping at very base, margins pinnatifid to fairly deeply lobed with several pairs of lobes, lobes denticulate, toothed or lobed, with broad rounded denticulate sinuses; cauline leaves similar but 4.5–12 × 0.8–2.4 cm, sessile and auriculate, margins less lobed to dentate-serrate in inflorescence leaves, apex obtuse to acute, glandular-pubescent. Capitula few to many in paniculate sub-corymbose terminal groups; calycular bracts very few; phyllaries 12–18, 7.5–12 × 0.4–0.8 mm and hairy with white hairs. Ray florets 8(12), corollas yellow, ray limb 4.5 mm long; disc florets 37–40, ± cylindrical (hardly widening upwards), 5.5 mm long. Achenes 2–4.5 mm long, ribbed, puberulous; pappus setae about 5.5 mm long.

**Zimbabwe**. S: Masvingo (Fort Victoria), Morgenster mission, 4.x.1949, *Wild* 3050 (K, SRGH). Also in Lesotho and South Africa. 'by Finger Rocks'; probably around 1250 m.

Conservation Status: Probably Least Concern in the other parts of its distribution area.

The identification is a tentative one, as the material lacks the basal parts; I still feel fairly sure about it.

19. **Senecio helminthioides** (Sch.Bip.) Hilliard, Notes Roy. Bot. Gard. Edinburgh **32**(3): 379 (1973). —Hilliard, Compositae Natal: 496 (1977). —Mapaura & Timberlake, Checkl. Zimb. Vasc. Pl.: 28 (2004). Type: South Africa, Natal Bay [Natalbai], vii.1839, *Krauss* 285 (P00122421 holotype, G, K, TUB).

    *Senecio quinquelobus* (Thunb.) DC. var. *helminthioides* Sch.Bip., Flora **27**(40): 700 (1844), as 'helminthoideus'.

Herbaceous twiner, stem and leaves slightly succulent, glabrous or with leaves slightly hairy below. Leaves ovate-deltate, petiole 1.5–6 cm long, lamina 2.5–6 × 4–6(7) cm, base cordate, margins shallowly and acutely 3–7-lobed, sinuses broad and often denticulate, apex acute; glabrous in ours or with a few hairs on lower surface near lamina base. Capitula discoid, solitary or up to 6 in very open corymbs terminating short side branches with reduced leaves; involucre narrowly campanulate; calycular bracts broadly elliptic, foliaceous, erect to spreading; phyllaries 11–13, 10–12 × 0.6–1.2 mm, keeled, veins resinous, margins pale. Florets 32–38, corollas yellow or orange, 8.5–9.7 mm long, lobe apex very papillate. Achenes cylindric, 2.8–4.5 in ours (reported as 5–6) mm long, ribbed, hispid between ribs; pappus setae 7–10.5 mm long.

**Zimbabwe**. C: Wedza Mt, 21.v.1968, *Mavi* 725 (K, SRGH). **Mozambique**. M: Inhaca, Barreira Vermelha, 26.vii.1960, *Mogg* 31471 (K).

Also from Madagascar, South Africa (Cape Province, Transvaal, Natal) and Eswatini. 'Kloof forest'; 80–1300 m? [no altitude given, but Wedza Mt is 1780 m; the scheelite mine collection site is just below the saddle, and possibly at 1300 m]. I have seen four collections for our area.

Conservation Status: While the species has a wide distribution, I have been unable to determine its habitat in South Africa, but probably Least Concern.

Much confused with *S. quinquelobus* (Thunb.) DC. from South Africa, but distinct in the leafy calycular bracts and the hispid achenes.

20. **Senecio hochstetteri** Sch.Bip. ex A.Rich., Tent. Fl. Abyss. **1**: 435 (1848). —Oliver & Hiern, F.T.A. **3**: 414 (1877). —Jeffrey in Kew Bull. **41**(4): 903 (1986). —Lisowski, (Asterac. Fl. Afr. Cent. 2) Fragm. Flor. Geobot. **36** Suppl. 1: 316, t. 69 (1991). —Mapaura & Timberlake, Checkl. Zimb. Vasc. Pl.: 28 (2004). —Jeffrey & Beentje in F.T.E.A., Compositae **3**: 666 (2005). —Burrows & Willis, Pl. Nyika Plateau: 105 (2005). Type: Ethiopia, Mt Kubbi [Koubi], 26.vi.1837, *Schimper* 268 (P00122445 holotype, BR836236, BR887866, ETH000000131, GOET002366, HBG505228, J7208, K00311446, K00311447, K00311448, M0105165, MO391701, MPU011968, MPU011971, NY259518, P00122446, P00122447, SS08-7291, WAG0000651, WAG0000652). FIGURE 6.4.**4**.

   *Senecio chlorocephalus* Muschl. in Mildbraed, Wiss. Ergebn. Deut. Zentr.-Afr. Exped. (1907-1908), Bot. **2**: 393 (1911). Type: D.R. Congo, Kivu, SE Karisimbi, W of Lake Karago, xi.1907, *Milbraed* 1637 (B† holotype, BR).

   *Senecio polygonoides* Muschl. in Mildbraed, Wiss. Ergebn. Deut. Zentr.-Afr. Exped. (1907-1908), Bot. **2**: 398 (1911). Type: Rwanda/D.R. Congo, E of Karisimbi, *Mildbraed* 1571 (B† holotype; BR).

   *Senecio rusisiensis* R.E.Fr. in Wiss. Ergebn. Schwed. Rhodesia–Kongo-Exped. 1911–1912, **1**: 344 (1911). Type: D.R. Congo [Deutsch Ost-Afrika], S of Lake Kivu along Rusisi R, *Fries* 1550 (UPS holotype).

   *Senecio tshitirungensis* De Wild., Pl. Bequaert. **5**: 125 (1929). Type: D.R. Congo, Tshitirunge, 5.x.1914, *Bequaert* 5999 (BR holotype).

   *Senecio tongoensis* De Wild., Pl. Bequaert. **5**: 122 (1929). Type: D.R. Congo, between Tongo and Mukule, 25.ix.1914, *Bequaert* 5864 (BR holotype).

   *Senecio acroleucus* Merxm., in Proc. Trans. Rhodesia Sci. Assoc. **43**: 68 (1951). Type: Zimbabwe, Marondera (Marandellas), 1.i.1941, *Dehn* 240 (M0105180 holotype).

   *Senecio greenwayi* C.Jeffrey in Kew Bull. **41**(4): 903 (1986). —Lisowski, (Asterac. Fl. Afr. Cent. 2) Fragm. Flor. Geobot. **36** Suppl. 1: 319 (1991). —Jeffrey & Beentje in F.T.E.A., Compositae **3**: 666 (2005). Type: Tanzania, Mbeya, 2.vii.1941 *Greenway* 6182 (K holotype).

Perennial herb 20–120 cm high, with a fleshy rootstock to 2 cm thick, rosette-forming; stems 1–several from rosette, glandular, viscid, purplish near base. Basal leaves sessile, often a little to one side of flowering stem, slightly fleshy, narrowly to broadly obovate to narrowly elliptic, (5)7–38 × 1–10 cm, base long-attenuate into a petioloid base, margins shallowly to deeply and coarsely pinnately or lyrately lobed, lobes obovate, ovate or oblong, sometimes sinuate-denticulate, apex obtuse to acute, apiculate; stem leaves sessile, ovate-oblong to oblanceolate, 3–14 × 0.4–3.7 cm, subauriculate, pinnately lobed or dentate to (especially uppermost) entire. All leaves sticky pubescent with twisty glandular hairs. Capitula discoid, many in lax to congested usually long-stalked terminal corymbose panicles; peduncles glandular; involucre 4.5–9 × 2.5–5 mm, glandular; calycular bracts 1–5, lax, narrow, lanceolate, 0.5–5 mm long; phyllaries 8–15(17), 4–9 × 0.4–0.8 mm, glandular, green, tips purple to black. Florets 15–50, corollas white, cream or pale yellow, infundibuliform with very narrow tube and widening upper part, (4)5–10 mm long, of which lobes 0.7–1.6 mm long, tips minutely papillose; anthers blue, mauve or purple. Achenes 2–4 mm long, ribbed, short-hairy between ribs; pappus setae 5–8 mm long, white.

**Zambia**. B: Kasuka, Sefula dambo, 14.x.1993, *Bingham* 9736 (K). N: Nyika Plateau, 2 m SW of Rest House, 2150 m, 21.x.1958, *Robson* 215 (K). E: Makutus, 28.x.1972, *Fanshawe* 11595 (K, NDO). **Zimbabwe**. C: Marondera, Looe, 22.xii.1948, *Wild* 2712 (K, SRGH). E: Odzani River valley, 1914, *Teague* 264 (BOL, K). S: Bikita, 15.xii.1953, *Wild* 4391 (K, SRGH). **Malawi**. N: Nyika Plateau, Chelinda bridge, 18.x.1975, *Pawek* 10282 (K, MAL, MO, SRGH, UC). C: Chongoni F.R., Chiwao Hill, 11.ix.1967, *Salubeni* 826 (K, SRGH). S: Mulanje Mt., skyline path to Chambe, 3.v.1989, *Pope et al.* 2279 (K, MAL).

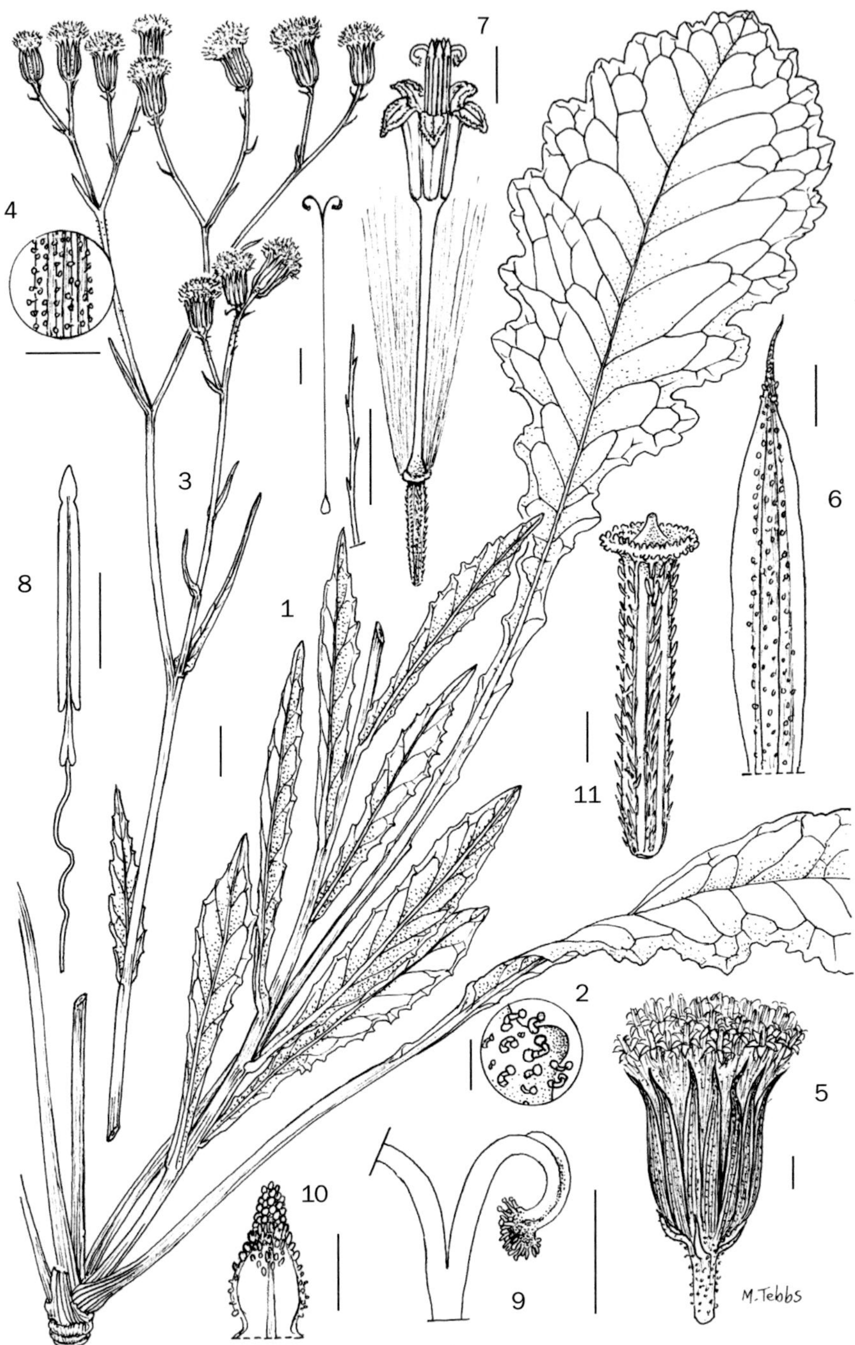

Fig. 6.4.4. SENECIO HOCHSTETTERI. 1, habit; 2, detail of leaf margin; 3, flowering branch; 4, detail of stem; 5, capitulum; 6, phyllary; 7, floret; 8, anther; 9, style branches; 10, detail of style branch apex; 11, achene. All from *Richards* 9926. Scale bars: 2, 9–11 = 0.5 mm; 4–8 = 1 mm; 1, 3 = 2 cm. Drawn by Margaret Tebbs.

Also known from W Africa from Sierra Leone to Cameroun, Sudan, D.R. Congo, Rwanda, Burundi, Ethiopia, Kenya, Tanzania, Uganda, possibly South Africa. Grassland, recently burned grassland, occasionally in forest margins or miombo, sometimes in dambos or pans or by lake shores and in bogs, in small colonies and never very common; 1000–2550 m.

Conservation Status: Due to a wide distribution in a range of common habitats, this should be Least Concern.

Two sheets from Zimbambwe, Chimanimani (*Phipps* 261, *Goldsmith* 36/71) are a bit on the dainty side; the Phipps specimen was growing in water in a pan.

21. **Senecio inornatus** DC., Prodr. **6**: 385 (1838). —Hilliard, Compositae Natal: 471 (1977). —Jeffrey in Kew Bull. **41**(4): 894 (1986). —Lisowski, (Asterac. Fl. Afr. Cent. 2) Fragm. Flor. Geobot. **36** Suppl. 1: 277, t. 59 (1991). —Mapaura & Timberlake, Checkl. Zimb. Vasc. Pl.: 28 (2004). —Jeffrey & Beentje in F.T.E.A., Compositae **3**: 633 (2005). —Gereau *et al.*, Lake Nyasa Climat. Reg. Florist. Checkl.: 31 (2012). —Burrows & Willis, Pl. Nyika Plateau: 105 (2005). Type: South Africa, Kei [Zwart-Key] et Buffelrivier, *Drège* s.n. [5151] (G-DC holotype).

    *Senecio caulopterus* DC., Prodr. **6**: 388 (1838). Type: South Africa, Zwarte Omsamwubo et Omsamculo, *Drège* s.n. [5148] (G-DC holotype, HAL, P).

    *Senecio serra* Sond. in Linnaea **23**(1): 68 (1850), nom. illeg., non Hook. (1834). Type: South Africa, Port Natal, 1847, *Gueinzius* 358 (G holotype, K, MEL, P).

    *Senecio sneeuwbergensis* Bolus in Hooker 's Icon. Pl. **11**: 54, t. 1067 (1870). Type: South Africa, Cape of Good Hope, near Graaff-Reinet, vi.1867, *Bolus 571* (K holotype, BOL, K, PRE).

    *Senecio lygodes* Hiern, Cat. Afr. Pl. **1**(3): 599 (1898). —Hilliard, Compositae Natal: 453 (1977). Type: Angola, Huilla, Quipumpunbime river, Humpata, 1860, *Welwitsch* 3676 (BM holotype, BR, K, P).

    *Senecio ommannei* S.Moore in J. Linn. Soc., Bot. **37**(260): 324 (1906). Type: South Africa, Steijns Farm, Johannesburg, 1903, *Ommanney* 111 (BM holotype).

    *Senecio diversidentatus* Muschl. in Bot. Jahrb. Syst. **43**(1): 59 (1909). Type: South Africa, 'Südostafrikanische Hochsteppe', i.1885, *Wilms* 802 (B† holotype, AMD, E, GOET, K).

    *Senecio serra* Sond. var. *longipedunculatus* Thell. in Vierteljahrsschr. Naturf. Ges. Zürich **66**: 245 (1921). Types: South Africa, Transvaal, Lydenburg, xi.1895, *Wilms* 800 (K, Z syntypes); between Middelburg and Crocodile River [Krokodilfluss], xii.1883, *Wilms* 800a (GOET, Z syntypes).

    *Senecio stolzii* Mattf. in Bot. Jahrb. Syst. **59** (Bleib 133): 37 (1924). Type: Tanzania, Kyimbila, Rungwe, 3.ii.1912, *Stolz* 1116 (B† holotype, JE, K, M, U, WAG).

    *Senecio fraudulentus* E.Phillips & C.A.Sm. in Rep. Director Veterin. Serv. Anim. Industr. S. Afr. **17**: 642 (1931), nom. nov. pro *S. serra* Sond.

    *Senecio macroalatus* M.D.Hend. in Bothalia **6**: 426 (1954). Type: South Africa, Natal, Bergville, Cathedral Peak, Organ Pipes Pass, 11.iii.1951, *Killick* 1486 (K, NH, NU, PRE syntypes).

Erect perennial herb 1–2.4 m high, with short creeping rhizome; stems unbranched, except for inflorescence, winged in lower part (with narrow wings running down from leaf base auricles), glabrous. Basal leaves preceding flowering stem, withering before flowering, sessile, elliptic-lanceolate, 20–73 × 1.5–10 cm (including 4–25 cm long petioloid base), gradually attenuate basally into a narrowly-winged petioloid base, margins dentate, apex acute; stem leaves sessile, ovate-lanceolate to oblanceolate, 6–33 × 1–7.5 cm, attenuate basally into an entire-winged, not or slightly expanded and decurrent petioloid base, margins denticulate, apex acute or shortly acuminate; all leaves bright green, with thick pallid midrib, glabrous. Capitula radiate, erect, many in a copious lax to rather congested terminal corymbs; stalk of individual capitula glabrous or shortly glandular-pubescent below capitula, 1–2.5 cm long; involucre shortly hairy at base, cylindrical, 6–7.5 × 2.5–4 mm; calycular bracts 2–6, lanceolate, 1.5–5 mm long; phyllaries 10–16, yellow-green, sometimes with black tips, 4–7.5 × 0.6–0.9 mm, glabrous or almost so, bearded-papillate at apex. Ray florets 5–9, corollas golden yellow to yellow-orange, corolla tube 2.5–4 mm long, glandular-hairy, rays limb 5–9.5 × 2–3.5 mm, 4–9-veined. Disc florets many (50–60), corollas bright yellow, with yellow styles and brown or yellow anthers, corolla 5–8 mm long, glabrous or almost so, tube gradually expanded above middle, lobes 1–1.5 mm long. Achenes ellipsoid, 1.5–4 mm long, angled, glabrous (or sometimes slightly hairy, *Milne-Redhead* 3221 or hairy, *Lewis* 6154); pappus setae 5–7.5 mm long.

**Zambia**. W: Solwezi, Lualuba P.F.A., 13.vi.1962, *Holmes* 1481 (K, NDO). **Zimbabwe**. E: Umtali, Vumba Mts, 10.v.1956, *Chase* 6105 (K, SRGH). S: Victoria, Muzero Farm, 30.iii.1971, *Chiparawasha* 379 (K, SRGH). **Malawi**. N: Viphya Plateau behind Nkalapya Mt., 9.xii.1973, *Pawek* 7589 (K, MAL, MO, SRGH, UC). C: Dedza, Chongoni Forest, 24.iii.1969, *Salubeni* 1284 (K, MAL, SRGH). S: Zomba Plateau, Chitinji Dambo, 3.i.1985, *Salubeni & Tawakali* 3920 (K, MAL). **Mozambique**. MS: Tsetserra, 3.iii.1954, *Wild* 4495 (K, SRGH).

Also known from D.R. Congo, Burundi, Tanzania, Angola and South Africa. Dambos, swamps, occasionally in drier sites in grassland and woodland; 1050–2350 m.

Conservation Status: Due to a wide distribution in a common habitat, this should be Least Concern.

Hilliard (1977: 453) recognizes *S. lygodes* as a separate species, differing from *S. inornatus* in the 8-rayed capitula (5 in *S. inornatus*) and the long-attenuate leaves with finely reticulate venation (Hilliard does not say how this differs from *S. inornatus*), as well as the habitat preferences: streamsides or marshes, rather than the open grassland of *S. inornatus* – but F.T.E.A. (2005: 634) and Lisowski (1991: 277) have it as a synonym of *S. inornatus*, and that is how I treat it here.

22. **Senecio katangensis** O.Hoffm. ex De Wild., Ann. Mus. Congo Belge, Bot. sér.4, **1**: 11 (1903). —Lisowski, (Asterac. Fl. Afr. Cent. 2) Fragm. Flor. Geobot. **36** Suppl. 1: 299 (1991). Type: D.R. Congo, Lukafu, xii.1899, *Verdick* 340 (BR holotype, BRLU).

Var. **latifolia** Lisowski, (Asterac. Fl. Afr. Cent. 2) Fragm. Flor. Geobot. **36** Suppl. 1: 300 (1991). Type: D.R. Congo, Katanga, Kundelungu Plateau 2 km E of Lutshipuka source, *Lisowski* 93748 (POZG holotype, BR).

Herb 40–100 cm high, branching from near base; small undergound stem; stem slightly ridged, glabrous. Leaves increasing in size up stem, broadly elliptic, 2.5–10.5 × 0.8–5.7 cm, base sessile and cuneate and half-clasping stem, margins entire, apex acute; with 5–7 roughly parallel veins; glabrous. Capitula discoid, 3–8 in lax terminal panicles; calycular bracts minute; phyllaries 10–13, 6 × 0.6–0.9 mm, with papillate apex. Ray florets absent. Florets ± 27, corollas pale yellow, infundibuliform, 5.7–7 mm long; lobes papillate at apex. Achenes 2.5–4 mm long, ribbed, pubescent; pappus setae white, 4–5.5 mm long.

**Zambia**. N: near Mbala [Abercorn], 14.xii.1954, *Richards* 3625 (K).
Also in D.R. Congo. On pan bank among grass, bracken and mixed vegetation, few trees, sandy loam; ?1670 m.

Conservation Status: in D.R. Congo restricted to the Kundelungu and Biano Plateaux, from where is known from seven collections; while the Kundelungu is a National Park, protection is nominal rather that real and threats such as conversion to agriculture are prevalent, as they are in the Mbala area. Probably Vulnerable (VU).

As Lisowski (1991: 301) notes, this rather resembles *S. ruwenzoriensis* but differs in the parallel-veined leaves and the lack of radiate florets.

23. **Senecio kayomborum** Beentje in Kew Bull. **58**(1): 233 (2003). —Jeffrey & Beentje in F.T.E.A., Compositae **3**: 626 (2005). Type: Tanzania, Iringa Distr., Mufindi, Igowole, 10.iii.1989, *Kayombo & Kayombo* 220 (K holotype, MO, NHT).

Erect or creeping herb with annual shoots up to 40 cm high or long from a thickened undergound part; stems in upper part reddish purple and sparsely pilose with multicellular hairs, otherwise glabrous. Leaves narrowly obovate to almost linear, 1.5–4.5 × 0.2–1.6 cm, base attenuate, margins entire or denticulate but slightly thickened and minutely revolute, apex obtuse and somewhat thickened; sparsely pilose-scabridulous on both surfaces, glabrescent (no basal leaves or root crown visible or indicated). Capitula solitary, erect, discoid; involucre cylindrical, dark purple-red, 8–12 × 6–9 mm, densely pilose-scabridulous at base; calycular bracts 5–10, narrowly lanceolate and up to 5 mm long, acute, pilose; phyllaries 15–20, 8–12 × 0.4–0.8 mm, attenuate, sparsely pilose-scabridulous, penicillate-bearded at apex. Ray florets absent; disc florets 30–40, corollas orange-yellow or yellow, 8.5–10.2 mm long, tube slightly and gradually widening in upper half, lobes 1.5–2.5 mm long. Achenes 2.4–3 mm long, 8–9-ribbed, pilose between ribs; pappus setae 7.5–8.5 mm long, white.

**Zambia**. C: Serenje Dist., Kundalila Falls, 13°09.3'S, 30°43.0'E, 1494 m, 14.i.2001, *Bingham & Zukas* 12311 (K, UZL).

Also known from Tanzania. The single Zambian specimen was from a wet dambo at 1494 m altitude.

Conservation Status: Known from three Tanzanian specimens (altitude: 1800—2100 m) from similar habitats; dambos are under threat from agricultural conversion, so the species is at least Vulnerable.

24. **Senecio latecorymbosus** Gilli in Ann. Naturhist. Mus. Wien **78**: 159 (1974). —Jeffrey in Kew Bull. **41**(4): 894 (1986). —Jeffrey & Beentje in F.T.E.A., Compositae **3**: 633 (2005). Type: Tanzania, Njombe Dist., Uwemba, 27.viii.1958, *Zerny* 634 (W holotype).

Coarse perennial herb 1.5–2.4 m high; stems erect, leafy, sparsely pubescent or glabrous. Stem leaves sessile, elliptic, narrowly elliptic or lanceolate, 14–39 × 3–8 cm, base rounded and semi-amplexicaul (especially in upper leaves) to tapered and slightly expanded (in lower leaves), very shortly narrowly decurrent, margins regularly dentate, apex attenuate, acute, glabrous above, sparsely and minutely pilose beneath. Capitula radiate, erect, many in a copiously branched terminal leafy corymbose panicle; involucre cylindrical, 9–12 × 5–10 mm, glabrous except for a few glandular hairs at base; calycular bracts 8–12, lanceolate, 3.5–8 mm long, pilose especially on margins; phyllaries 12–16, 8.3–11 × 1–1.3 mm long, glabrous, bearded at apex. Ray florets 7–8, corollas yellow, tube 5–7 mm long, glandular-hairy or glabrous, ray limb 7–13 × 2–5 mm, 4–6-veined. Disc florets 18–25, corollas yellow, 9–10 mm long, tube glabrous, expanded above middle, lobes 1.5–2 mm long. Achenes 5–5.5 mm long, very sparsely hairy, many-ribbed; pappus setae 8–9 mm long, white.

**Zambia**. E: Nyika, 26.vi.1966, *Fanshawe* 9765 (K, NDO). **Malawi**. N: Nyika, S of Kasaramba view point, 8.vii.1970, *Brummitt* 11897 (K, MAL).

Also known from Tanzania. Upland forest, in semi-shade, in gaps or on edges; 2200–2350 m.

Conservation Status: There are only three specimens from our area, two from the Malawi side of the Nyika Plateau, and one from the Zambian side; only the Malawi side is well-protected. The six Tanzanian specimens are all from the S of the country, a few from protected areas. Data Deficient until a proper assessment is done; probably Vulnerable.

The original spelling by Gilli was *latecorymbosus*, but Jeffrey changed this to *laticorymbosus*. I am unable to find a rule regarding this in the ICBN, so I am using the original spelling.

25. **Senecio latifolius** DC., Prodr. **6**: 387 (1837). —Hilliard, Compositae Natal: 486 (1977). —Da Silva *et al.*, Prelim. Checkl. Vasc. Pl. Mozamb.: 34 (2004) —Mapaura & Timberlake, Checkl. Zimb. Vasc. Pl.: 28 (2004). —Burrows & Willis, Pl. Nyika Plateau: 105 (2005). —Phiri, Checkl. Zamb. Vasc. Pl.: 33 (2005). —Setshogo, Prelim. Checkl. Pl. Botsw.: 38 (2005). —Bandeira *et al.*, Wild Flowers South. Mozamb.: 214 (2007). Type: South Africa, Umsikaba [Omsamcaba], *Drège* s.n. [5140] (G-DC holotype, E, HBG, K, P, TUB).

*Senecio sceleratus* Schweik. in Onderstepoort Journ. **16**: 130 (1941). —Da Silva *et al.*, Prelim. Checkl. Vasc. Pl. Mozamb.: 299 (1993). Type: South Africa, Zoutpansberg, Tzaneen, 1.xi.1921, *Phillips 3265* (PRE0226671-0 holotype).

*Senecio pergamentaceus* Baker in Bull. Misc. Inform. Kew **1898**(139): 154 (1898). —Jeffrey in Kew Bull. **41**(4): 896 (1986). —Mapaura & Timberlake, Checkl. Zimb. Vasc. Pl.: 28 (2004). —Jeffrey & Beentje in F.T.E.A., Compositae **3**: 643 (2005). —Burrows & Willis, Pl. Nyika Plateau: 105 (2005). —Phiri, Checkl. Zamb. Vasc. Pl.: 33 (2005). —Gereau *et al.*, Lake Nyasa Checkl.: 31 (2012). Type: Malawi, Zomba, *Whyte* s.n. (K holotype, Z000003961).

Perennial herb from stout woody stock, glabrous except for woolly root crown (and some hidden wool in leaf axils); flowering stems 1–2 from a crown, 0.4–1.7 m tall, usually unbranched below inflorescence branches, leafy throughout, glabrous. Leaves leathery, sessile, oblong to elliptic to narrowly ovate or obovate, 6–21 × 0.7–8 cm, decreasing only slightly upwards, base ± cordate, half-clasping and sometimes slightly decurrent, margins thickened and distantly callose-denticulate or entire, apex (sub)acute; glabrous, minutely glandular, and quite often glaucous. Capitula radiate, many in large lax corymbose panicles to 20 cm across; peduncles of individual capitula 6–40 mm long; calycular bracts few, 1–2 mm long; involucre turbinate to cylindrical, 5.5–6 × 3.5 mm; phyllaries 6–8, 4–7 × 0.8–1.6 mm, flat with resinous veins and some with pale membranous margins, glabrous but for a little tuft of hairs at tip. Ray florets usually 5 (occasionally 1–4), corollas bright yellow, tube 2.8–4.5 mm long, limb 4–13 × 2–3 mm; disc florets 9–15, corollas bright yellow to orange (this last might just be anther colour), narrowly infundibuliform, 5–6.5 mm long of which lobes 1–2.6 mm, papillose at tip. Achenes 2.2–5 mm long, slightly ribbed, glabrous; pappus setae white, 4.2–6.5 mm long.

**Botswana**. SE: Kanye, Stony Hill, 21.x.1978, *Hansen* 3497 (C, GAB, K, PRE, SRGH, WAG). **Zambia**. N: Nkali dambo, 5.i.1952, *Richards* 368 (K). W: Ichimpi, Kitwe, 8.iii.1968, *Mutimushi* 2527 (NDO, K). C: Serenje, Kundalila Falls, 4.ii.1973, *Kornas* 3208 (K, UZL). S: Choma, 14.i.1983, *Chisumpa* 728 (NDO, K). **Zimbabwe**. N: Msasa Farm, Tatagura Valley, 9.xi.1962, *Angus* 3393 (FHO, K, SRGH). W: Matobo, Hope Fountain School, 23.vii.1973, *Norrgrann* 349 (K, SRGH). C: Helensvale, 31.x.1965, *Simon* 491 (K, SRGH). E: Mutare [Umtali] North, 30.ix.1964, *Cleghorn* 984 (K, SRGH). **Malawi**. N: Chitipa, Jembya F.R., 14.i.1989, *Thompson & Rawlins* 5982 (CM, K, MAL). C: Dedza Mt, 23.x.1956, *Banda* 288 (K, MAL, SRGH). S: Zomba, Chitinji Valley, 26.x.1984, *Salubeni & Tawakali* 3909 (K, MAL). **Mozambique**. N: Metonia, no date, *Hornby* 3754 (K, PRE). Z: Mt Namuli, Muretha Plateau, 15.xi.2007, *Harris et al.* 313 (K, LMA). T: Angonia, Chitambe, 17.xii.1980, *Macuacua* 1456 (K, LISC). M: Namaacha, 19.x.1983, *Groenendijk & de Koning* 737 (K, LISU, WAG).

Also known from Tanzania, Angola, South Africa (Natal, Transvaal) and Eswatini. Grassland on shallow soils, bracken stands, rocky miombo, pioneer on burnt ground and one of the earliest plants to flower after burning; 1000–2250 m.

Conservation Status: Due to a wide distribution in a range of common habitats and with a wide altitude range, this should be Least Concern.

Toxic to cattle, fide Bandeira *et al.* (2007).

*Mutimushi* 2372 from Kitwe (Zambia W) looks very much like this taxon, but has 14–15 phyllaries. The heads are rather young but probably have few rays, and otherwise the specimen looks like *S. latifolius*.

The name *S. venosus* Harv. has been used extensively in our area; the type of this taxon is from the Cape and it is possible the two are synonymous. They are certainly very close, the differences reported seem to be restricted to whether the leaf is thicker or thinner.

26. **Senecio lisowskii** Long Wang & Beentje in Phytotaxa **522**(1): 73 (2021). Type: D.R. Congo, Katanga, W of Lubumbashi, Natwebo bank, *Lisowski* 57937 (POZG-V-0100059 holotype, BR00000014982828, BR00000014982835, POZG-V-0004882, POZG-V-0004883, POZG-V-0004884).

    *Senecio jeffreyanus* Lisowski, (Asterac. Fl. Afr. Cent. 2) Fragm. Flor. Geobot. **36** Suppl. 1: 321 (1991), nom. illeg., non Diels in Notes Roy. Bot. Gard. Edinburgh **5**: 192 (1912).

Annual or short-lived perennial herb to 70 cm high; stem erect, branched, puberulous but becoming glabrous. Leaves obovate in outline and pinnatipartite to pinnatisect, 4.5–8 × 0.8–3 cm (larger in D.R. Congo), base half-clasping stem, lobes narrow to ovate and sparsely dentate, apex obtuse; puberulous. Capitula discoid, in terminal corymbose panicles; involucre cylindrical at anthesis; calycular bracts 8–10, small; phyllaries ± 15, 7–8 × 0.4–0.6 mm, margins pale, apex acute, glabrous except for very apex. Florets 40–56, corollas yellow, narrowly infundibuliform, 4–4.5 mm long, lobes 0.3–0.6 mm. Achenes cylindrical, 4–5 mm long, angular and puberulous; pappus setae white, 4–5 mm long.

**Zambia**. W: Busanga Flats, Lufupa drainage, 19.x.1962, *Vesey-Fitzgerald* 3772 (K, UZL). S: E of Ngoma, 1.ix.1964, *van Rensburg* 2951 (K).

Also known from the D.R. Congo. Along a stream coming from hot springs; ± 1000 m.

Conservation Status: This taxon is known from three collections, one from D.R. Congo and the two cited above; a proper assessment needs to be done based on threats to the habitat in these sites, but here I assume it is at least Vulnerable and possibly worse.

27. **Senecio madagascariensis** Poir., Encycl., Suppl. **5**: 130 (1817). —Hilliard, Compositae Natal: 404 (1977). —Mapaura & Timberlake, Checkl. Zimb. Vasc. Pl.: 28 (2004). — Jeffrey & Beentje in F.T.E.A., Compositae **3**: 669 (2005). —Bandeira *et al.*, Wild Flowers South. Mozamb.: 214 (2007). Type: Madagascar, *Commerson* s.n. (?P- herb. Desfontaines holotype, MPU011695, P00557664).

    *Senecio ruderalis* Harv. in Harvey & Sonder, Fl. Cap. **3**: 355 (1865). Type: South Africa, near Natal, *Williamson* s.n. (TCD syntype); *Sanderson* 326 (TCD syntype); 1861, *Gerrard & McKen 324* (NH, TCD syntypes); *Grant* s.n. (K syntype); Delagoa Bay, near Lorenzo Marquas, 1860, *Speke* s.n. (?K syntype).

    *Senecio bakeri* Scott Elliot in J. Linn. Soc., Bot. **29**(197): 30 (1891). Types: Madagascar, Fort Dauphin, *Scott-Elliot* 2307 (E, P00557663 syntypes); '2955' (probably in error for 2956, K syntype, annotated by Scott Elliot).

    *Senecio junodianus* O.Hoffm. in Schinz & Junod, Mem. Herb. Boiss. **10**: 74 (1900). Type: Mozambique, Delagoa Bay, *Junod* 79 (Z holotype).

Annual or short-lived perennial herb 15–90 cm, glabrous or very sparsely hairy; stem unbranched below or branched from base, branching above, leafy throughout. Leaves membranous, very variable, linear-lanceolate to narrowly elliptic, 2–12.5 × 0.3–3 cm, base attenuate and half-clasping, margin denticulate to dentate to lobed, apex acute; glabrous or sparsely puberulous. Capitula radiate, in corymbose panicles, few to many; involucre cylindrical to turbinate; calycular bracts 8–12 and small, ciliate; phyllaries 19–21, 4–7 × 0.6–0.8 mm, margins pale, apex attenuate, glabrous. Ray florets 12–15, corollas canary yellow to golden, tube 2.5–3 mm long, ray limb 4.5–6 × 1.4–2.5 mm, often inrolled. Disc florets (40)80–110, corollas yellow, very narrowly infundibuliform, 3.5–5.3 mm long of which lobes 0.2–0.5 mm, with a few small glands near apex. Achenes cylindrical, 1.7–2.5 mm long, densely hairy between ribs; pappus setae white, 4.5–7 mm long.

Reported to occur in Zimbabwe (fide Mapaura & Timberlake 2004) but I have not seen any specimens; similarly reported from Zambia, but the taxon seems to be restricted to low-lying coastal areas.

**Mozambique**. GI: Pomene, 24.ix.1980, *Jansen et al.* 7531 (K, LMU, WAG). M: Inhaca Is., 20.vii.1980, *de Koning & Nuvunga* 8311 (K, LMU, WAG).

Also known from Kenya, Madagascar, Mascarenes, South Africa. Open places in moist sites on sand or alluvium, in abandoned cultivations; can be locally common; 0–150 m.

Conservation Status: Due to a wide distribution in a common habitat, this should be Least Concern.

*Senecio pellucidus* DC. was listed in Da Silva *et al.* (2004: 34); this is a species of the southern and western Cape and part of the *Senecio madagascariensis* complex – I believe the basis of the inclusion in the checklist was a misidentified specimen of *S. madagascariensis* (such as *Jansen* 7359).

The South African *Senecio skirrhodon* DC., Prodr. **6**: 401 (1838). —Hilliard, Compositae Natal: 405 (1977). Type: South Africa, between Umzimvubu and Umzimkulu rivers [Omsamwubo & Omsamculo], *Drège* [5158] (G-DC holotype, P0000353, P0000354, P0000355). Synonym: *Senecio madagascariensis* var. *crassifolius* Humbert, Not. Syst. **15**: 372 (1959). —Humbert, Fl. Madagasc. Composées **189**: 736 (1963). Type: Madagascar, Ambovombe (Sud), 7.vi.1931, *Decary* 8968 (P00557662 holotype) – is known from Madagascar and in southern Africa from Natal S to the southern Cape Province, growing on the seashore, on sand in reach of the salt spray; it flowers throughout the year. According to Hilliard (1977: 405): "specimens of *S. madagascariensis* from near the sea grade into *S. skirrodon* which is possibly no more than a maritime form of *S. madagascariensis* and better placed as a variety of that species".

28. **Senecio maranguensis** O.Hoffm. in Engler, Pflanzenw. Ost-Afrikas C: 418 (1895). —Jeffrey in Kew Bull. **41**(4): 885 (1986). —Lisowski, (Asterac. Fl. Afr. Cent. 2) Fragm. Flor. Geobot. **36** Suppl. 1: 262 (1991). —Jeffrey & Beentje in F.T.E.A., Compositae **3**: 623 (2005). Type: Tanzania, 'Kiboscho to Marangu and on the Kiginika', *Volkens* 1127 (B† syntype);1491 (B†, BM syntypes).

   *Senecio scrophulariifolius* O.Hoffm. in Bot. Jahrb. Syst. **24**(3): 474 (1898). Type: Tanzania, Uluguru Mts, 6.xi.1894, *Stuhlmann* 9172 (B† holotype).

   *Senecio psiadioides* O.Hoffm. Bot. Jahrb. Syst. **30**(3): 436 (1901). Type: Tanzania, Kinga Mts, Bulongwa, 9.ix.1899, *Goetze* 1201 (B†, BR, P syntypes); Ussangu, Pikurugwe crest, 16.ix.1899, *Goetze* 1259 (B†, BM, BR, E, P syntypes).

   *Senecio subcarnosulus* De Wild., Pl. Bequaert. **5**: 118 (1929). Type: D.R. Congo, Ruwenzori, Butagu valley, 15.iv.1914, *Bequaert* 3728 (BR holotype).

Woody herb or shrub to 3 m or semi-scandent to 6 m; stems leafy, long and whippy, sometimes tinged purplish or red, finely pubescent or thinly arachnoid, glabrescent. Leaves petiolate, leathery, lanceolate, ovate or oblong, 3–16.5 × 1–6 cm, base cuneate, rounded, truncate or weakly cordate, margins finely to coarsely sinuate-serrate, dentate or bidentate, apex obtuse to acute; glabrous to finely pubescent especially on veins beneath, glabrous above or glabrous except for midrib and also sometimes main veins and with impressed veins; somewhat marcescent when old; petioles sparsely pubescent or arachnoid above to glabrous, 0.2–3 cm long, very narrowly winged, exauriculate or with minute to moderately conspicuous coarsely dentate stipuliform auricles at base. Capitula radiate (and in our material obscurely so!), many in copious rather congested to rather lax spreading terminal compound corymbs; stalk of individual capitula finely pubescent or arachnoid; involucre 3–5 × 1.7–2.5 mm; calycular bracts 3–6, 1–3 mm long, arachnoid or pubescent at least on margins and apex; phyllaries 7–8, green with brown tips, 2.5–5 × 0.7–1.1 mm , glabrous or sparsely arachnoid. Ray florets 5–8, corollas pale to bright yellow, tube 2.5–4 mm long, hairy at apex or glabrous, limbs 1.5 × 0.7 mm (3–5 × 1–2.2 mm elsewhere), 4-veined, spreading. Disc florets 16–22, corollas yellow turning red-brown, 4–6.2 mm long, tube glabrous, expanded above middle, lobes 0.5–0.7 mm long. Achenes ribbed, 1.5–2 mm long, glabrous; pappus setae 3.5–6 mm long.

**Malawi**. S: Mulanje Mountain, Nayawani shelf near top of Madzeka basin path, 28.vii.1970, *Brummitt* 12293 (K, MAL).

Also known from Burundi, D.R. Congo, Rwanda, Kenya, Uganda and Tanzania. Montane forest or forest margins, secondary bushland replacing forest; 1800–2100 m.

Conservation Status: Due to a wide distribution in a common habitat, this should be Least Concern.

Ray florets were reported to be absent in *Brummitt* 12293, but are there – albeit tiny!

29. **Senecio milanjianus** S.Moore in J. Linn. Soc., Bot. **35**(245): 359 (1902). —Brenan *et al.* in Mem. NY Bot. Gard. **8**(5): 484 (1954). —Jeffrey in Kew Bull. **41**(4): 897 (1986). —Mapaura & Timberlake, Checkl. Zimb. Vasc. Pl.: 28 (2004). —Jeffrey & Beentje in F.T.E.A., Compositae **3**: 644 (2005). Type: Malawi, Mt Milanji, *Whyte* s.n. (BM holotype).

   *Senecio tropaeolifolius* O.Hoffm. in Bot. Jahrb. Syst. **30**(3): 437, t. 21 (1901), nom. illeg., non MacOwan ex F.Muell. (1867). Type: Tanzania, Kinga Mts, Kipengere crest, 28.v.1899, *Goetze* 964 (B† holotype, BM, P).

   *Senecio conradii* Muschl. in Bot. Jahrb. Syst. **43**(1): 43 (1909), nom. nov. pro *S. tropaeolifolius* O.Hoffm.

Perennial herb from large underground rootstock to 2.5 cm across (and longer) from which arise erect leafy flowering stems 18–120 cm high; stem semi-succulent, pale green, purple- or red-tinged, glabrous. Basal and median stem leaves succulent, dull green with purplish margins or lower surface purple, paler and slightly glaucous beneath, petiolate, peltate and tending to fold in half down middle, circular, 3.5–9 cm in diameter, shallowly and regularly sinuate to almost lobed; petiole 2–10 cm long, purplish, exauriculate; upper stem leaves few, sessile, oblong to ovate, 3.5–5 × 1.2–2.2 cm, base semi-amplexicaul, margins entire, apex obtuse. Capitula radiate, erect in rather congested long-stalked terminal corymbose panicles with auriculate bracts/sessile leaves; peduncles of individual capitula less than 10 mm; involucre cylindrical, 5.5–9 × 3–5 mm; calycular bracts 6–7, linear, 1.3–7 mm long; phyllaries 10–13, yellow-green with purple tips, 6.5–9 × 0.4–1.1 mm . Ray florets 5–8, tube 3–5.5 mm long, glabrous, ray limbs bright yellow, 4.5–7 × 1.5–3 mm, 4–5-veined. Disc florets less than 20, corollas yellow, 4.5–7.2 mm long, tube glabrous, expanded from below middle, lobes 0.7–1.2 mm long. Achenes narrowly oblong, angled, 1.5–2.7 mm long, pubescent; pappus setae 4.5–6 mm long.

**Zimbabwe**. E: Chimanimani, Makurupini, 27.viii.1969, *Wild* A23 (K, SRGH). **Malawi**. S: Mt Mulanje, Chinzama–Sombani path, 3.iii.2005, *Bisson et al.* 22 (K, MAL). **Mozambique**. N: Mt Ingao, Namuruquele, 8.v.2009, *Matimele et al.* 96 (IIAM, K). Z: Mt Namuli, Muretha Plateau, 28.v.2007, *Patel et al.* 7358 (K, IIAM).

Also known from Tanzania. In rock crevices or on stone pavements with water seep, stems creeping between rocks; also in stony grassland or in *Widdringtonia* forest margins; 1550–2250 m.

Conservation Status: Due to a wide distribution in a common habitat and with a fairly wide altitude range, this should be Least Concern.

30. **Senecio ngandae** Beentje in Kew Bull. **74**-67: 2 (2019). Type: Malawi, Nyika Plateau, Nganda Peak, 5.iii.1979, *Pawek* 12476 (K holotype, MAL, MO, SRGH, UC). FIGURE 6.4.**5**.

Erect herb 40–90 cm high with basal leaves and fewer cauline leaves; stems unbranched except for inflorescence, glabrous except for inflorescence axes. Basal leaves narrowly elliptic, 20–35 × 1–2.7 cm, base long-attenuate, margins dentate, apex attenuate; glabrous except for some minute hairs on midrib and margins; cauline leaves similar but smaller, and more distal ones with cuneate to almost rounded base. Capitula discoid, 4–13 together in erect sub-corymbose panicles, with small bracts, inflorescence axes slightly hairy and with small bracts; peduncles of individual capitula 1–7.5 cm long; involucre campanulate; calycular bracts several, linear and to 5 mm long; phyllaries 20–21, 5.3–5.5 × 0.8–1 mm, sparsely hairy at base and with little tuft at apex, darker at apex. Florets 34–38, corollas maroon or crimson-brown, ± 4.5–6 mm long, papillate at lobe apices. Achenes 1.2 mm long, slightly 10-ribbed, glabrous; pappus setae white, 4.5–6 mm long.

**Malawi**. N: Nyika Plateau, 14.iii.1961, *Robinson* 4513 (K); idem, Nganda Peak, 5.iii.1979, *Pawek* 12476 (K, MAL, MO, SRGH, UC); idem, *Chapama et al.* 803 (K, MAL); idem, *Patel & Ludlow* 4995 (K, MAL).

Endemic to the Nyika plateau. Grassland; 2200–2400 m.

Conservation Status: This taxon has an EOO of less than 1 km² and an AOO of 4 km². As a strict and narrow endemic to the Nyika National Park, this is a clear example of Least Concern (Conservation Dependent), as it occurs in a well-protected area; any negative change in the status of the protected area would very quickly change the assessment.

31. **Senecio ngoyanus** Hilliard in Notes Roy. Bot. Gard. Edinburgh **32**(3): 382 (1973). —Hilliard, Compositae Natal: 429 (1977). —Da Silva *et al.*, Prelim. Checkl. Vasc. Pl. Mozamb.: 34 (2004). Type: South Africa, Natal, Mtunzini Dist., Ngoye, 14.xii.1968, *Hilliard & Burtt* 5632 (E holotype, K, NH, NU).

Perennial herb to 60 cm, from a stout woody stock to 1 cm across; stem solitary, erect, branched sub-dichotomously into a very open few-headed corymbose panicle, glabrous except for sparse short glandular pubescence on inflorescence branches. Leaves mostly radical, narrowly elliptic, 12–17 × 1–1.9 cm, half of this petiole, base attenuate, margins entire to denticulate, apex subacute, glabrous; cauline leaves much smaller and narrow, gradually passing into bracts. Capitula discoid, few in open corymbose panicles, campanulate, 6–7 × 4–8 mm; calycular bracts 2–3, 1–2 mm long and slender, often tinged purple; phyllaries ± 14, 4.5–5 × 0.3–0.5 mm, glabrous. Florets 30–50, corollas purple, 3.8–4.5 mm long, including 0.5–0.6 mm lobes; lobe tips minutely papillate. Achenes cylindric, 2.8–3 mm long, ribbed, glabrous; pappus setae white, 5–6 mm long.

**Mozambique**. M: Inhaca Island, Hlangwini swamp, 26.ix.1959, *Mogg* 27549 (K).

Also known from South Africa (Natal, Zululand). Swamp; low altitude.

Conservation Status: Assessed as Vulnerable on the Red List of South African Plants (2007).

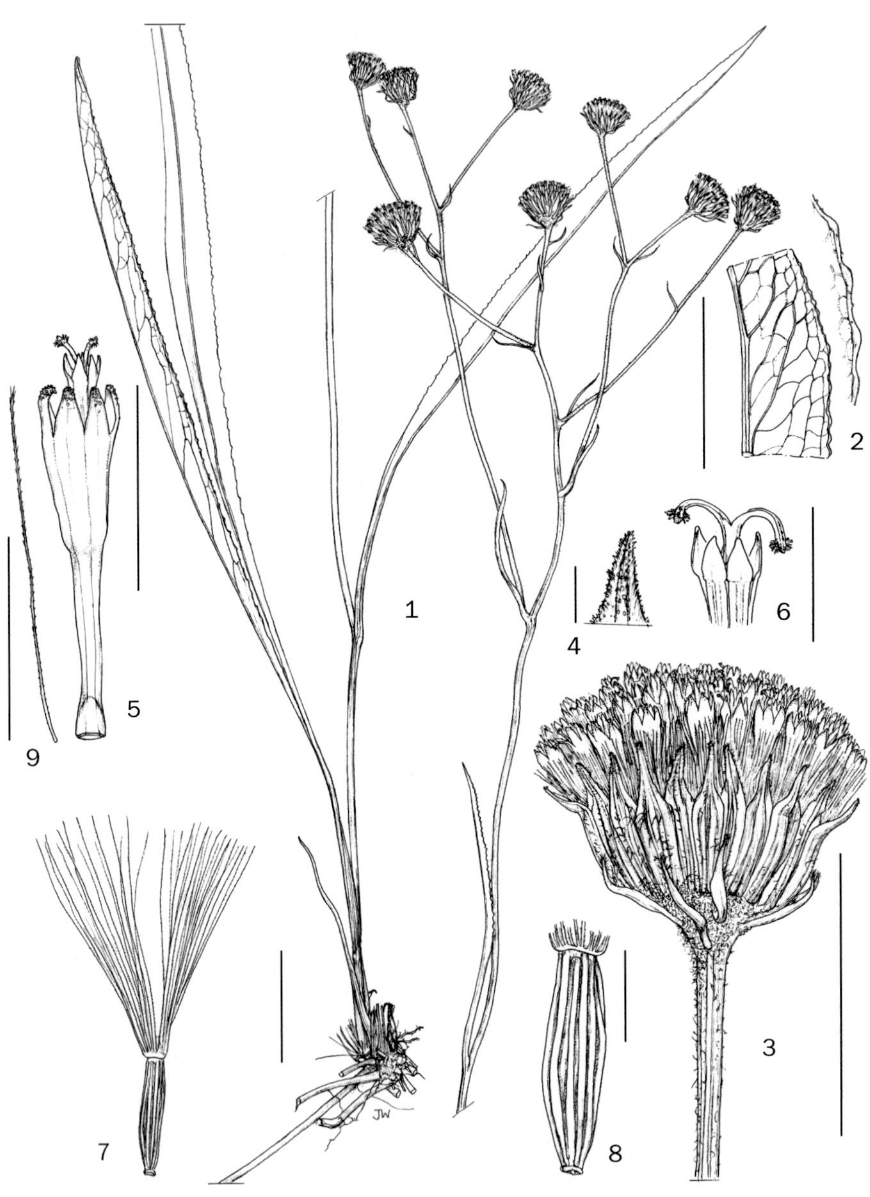

Fig. 6.4.5. SENECIO NGANDAE. 1, habit; 2, detail of abaxial leaf surface, with leaf margin, enlarged; 3, capitulum; 4, apex of phyllary; 5, floret corolla; 6, floret detail; 7, achene with pappus setae; 8, achene; 9, pappus seta. 1, 3–9 from *Chapama et al.* 803; 2 from *Pawek* 12476. Scale bars: 4, 6, 8 = 1 mm; 5, 7, 9 = 3 mm; 3 = 1 cm; 1, 2 = 3 cm. Drawn by Juliet Williamson. Reproduced from Kew Bulletin (2019) with permission of the artist.

32. **Senecio nyangani** Beentje in Kew Bull. **74**-67: 3 (2019). Type: Zimbabwe, Nyanga, Mtenderere source, *Wild* 4593 (K holotype, SRGH). FIGURE 6.4.**6**.

Perennial herb 60–120 cm high, with stem woody near base, branching only at inflorescence level; stem glabrous. Leaves densely set just below inflorescence, with marcescent leaves hanging down, upper leaves grading into inflorescence bracts; leaves narrowly elliptic, 4–9.5 × 0.4–0.7 cm, base sessile and cuneate, margins thickened and slightly revolute when dry, apex acute to an obtuse tip, glabrous. Capitula radiate, 5–16 together in terminal corymbose panicles; phyllaries 13–16, 7–7.5 × 0.7–1.2 mm, glabrous but for papillate apex. Ray florets 8, corollas yellow, limb 12–13.5 × 5 mm; disc florets ± 25, 7.5–8.3 mm long. Achenes 3.3–3.7 mm long, puberulous; pappus setae white, 8–9 mm long.

**Zimbabwe**. E: Nyanga (Inyanga), Mtenderere source, 5.ix.1954, *Wild* 4593 (K, SRGH); below summit of Inyangani, 22.ix.1958, *Drummond & Robson* 5833 (K, SRGH).

Not known elsewhere. In heath scrub and/or among boulders; 2100–2600 m.

Conservation Status: This taxon has an EOO of less than 1 km$^2$ and an AOO of 8 km$^2$. As a narrow and strict endemic to the Nyanga National Park, this is a clear example of Least Concern (Conservation Dependent), as it occurs in a well-protected area; any negative change in the status of the protected area would very quickly change the assessment.

One of the labels says *Goodier & Phipps* 206 (SRGH) is similar; M. Hyde (pers. comm.) confirms this looks very like *S. nyangani*.

33. **Senecio oxyriifolius** DC., Prodr. **6**: 405 (1838), as '*oxyriaefolius*'. —Hilliard, Compositae Natal: 473 (1977). —Jeffrey in Kew Bull. **41**(4): 897 (1986). —Lisowski, (Asterac. Fl. Afr. Cent. 2) Fragm. Flor. Geobot. **36** Suppl. 1: 303, t. 66 (1991). —Da Silva *et al.*, Prelim. Checkl. Vasc. Pl. Mozamb.: 34 (2004) —Mapaura & Timberlake, Checkl. Zimb. Vasc. Pl.: 28 (2004). —Jeffrey & Beentje in F.T.E.A., Compositae **3**: 646 (2005). —Burrows & Willis, Pl. Nyika Plateau: 105 (2005). —Phiri, Checkl. Zamb. Vasc. Pl.: 33 (2005). —Setshogo, Prelim. Checkl. Pl. Botsw.: 38 (2005). Type: South Africa: Cape Province, Vischrivier, *Drège* [5653] (G-DC holotype, P).

    *Senecio peltatus* DC., Prodr. **6**: 405 (1838). Type: South Africa: Zwarte Omsamculo & Omtendo, 25.ii.1832, *Drège* s.n. [5029] (G-DC holotype, P).

    *Senecio subnudus* DC., Prodr. **6**: 405 (1838). Type: South Africa: Stellenbosch, *Ecklon* s.n. [787] (G-DC holotype).

    *Senecio peltiformis* DC., Prodr. **6**: 405 (1838). Type: South Africa: Zwarte-Key & Buffelrivier, *Drège* s.n. [3776] (G-DC holotype, K)

    *Senecio subpeltatus* Steud., Nomencl. Bot., ed. 2, **2**: 565 (1841), nom. illeg. superfl. pro *S. subnudus* DC., Prodr. **6**: 405.

    *Senecio orbicularis* Sond. ex Harv. in Harvey & Sonder, Fl. Cap. **3**: 376 (1865). —Da Silva et al., Checkl. Vasc. Pl. Mozamb.: 299 (1993). Type: South Africa, Magalisberg, *Burke s.n.* (K, TCD syntypes); *Zeyher* 931 (BM, S syntypes).[20]

Perennial herb 40–130 cm high (Chase calls it an annual) with horizontal, fleshy tuberous rhizome (to 7 cm long and 2.5 cm diam.) and erect flowering stems; stems glabrous, succulent or rather succulent. Lower and median stem leaves succulent or somewhat succulent, pale green or glaucous blue-green, lowermost small, soon disappearing, median functional, 4–8, petiolate with petiole 2.5–10.5 cm long, lamina peltate, circular, subreniform-circular to almost rounded-triangular or rounded-subhastate, 2.5–9 × 3–10.5 cm, margins rather closely sinuate-dentate to broadly sinuate-apiculate-dentate, occasionally

---

[20] There is a handwritten 'determinavit' note label in C.A. Smith 's hand saying that the K syntype matches the *Zeyher* 931 syntype in the BM; this suggests that two collections were involved. The K sheet is marked 'Syntype' in Smith's hand, and is a *Burke* s.n. collection. Smith's note goes further suggesting that the BM material has no inflorescence, the one mounted on the sheet belonging to another species. Lectotypification, if necessary, would be better carried out designating the K syntype, which is also heavily annotated with sketches of a floret, achene and capitulum, and other descriptive notes.

slightly lobed, apex rounded to obtuse and apiculate, upper leaves few, remote, small, lanceolate, scale-like. Inflorescences of rather dense to spreading or lax many-headed terminal corymbose panicles. Capitula discoid; involucre cylindrical, pale green or glaucous, 6.5–8.5 mm long; calycular bracts 0–2, lanceolate, 1.5–5.5 mm long; phyllaries 7–12, 6–8 × 0.5–1 mm, pale green or glaucous. Florets 11–20, corollas pale to golden yellow, 4.8–6 mm long, tube narrowly infundibuliform and 4–5 mm long, lobes 0.7–1.6 mm long, glabrous or with a few papillae. Achenes 2–3.2 mm long, ribbed, beset with short hairs; pappus setae 5–6.5 mm long, barbellate, white.

Fig. 6.4.**6**. SENECIO NYANGANI. 1, habit; 2, capitulum; 3, ray floret corolla; 4, disc floret corolla; 5, disc floret detail; 6, disc floret achene and pappus setae; 7, achene; 8, pappus seta. 1 from *Drummond & Robson* 5833; 2–8 from *Wild* 48153. Scale bars: 5, 7, 8 = 3 mm; 2–4, 6 = 5 mm; 1 = 3 cm. Drawn by Juliet Williamson. Reproduced from Kew Bulletin (2019) with permission of the artist.

**Botswana**. SE: old road Hildavale–Digawane, 1312 m, 30.xi.2004, *Farrington et al.* MSB 129 (K, NPGRC). **Zambia**. N: road to Kasulu, 5000′, 1.i.1952, *Richards* 220 (K). W: 0.5 mile S of Matonchi Farm, 6.xii.1937, *Milner-Redhead* 3520 (K). **Zimbabwe**. N: Guruve (Sipolilo), Nyamunyeche Estate, 15.xii.1978, *Nyariri* 582 (SRGH, fide M. Hyde). W: Bulawayo, i.1914, *Rogers* 3563 (K). C: Mount Wedza, Dangamvuri, 1707 m, 27.ii.1964, *Wild* 6348 (K, SRGH). E: Chimanimani Mts, 3 mile S of Outward Bound school, 10.xi.1965, *Chase* 6338 (K, SRGH). S: Mberengwa (Belingwe), 17.iii.1964, *Wild* 6391 (SRGH, fide M. Hyde). **Malawi**. N: Vipya Plateau, 37 miles SW of Mzuzu, 5500′, 3.ii.1974, *Pawek* 8043 (K, MAL, SRGH). C: Dedza Mountain, 3.i.1967, *Jeke* 45 (K, SRGH). S: Mt Chiradzulu above Lisao Forest, 13.iii.1977, *Brummitt, Seyani & Patel* 14851 (K, MAL). **Mozambique**. no specimens seen, but reported in Da Silva *et al.* (2004).

Also known from D.R. Congo, Angola, Tanzania and South Africa (Cape Province, Transvaal). Grassland or bushed grassland, often in rocky sites, sometimes in miombo; 900–2300 m.

Conservation Status: Assessed as Least Concern on the Red List of South African Plants (2020).

34. **Senecio pachyrhizus** O.Hoffm. in Bot. Jahrb. Syst. **30**(3): 435, t. 19/a–g (1901). — Brenan *et al.* in Mem. NY Bot. Gard. **8**(5): 486 (1954). —Jeffrey in Kew Bull. **41**(4): 887 (1986). —Lisowski, (Asterac. Fl. Afr. Cent. 2) Fragm. Flor. Geobot. **36** Suppl. 1: 270 (1991). —Jeffrey & Beentje in F.T.E.A., Compositae **3**: 626 (2005). —Burrows & Willis, Pl. Nyika Plateau: 105 (2005). —Phiri, Checklist Zambian Vasc. Pl.: 33 (2005). Type: Tanzania, Usafwa [Usafua], 4.viii.1899, *Goetze* 1121 (B† holotype).

 *Senecio dekindtianus* Volkens & O.Hoffm. in Bot. Jahrb. Syst. **32**(1): 151 (1902). Type: Angola, Benguella, Tyindigiro, Lualo River source, *Dekindt* 880 (B† holotype, MPU).

 *Senecio adustus* S.Moore in J. Bot. **56**: 228 (1918). Type: Angola, Benguella, East of Keiando, along path to Boea Varaquanha, and near the Cubal rivulet, 14.vii.1905, *Gossweiler* 1749 (BM000924790 holotype, BR8878076, LISC014905).

Dwarf perennial herb with woody rootstock, inflorescences appearing before leaves; rootstock 1–2.5 cm in diameter, crown often woolly and with conspicuous fibrous remnants of old leaf-bases. Stems several from each crown, forming small cushions 5–16 cm high, pale green, minutely glandular-pubescent and sometimes also slightly arachnoid, glabrescent. Leaves (sometimes more than 20 from rootstock) all basal, oblong to elliptic, 13–35(–45) × 1–4.5 cm, long-attenuate into petioloid base, margins crenate-serrate, apex acute to attenuate, arachnoid-pubescent above and beneath but grey-green above and silvery or whitish beneath; stem-leaves scale-like, sessile, concave, pale green or brown, oblanceolate or lanceolate, 1–2.5 × 0.2– 0.5 cm, margins erose-denticulate, apex acuminate or acute, glabrous or sparsely arachnoid beneath, uppermost ones merging into calyculus. Capitula discoid, erect, terminal on stems 2–14 cm long, usually solitary; involucre broadly cylindrical, brownish, 15–22 × 11–18 mm, glabrous or sparsely arachnoid at base; calycular bracts 5–8, broadly ovate to lanceolate, 10–20 × 1.2–8 mm, acuminate or acute, shortly erose-denticulate; phyllaries 16–25, 15–22 mm long, long-attenuate, margins short-ciliate near apex but otherwise glabrous. Florets 30–60, corollas white or cream, anthers fully exserted, 15.5–20 mm long, tube shortly hairy on narrow cylindrical lower part, rather abruptly expanded and campanulate in upper one-sixth, lobes 2–2.5 mm long. Achenes (mature) 11 mm long, densely long-setuliferous; pappus setae 8–22 mm long, white.

**Zambia**. N: Mpika, Mutinondo Wilderness Area, 31.viii.2002, *Bingham* 12554 (K). **Malawi**. N: Nyika National Park, Sawi Valley, 11.ix.1972, *Synge* 434 (K, MAL, SRGH). C: Chenga Hill, 9.ix.1946, *Brass* 17602 (K, NY).

Also known from Cameroon, D.R. Congo, Burundi, Uganda, Tanzania and Angola. Pyrophytic, in upland grassland and miombo, flowering about a month after fire; in large clumps or cushion-forming; 1250–1750 m.

Conservation Status: Widespread in common habitats, Least Concern.

35. **Senecio peltophorus** Brenan in Brenan *et al.* Mem. New York Bot. Gard. **8**(5): 484 (1954). Type: Malawi, Mlanje, Tuchila Plateau, v.1901, *Purves* 10 (K holotype).

Herb 10–40 cm; rhizome thick to tuberous, 6–10 mm across; stem pale green, glabrous, leafy only near base. Leaves fleshy, peltate (petiole 1.7–5.7 cm long), pale green, sometimes mauve or grey below, ovate to suborbicular, 1.5–3 cm in diameter, margins sinuate to almost 10-lobed, glabrous. Inflorescences of dense many-headed long-stalked corymbose panicles to 12 cm, with small auriculate sessile bracts 1–5 mm long; peduncles of individual heads 4–10 mm long. Capitula radiate; involucre 4–5.5 mm long; calycular bracts few, narrowly ovate to linear, 1–2.5 mm; phyllaries 7–11, linear to oblong, 4–5.5 × 0.4–0.6 mm, at apex purplish and with a small tuft of hairs. Ray florets 4–9, corollas yellow, tube 2.2–3.5 mm long, lamina 4.5–5 × 1.5–3.5 mm. Disc florets 10–13, corollas yellow, 3–3.9 mm long. Achenes oblong-fusiform, 1.7–2 mm long, angled, pubescent; pappus setae 3–3.5 mm.

**Malawi**. S: Mt Mulanje, Litchenya Plateau, 30.iv.2008, *Patel et al.* 874 (FRIM, K). **Mozambique**. Z: Mt Namuli, Muretha Plateau, 31.v.2007, *Timberlake* 5141 (K, LMA).

Not known elsewhere. Thin soil over rock, seepage; 1600–2200 (?2500) m.

Conservation Status: This taxon is known from ten specimens, but occurs in a fairly common habitat with a wide altitude range. A careful check needs to be made of occurrence in protected areas, and threats to unprotected sites, to decide whether the taxon is Least Concern or Vulnerable.

The protologue says this species is very close to *S. milanjianus* but *S. peltophorus* is terrestrial and smaller in all parts; I feel certain these are two good spp.

36. **Senecio polyanthemoides** Sch.Bip. in Flora **27**(40): 697 (1844). —Hilliard, Compositae Natal: 407 (1977). —Da Silva *et al.*, Prelim. Checkl. Vasc. Pl. Mozamb.: 34 (2004). Type: South Africa, Natal Bay [Natalbai] and near Umlaas River, x.1839, *Krauss* 424 (?P holotype, FR, K, M, TUB).

Annual herb 1–1.8 m tall, bushy or simple-stemmed; stems woody at base, sometimes arachnoid-pubescent, minutely glandular, leafy throughout. Leaves lanceolate, 6–15 × 0.7–4 cm, becoming smaller upwards, base narrowed, petiole-like, sometimes auriculate, sometimes slightly decurrent on stem, margins revolute, lower leaves sharply serrate or more deeply pinnately divided, upper leaves entire, apices acute to acuminate, smooth or scabrid above, white- or grey-felted beneath. Inflorescence corymbose-paniculate, many-headed and up to 20 cm across. Capitula radiate, in corymose panicles; involucre campanulate, 3.7–5 × 3–4 mm; calycular bracts 1–1.5 mm long; phyllaries 14–20, 3.5–4.7 × 0.3–0.7 mm, glabrous, keeled, 1–3-veined, resinous. Ray florets ± 8, corollas canary yellow, tube very narrow and 3–3.5 mm, ray limbs 3–4 × 0.8–1.6 mm. Disc florets 30–42, corollas canary yellow, narrowly infundibuliform, 3–4.5 mm long of which lobes 0.6–0.8 mm. Achenes 1.2–1.7 mm long, cylindrical, striate, slightly angled, sparsely pilose; pappus setae white, 4–4.8 mm long.

**Mozambique**. GI: Gaza, Vile de João Belo Cotton Experimental Fields, 8.x.1945, *Pedro* 276 (K, PRE). M: Maputo, Namaacha, 25.viii.1967, *Gomes e Souza & Balsinhas* 4941 (K).

Also known from South Africa (Natal, Transvaal), Eswatini. On alluvium (clay soils) and abandoned cultivated areas, sometimes in colonies; visited by bees; near sea level.

Conservation Status: Based on distribution and preference for common and ruderal habitat, Least Concern.

Very similar to *S. polyanthemoides* is *S. pterophorus* DC., Prodr. **6**: 389 (1838). —Hilliard, Compositae Natal: 407 (1977). —Da Silva *et al.*, Prelim. Checkl. Vasc. Pl. Mozamb.: 34 (2004). Type: South Africa, between Umsikaba and Umzimkulu rivers [Omsamcaba & Omsamculo], *Drège* s.n. [5154] (G-DC holotype, HAL, HBG, K, P); the only difference seems to be that *S. pterophorus* has bigger capitula with 13 ray florets. Da Silva *et al.* (2004: 34) report this from Mozambique, but I have seen no specimens from our area. *S. pterophorus* is known from South Africa (Cape Province, Natal). Hilliard (1977: 407–408) suggests this only just gets into S Natal – which implies Mozambique occurrence is unlikely.

37. **Senecio polyodon** DC., Prodr. **6**: 386 (1838). —Hilliard, Compositae Natal: 430 (1977). —Mapaura & Timberlake, Checkl. Zimbab. Vasc. Pl.: 28 (2004). Type: South Africa: 'in Africâ australi', x.1829, *Drège* s.n. [6359] (G-DC holotype, P).

 *Senecio versicolor* Hiern, Cat. Afr. Pl. **1**(3): 597 (1898). Type: Angola, Huila, xii.1859, *Welwitsch* 3677 (BM, LISU syntypes); Cacolovar river between Ivant âla and Quilengues, ii.1860, *Welwitsch* 3678 (BM, K, LISU syntypes).

 *Senecio metallicorum* S.Moore, J. Bot. **41**: 134 (1903). Type: South Africa, North of Johannesburg, x.1902, *Rand* 979 (BM holotype).

 *Senecio breyeri* S.Moore, J. Bot. **59**: 230 (1921). Type: South Africa, Transvaal, Nelspruit, Barberton, x.1917, *Breyer* s.n. [in herb. Rogers 24019] (BM holotype, PRE).

## Var. **polyodon**

Perennial herb 20–100 cm; stems solitary or several from root crown, branching only in inflorescence, often decumbent at base and rooting, glandular-pubescent. Radical leaves in a rosette, spatulate to elliptic, 3–7 × 0.8–2 cm including petiole, base attenuate into petiole, margins coarsely dentate to sinuate-denticulate, apex acute to obtuse; glandular-pubescent on both surfaces; cauline leaves narrowly spatulate, 2–8 × 0.2–0.8 cm, base broad and clasping, margins callose-dentate, almost entire to subpinnatifid, glandular-pubescent on both surfaces, called 'aromatic' by some collectors. Capitula in lax panicles, discoid; involucre campanulate and 6 × 7 mm; calycular bracts few, burgundy-red, 1.5–2 × 0.4–1.2 mm; phyllaries green with burgundy-red tips, 14–20, 4.8–9 mm long, glandular-pubescent. Florets 40–80, corollas white or dull yellow, narrowly infundibuliform, 4.2–6.2 mm long including 0.5–0.6 mm long lobes; anthers purple or blue. Achenes cylindric, 2–2.2 mm long, ribbed, hispid between ribs; pappus setae white, 4.5–6 mm long.

**Zambia**. N: Mutinondo Wilderness area, Big Kabasano dambo, 20.xii.2012, *Merrett* 1095 (K). **Zimbabwe**. C: Marondera (Marandellas), Hunyani source, 27.ix.1949, *Wild* 2978 (K, SRGH). E: Nyanga (Inyanga) Mare River, 21.x.1946, *Wild* 1517 (K, SRGH). **Malawi**. N: Nyika, Chelinda Dam, 7500ʹ, 5.xi.1977, *Pawek* 13160 (K, MA, MO, SRGH, UC).

Also known from Angola, Lesotho, South Africa (Transvaal, Natal, Orange Free State) and Eswatini. Streamside grassland, grassland on peaty soil or in dambo, may be locally common; 1400–2350 m.

Conservation Status: Based on the wide distribution and preference for common habitat and large altitude range, Least Concern.

Some specimens at Kew were named *S. erubescens*, but Wild remarks (on his 1370) that while the flowers had a purplish tinge, this was due to purple anthers (or, I might add, the violet stigmas); the florets were consistently dull yellow in that population.

Our material belongs to var. *polyodon*; the other variety, *S. polyodon* var. *subglaber* (O.Hoffm. ex Kuntze) Hilliard & B.L.Burtt, is endemic to South Africa and Lesotho and is distinguished by the almost glabrous stems, leaves and inflorescence.

After this account was written two more specimens from Zambia turned up: C: Mutinondo Wilderness area, Big Kabasano dambo, 24.x1.2015, *Merrett* 2000 & idem, 19.x.2016, *Merrett* 2139. These are very much like *S. polyodon*, which is known from the same area (e.g. *Merrett* 1095, 1122, 1709) – but differ in slightly larger leaves (to 12.5 × 3 cm) and phyllaries (9.3–10.5 mm), larger florets (4.8–7.8 mm) and achenes (2.5–3.3 mm), and yellow florets with yellow anthers. The other *Merrett* specimens cited have measurements that agree with the above description for *S. polyodon*, and in the one specimen that has a flower description, white florets with blue-black anthers. *Merrett* 2139 also lacks glandular hairs on the leaves, and has multicellular hairs instead. I am not sure what to make of these differences – the general look is very much that of *S. polyodon*.

38. **Senecio proprior** S.Moore in J. Linn. Soc. Bot. **40**: 118 (1911), as '*propior*'. Type: Zimbabwe, Chimanimani Mts at 7000ʹ, 26.ix.1906, *Swynnerton* 1879 (BM holotype, K).

Erect perennial herb, 45–90 cm; rootstock 3–6 × 1.5–3.5 cm, with fibrous roots; root crown often (in burnt areas?) with a dense mass of fibres surrounding lower stem and leaf bases; stems 1–2 from roots, nearly always unbranched except for inflorescence, thinly pubescent (with dark-tipped light hairs) but soon becoming glabrous. Leaves near base narrowly elliptic to almost linear, 16–50 × 0.5–2.5(3.2) cm, base long-attenuate with petiole-like part about ⅓ of its length but very base almost sheathing, margins denticulate or serrulate with slightly callose teeth, apex acute, hairy; gradually changing to upper leaves which are sessile with a sub-auriculate base, and sometimes more strongly dentate, 2–17 × 0.4–2(3.9) cm; densely puberulous-pubescent when very young but soon glabrous or nearly so. Inflorescences lax with relatively few (4–13) capitula in slightly corymbose panicles; axes thinly pubescent to glabrous, parts near small dentate bracts often remaining hairy longest. Involucre cylindrical; calycular bracts several, linear, 1.8–4 mm long and often pubescent at base; phyllaries 11–14(17), light green, 5.3–7.5 × 0.3–1 mm with pale membranous margins and with darker papillate tips, puberulous or glabrous. Capitula radiate, or rarely discoid only (*Hilliard* 4307). Ray florets 7–9, corollas bright yellow, tube 4–4.8 mm and lamina 5–11 × 1.5–3.4 mm. Disc florets 30–65, corollas yellow, narrowly infundibuliform, 5.6–7 mm long of which lobes 0.5–0.7 mm. Achenes oblong, pale brown, 1.5–3(4.9) mm, angular, hairy on angles (sometimes black and glabrous, *Juwawo* 437, rarely pale and glabrous, *Hilliard* 4307); pappus setae white, 5–7.5 mm.

**Zambia**. N: Mutinondo Wilderness area, Mayense, xii.2012, *Merrett* 1117 (K). **Zimbabwe**. C: Wedza, Mt Wedza, 27.ii.1964, *Wild* 6351 (K, SRGH). E: Nyanga Downs, 2100 m, 7.xii.1959, *Wild* 4892 (K, SRGH). **Malawi**. N: Nkhata Bay Dist., Vipya Link road, 11.i.1975, *Pawek* 8943 (K, MAL, MO, SRGH, UC). C: Dedza, Chongoni Forest, 27.x.1965, *Banda* 700 (K, SRGH). S: Mt Mulanje, Sombani Hut, 2115 m, 18.ii.2006, *Juwawo et al.* 437 (FRIM, K, MAL). **Mozambique**. T: Tete, Ulongue, 21.xi.1980, Macuacua 1294 (K, LMA); MS: Chimanimani Mts, 8.vi.1949, *Wild* 2918 (K, SRGH).

Endemic to the FZ area. Grassland, miombo woodland, sometimes in moist sites; can be locally common; (850)1400–2350 m.

Conservation Status: This endemic has a moderately wide distribution, in common habitats and with a wide altitude range; Least Concern.

Several of the collections from Mutinondo (Zambia) have irregular numers of ray florets, varying from 4–6 per head; these, and others, also have cauline leaves that are wider than is usual for this taxon.

39. **Senecio purpureus** L., Syst. Nat., ed. 10, **2**: 1214 (1759). —Hilliard & Burtt in Notes Roy. Bot. Gard. Edinburgh **34**(1): 96 (1975). —Hilliard, Compositae Natal: 419 (1977). —Jeffrey in Kew Bull. **41**(4): 904 (1986). —Lisowski, (Asterac. Fl. Afr. Cent. 2) Fragm. Flor. Geobot. **36** Suppl. 1: 320 (1991). —Da Silva *et al.*, Prelim. Checkl. Vasc. Pl. Mozamb.: 34 (2004). —Mapaura & Timberlake, Checkl. Zimb. Vasc. Pl.: 28 (2004). — Jeffrey & Beentje in F.T.E.A., Compositae **3**: 668 (2005). —Phiri, Checkl. Zamb. Vasc. Pl.: 33 (2005). —Gereau *et al.*, Lake Nyasa Climat. Reg. Florist. Checkl.: 31 (2012). Type: none mentioned in protologue; '*Senecio viscosus aethiopicus, flore purpureo*' Breyne, Exot. Pl. Cent. t. 67 (1678), iconotype, lectotypified by Hilliard (1977: 419).

 *Cacalia villosa* Jacq., Icon. Pl. Rar. **3**: 14, t. 580 (1789), nom. illeg. based on *S. purpureus* L. (1759).
 *Senecio mucronulatus* Sch.Bip. in Flora **27**(40): 701 (1844). Type: South Africa, all of Zwellendam district, xi.1838, *Krauss* 555 (P0000062 holotype, FR, G).
 *Senecio bussei* Muschl., Bot. Jahrb. Syst. **43**(1): 67 (1909). Type: Tanzania, SE Usagara, upper Mgalca Valley, *Busse* 1315 (B† holotype, EA).
 *Senecio lactucifolius* S.Moore in J. Bot. **54**: 283 (1916). —Mapaura & Timberlake, Checkl. Zimb. Vasc. Pl.: 28 (2004), as '*lactucaefolius*'. Type: Zimbabwe, Melsetter, *Swynnerton* s.n. (BM holotype).
 *Senecio lubumbashiensis* De Wild., Pl. Bequaert. **5**: 111 (1929). Type: D.R. Congo, Lubumbashi R source, *De Giorgi* 299 (BR holotype).

Erect perennial herb or woody herb with many stems from woody base, 0.6–2.4 m tall but usually 1.2–1.5 m; stems thick, hollow, pale green, with twisty glandular hairs. Basal leaves sessile, oblong-elliptic or lyrate, 35–73 × 9–12.5(15) cm, tapered into a petioloid base, margins usually pinnately lobed with large, oblanceolate, bidentate-lobulate terminal lobe and smaller ovate or broadly ovate dentate lateral lobes with rounded and apiculate apex (less often leaves almost entire); median and upper leaves sessile, lanceolate, oblong, oblong-elliptic or broadly elliptic in outline, 5.5–18 × 2.2–9 cm, base auriculate, margins pinnately lobed especially in lower part with ovate denticulate lobes, in uppermost leaves often simply dentate, apex obtuse to acute, acuminate-apiculate; all leaves sparsely shortly arachnoid above, ± finely arachnoid-tomentose beneath, with small glandular hairs. Inflorescences of many-headed corymbose-panicles, peduncles densely glandular-pubescent. Capitula discoid; involucre turbinate to campanulate, 7–9 × 3–3.5 mm; calycular bracts 4–9, lax, lanceolate, 2–5 mm long, with stalked glands; phyllaries 10–14, 5–8.5 × 0.4–0.9 mm, green with purplish tips to entirely purple, with short-stalked glands. Florets 20–70, corollas purple or mauve, rarely blue or white, thin lower tube and wider upper part, 4.7–7.2 mm long including lobes 1–1.2 mm long. Achenes narrowly cylindrical, 3–5 mm long, ribbed, glabrous or rarely pilose between ribs; pappus setae 5–7 mm long.

**Zambia**. N: Mutinondo, 14.i.2013, *Merrett* 1140 (K). W: Solwezi, 8.i.1969, *Mutimushi* 2929 (K, NDO). **Zimbabwe**. E: Mutare, Engwa, 2.iii.1954, *Wild* 4434 (K, SRGH). **Malawi**. N: Misuku Hills, Mugesse Forest, 14.vii.1970, *Brummitt* 12127 (K, MAL). S: Zomba Plateau, Chitinji Valey, 26.x.1984, *Salubeni & Tawakali* 3911 (K, MAL). **Mozambique**. T: Ulongue, Chitambe, 17.xii.1980, *Macuacua* 1458 (K, LMA). MS: Mossurize, 22.ii.1907, *Johnson* 144 (K).

Also known from D.R. Congo, Angola, South Africa (Cape Province, Natal). Boggy or marshy ground, grassland, sometimes in ruderal places; 1050–2100 m.

Conservation Status: With a wide distribution, common habitats and a wide altitude range, this must be Least Concern.

*Johnson* 144 says the whole plant smells of *Mentha* (mint).


40. **Senecio randii** S.Moore in J. Bot. **37**(422): 402 (1899). —Lisowski, (Asterac. Fl. Afr. Cent. 2) Fragm. Flor. Geobot. **36** Suppl. 1: 267 (1991). —Mapaura & Timberlake, Checkl. Zimb. Vasc. Pl.: 28 (2004). Type: Zimbabwe, Harare [Salisbury], ix.1898, *Rand* 625 (BM holotype, MO).[21]

Short-lived herb from a slightly thicker base emitting many erect or upwards-curving stems; these 15–40 cm high, glabrous. Leaves with short petiole in lower ones, upper sessile; blade narrowly obovate to oblong, 0.8–4.5(7.5) × 0.2–0.6 cm, base attenuate, margins subentire to remotely callose-dentate, apex obtuse; glabrous. Capitula solitary on branches or in leafy branched few-headed inflorescences, causing whole plant to look many-headed, radiate, 5–8 mm long, long-stalked; calycular bracts 7–8, minuscule; phyllaries 20–23, very narrowly elliptic, 4–6 × 0.6–1.3 mm, acute to narrowly acuminate, with a little tuft of hairs near apex, otherwise glabrous, inner with short glands on darker part between wide midrib and pale margins. Ray florets 10–15, corollas yellow, tube 2–3.5 mm long, ray limbs 4.5–6 × 1.3–2.5 mm. Disc florets 56–110, tube 4–5 mm long, lobes 0.5–0.8 mm, acute. Achenes narrowly ellipsoid to narrowly obovoid, 2–2.5 mm, pubescent; pappus setae 5–5.2 mm long, white.

**Zambia**. W: Mpongwe, 2.ix.1963, *Fanshawe* 7965 (K, NDO). C: Mumbwa, Kafue Park entrance, 26.ix.1964, *van Rensburg* 2971 (K). **Zimbabwe**. N: Mazowe [Mazoe], 10.ix.1971, *Biegel* 3603 (K, SRGH). C: Selukwe, Umbetekwana River, 10.ix.1975, *Wild* 8014 (K, SRGH). E: Nyangani [Inyangani], iv.1935, *Gilliland* 1890 (K). S: Masvingo (Victoria) Dist., kyle National Park, 7.v.1973, *Basera* 755 (SRGH, fide M. Hyde).

Also in D.R. Congo. In moist clayey dambos, along drainage lines, vleis, seasonally waterlogged black soil where it may form large patches; also a colonist of cultivated areas (fide Flora of Zimbabwe); (1050)1350–2550 m.

Conservation Status: Despite its rather restricted range, this has to be assessed as Least Concern due to its habitat preferences and altitude range.


41. **Senecio ruwenzoriensis** S.Moore in J. Linn. Soc. Bot. **35**: 355 (1902). —Hilliard & Burtt in Notes Roy. Bot. Gard. Edinburgh **34**(3): 280–281 (1976). —Hilliard, Compositae Natal: 475 (1977). —Jeffrey in Kew Bull. **41**(4): 897 (1986). —Lisowski, (Asterac. Fl. Afr. Cent. 2) Fragm. Flor. Geobot. **36** Suppl. 1: 296, t. 63 (1991). —Mapaura & Timberlake,

---

[21] Lisowski (1991: 267) says the type is *Frank* 625 – but Frank was Rand's first name!

Checkl. Zimb. Vasc. Pl.: 28 (2004). —Jeffrey & Beentje in F.T.E.A., Compositae **3**: 643 (2005). —Burrows & Willis, Pl. Nyika Plateau: 105 (2005). —Phiri, Checkl. Zamb. Vasc. Pl.: 33 (2005). —Gereau *et al.*, Lake Nyasa Climat. Reg. Florist. Checkl.: 31 (2012). Type: Uganda, Ruwenzori, 1893–1894, *Scott Elliot* 8043 (BM holotype, K).

*Senecio paucifolius* DC., Prodr. **6**: 403 (1838), nom. illeg., non S.G.Gmel. (1770–1774). —Oliver & Hiern, F.T.A. **3**: 414 (1877). Types: South Africa, Uitenhagen et Tambukiland, *Ecklon* [1253 & 1408] (G-DC, ?HAL, ?P syntypes), Zeeurobergen, 11.xi.1829, *Drège* 5847 (G-DC, P syntypes).

*Senecio othonniformis* Fourc., Trans. Roy. Soc. South Africa **21**(1): 89 (1932). —Da Silva *et al.*, Prelim. Checkl. Vasc. Pl. Mozamb.: 34 (2004). Type as for *S. paucifolius* DC.

Perennial herb, 15–100 cm high, with creeping fleshy rhizome (to 2 cm thick, sometimes over 7 cm long) sending out short-lived stems; stems leafy, erect, pale green, sometimes purplish at base, glabrous. Leaves cauline, succulent, light green, usually with glaucous bloom beneath, sessile, obovate to almost rhombic, oblanceolate, elliptic or lanceolate, 2.5–12 × 1–5.3 cm, base rounded to cuneate, margins entire or few-denticulate (strongly dentate to sub-lobed in *Chase* 6757 and *Pawek* 7920; almost 3-lobed in *Chase* 7241), apex rounded to obtuse and shortly apiculate, glabrous; pinnately veined but veins strongly ascending and often with two especially prominent ascending lateral veins from near base. Capitula radiate, 1–20 in a lax, long-stalked leafless terminal corymb; involucre cylindrical, slightly swollen at base, 7–10 × 4–8 mm, glabrous; calycular bracts 1–2, lanceolate, 1–2 mm long, glabrous; phyllaries 10–14, pale green and sometimes with darker tips, 6–9 × 0.6–1.4 mm long, glabrous, shortly tapered and bearded at apex. Ray florets (5)8(9), corollas pale to bright yellow, tube 3.5–4.5 mm long, glabrous or with a few hairs in upper part, ray limbs 5–11 × 1.5–3.5 mm, 4-veined. Disc florets 9–35, corollas yellow, 4.5–7.5 mm long, gradually expanded above middle, lobes 0.5–1 mm long; anthers brown; style yellow. Achenes 2–4 mm long, ribbed, glabrous or sometimes sparsely hairy; pappus setae 4–7 mm long, white.

**Zambia**. C: Mumbwa, no further data, *Macaulay* 1065 (K). E: Makutus, 29.x.1972, *Fanshawe* 11633 (K, NDO). **Zimbabwe**. W: Matobo, Besna Kobila, i.1961, *Miller* 7639 (K, SRGH). C: Marondera (Marandellas), Looe, 22.xii.1948, *Wild* 2709 (K, SRGH). S: Zimbabwe ruins, 1.vii.1930, *Hutchinson & Gillett* 3325 (K, PRE). **Malawi**. N: Mzimba, road to Mazamba Forest, 27.iv.1967, *Salubeni* 680 (K, MAL, SRGH). E: Mutare (Umtali), Tsetsera Mts, 450ʹ, 20.xi.1957, *Chase* 6757 (K, SRGH). S: Chiradzulu Mountain above Lisao Forest, 13.iii.1977, *Brummitt, Seyani & Patel* 14860 (K, MAL). **Mozambique**. Z: Moramballa, 18.i.1863, *Kirk* s.n. (K).

Also known from Nigeria to the Sudan, D.R. Congo, Rwanda, Burundi, Kenya, Uganda, Tanzania, South Africa (Cape Province, Natal, Orange Free State, Transvaal) and Lesotho. Shallow soil among rocks, rocky grassland or bushland, occasionally in wooded grassland, sometimes by streams or a weed of cultivation; (250)750–2550 m.

Conservation Status: Due to a wide distribution, a range of habitats and a wide altitude range, this must be Least Concern.

Comes up after the first rains, also flowers after fires. Hilliard (1977: 475) says widespread but seemingly highly localized.

42. **Senecio speciosus** Willd., Sp. Pl., ed. 4, **3**(3): 1991 (1803). — Ker in Edwards, Bot. Reg. **1**: t. 41 (1815). —Wood, Natal Pl. **6**(2): t. 550 (1910). —Hilliard & Burtt, Notes Roy. Bot. Gard. Edinburgh **34**(1): 98 (1975). —Hilliard, Compositae Natal: 433 (1977). —Bandeira *et al.*, Wild Flowers South. Mozamb.: 215 (2007). Type: Andrews, Bot. Repos. **5**: 291, t. 291 (iconotype).[22]

*Senecio pseudo-china* sensu Andrews, Bot. Repos. **5**: t.291 (1803), non L. (1753)[= *Gynura pseudochina* (L.) DC.].

---

[22] It is probable that Willdenow merely based his description on the description and plate in Andrews Botanical Repository, so Andrews' original plate of 'Senecio pseudo-china' should probably be considered the type. No herbarium material is known to exist of Andrews plants' which were all drawn from living material. It was suspected that the material came from China, although this has largely been discredited; it is highly probable that material came back onboard ships homeward bound from China, via the Cape, to England.

*Senecio concolor* DC., Prodr. **6**: 407 (1838). —Hooker, Bot. Mag. **109**: t. 6713 (1883). Types: South Africa, Cape, Tulbagh, *Drège* 5914 (G-DC lectotype), lectotypified by Hilliard & Burtt (1975: 98); 'Zneeurobergen' [Zuurebergen], *Drège* s.n. (G-DC syntype); Katriviersberg, *Drège* s.n. (G-DC syntype).

*Senecio concolor* [var.] β *hispido-scaber* DC., Prodr. **6**: 407 (1838). Types: South Africa, 'Between the Kei and Buffalo Rivers, *Drège* 5892' (G-DC lectotype, E, K, P), lectotypified by Hilliard & Burtt (1975: 98); Albany & Uitenhagen, *Ecklon* s.n. (G-DC syntype); Zwarte Key et Buffelrivier, *Drège* s.n. (G-DC syntype).

*Senecio concolor* var. *hispidus* Harv. in Harvey & Sonder, Fl. Cap. **3**: 363 (1865). Types: South Africa, Uitenhage and Albany 'in Br. Caffraria', *Ecklon* s.n.; *D' Urban* s.n.; *Hutton* s.n.; *Genl. Bolton* s.n.; Natal, *Gueinzius* s.n.; Sanderson *s.n.* (K, S, TCD syntypes).

*Senecio concolor* var. *lyratus* Harv. in Harvey & Sonder, Fl. Cap. **3**: 363 (1865). Types: South Africa, Simonstown, *Wright* 309 (K lectotype), lectotypified by Hilliard & Burtt (1975: 98); Modderlagd (*sic*), *Zeyher 941* (?K, ?S, ?TCD syntypes).

Perennial herb 30–100 cm tall; rootstock horizontal and woody to subfleshy, ± 15 mm diameter. Leaves mostly radical and looking like a rosette, ± fleshy, spatulate or elliptic, 10–40 × 2–5(11) cm, gradually attenuate into flat pseudopetiolate broadly-winged base 7–15 mm wide, base tightly clasping, margins pinnately lobed, sinuately lobed, lobulate or coarsely deltate-toothed, margins of lobes appearing ciliate (cilia of long uniseriate, multicellular eglandular hairs), apex subacute to obtuse, lamina glabrous or very sparsely pubescent over midrib sometimes (in some South African material) appearing 'lightly cobwebby' pubescent near margins, hairs uniseriate, multicellular, eglandular or with amber-coloured glandular tips. Flowering stems erect, striate, simple at base, branched higher up, glabrous or (not in our material) long-pilose or glandular-pilose. Inflorescence corymbose, few-bracted, lower bracts reduced and leaf-like, upper bracts much-reduced and ± scale-like, oblong to lanceolate, 7–70 × 2–10 mm, margins long-ciliate; capitula stalked for 15–60 mm. Capitula radiate, fragrant; calycular bracts few, 2–4 mm long, glabrous; phyllaries 16–20, (9)10–14 × 1.5–2 mm, glabrous (in ours), sometimes sparsely or moderately to densely pubescent (in South African material), hairs eglandular (in some South African material with distinctive amber-coloured glandular tips), margins scarious. Ray florets 8–12 (14–20), tube 6–6.5 mm long, ray limbs deep pink to purple, ± 8 × 2–2.5 mm, 4–7-veined. Disc florets many, corollas deep pink to purple, tube ± 4 mm long, widening to throat ± 3.5 mm long. Achenes cylindrical, 3.5–3.9 mm long, ribbed, sparsely to moderately pilose between ribs with twin-hairs; pappus setae white, 9–10 mm long, barbellate.

**Mozambique**. Z: Zambezia, entre o Lioma e Gerué, 41.7 km do Lioma, 15.ix.1949, *Grandvaux Barbosa & Carvalho* 4092 (K, LMJ). M: Maputo, Reserva de Caça de Maputo, 5.ix.1979, *de Koning* 7589 (K, LMU).

Also in South Africa (Cape Province, Natal, Transkei, Transvaal) and Eswatini. Damp grassland, marshy areas, lakesides; low altitude in ours (none of the four specimens gives an altitude), but in South Africa at 0 –750 m.

Conservation Status: As this taxon is fairly widespread in South Africa, the habitat is a common one and the altitude range is large, Least Concern seems the likely assessment.

Associated material: two specimens from Mt Namuli are similar, but come from much higher altitude, around 1950 m. The stems are sparsely glandular-pilose (against glabrous in *S. speciosus*); leaves are shorter than in typical *S. speciosus* (7–16 × 1.5–4 cm in Namuli specimens); calycular bracts are only 1 mm long; phyllaries are 7.8–9 mm long (also shorter than typical *S. speciosus*) and have scattered hairs; and achenes and pappus setae are slightly on the short side for *S. speciosus*. All in all there are not enough characters to convince me this is a new taxon; but the two specimens differ in some characters which makes me call attention to them.

**Mozambique**. Z: Mt Namuli, northern plateau, 16.xi.2007, *Harris et al.* 337 (K); Namuli, 1887, *Last* s.n. (K). Grassland; 1955 m.

During the writing of this treatment another interesting specimen was collected: **Mozambique**. Z: Ribaue, 15.x.2017, *Darbyshire & Agostenho* 1119 (K, LMA, LMU). This is closer to the Mt Namuli specimens in altitude (collected at 1442 m), leaf size and phyllary size; but differs from both the Mt Namuli specimens and *S. speciosus* in the mostly cauline leaves, which are ellipsoid, 5.5–7.5 × 1.4–2.2 cm with a cordate and semi-amplexicaul base, dentate margins and acute-apiculate apex. A second difference is the presence of short glandular hairs on the upper inflorescence axes and phyllaries, verging to (but different from) the long-stipitate glandular hairs mentioned for South African *S. speciosus* by Hilliard (1977: 434). Leaves, phyllaries and achenes are all slightly shorter than in the Mt Namuli specimens. The specimen came from rocks underneath a *Syzygium* on top of a granite inselberg.

43. **Senecio striatifolius** DC., Prodr. **6**: 387 (1838). —Hilliard, Compositae Natal: 456 (1977). —Mapaura & Timberlake, Checkl. Zimb. Vasc. Pl.: 28 (2004). Type: South Africa, Willbergen or Witbergen, *Drège* s.n. [5851] (G-DC holotype, HAL, P).

Perennial herb to 60 cm; rootstock woody, roots narrowly fusiform, crowned with old fibrous leaf bases and with woolly root crown; flowering stem solitary, unbranched below inflorescence, lightly cobwebby; leafy throughout. Radical leaves linear, 8–40 × 0.2–0.5(0.8) cm, base attenuate, clasping stem, margins closely minutely callose-denticulate and often revolute, apex attenuate; cauline leaves similar but quickly decreasing in size soon sessile, base often auriculate and half-clasping; sometimes with traces of wool, otherwise glabrous. Capitula discoid or radiate, up to 30 in corymbose panicles; all bracts thinly woolly; involucre campanulate; calyculus bracts several, short; phyllaries 16–20, 6–6.5 × 0.9–1 mm, keeled, with an apical tuft of hairs. Ray florets (when present, absent in ours) 8, corollas bright yellow. Disc florets ± 60, 6–6.5 mm long. Achenes cylindric, 3-4 mm long, ribbed, puberulous or glabrous (in ours); pappus setae 5–6 m long.

**Zambia**. SW of Dobeka Bridge, 13.x.1937, *Milne-Redhead* 2751 (K).

Mentioned in the checklist for Zimbabwe (Mapaura & Timberlake 2004), but without voucher specimens; I have not seen any specimens. M. Hyde states (pers. comm.) there is material as this taxon at SRGH but at least some of it is misidentified.

Also known from South Africa (Hilliard 1977 says 'widely distributed but rarely collected'). In boggy grassland; probably 1400–1500 m.

Conservation Status: Probably Least Concern.

44. **Senecio strictifolius** Hiern, Cat. Afr. Pl. **1**(3): 600 (1895). —Merxmüller, Prodr. Südwestafr. **139** Asteraceae: 165 (1967). —Jeffrey in Kew Bull. **41**(4): 895 (1986). — Lisowski, (Asterac. Fl. Afr. Cent. 2) Fragm. Flor. Geobot. **36** Suppl. 1: 284 (1991). —Da Silva *et al.*, Prelim. Checkl. Vasc. Pl. Mozamb.: 34 (2004). —Mapaura & Timberlake, Checkl. Zimb. Vasc. Pl.: 28 (2004). —Jeffrey & Beentje in F.T.E.A., Compositae **3**: 636 (2005). —Phiri, Checkl. Zamb. Vasc. Pl.: 33 (2005). —Setshogo, Prelim. Checkl. Pl. Botsw.: 38 (2005). —Heath & Heath, Field Guide Pl. North. Botsw.: 223 (2009). Type: Angola, near Lopollo and Lake Ivantala, 14.iii.1860, *Welwitsch* 3675 (BM holotype, BR, G, K, LISU, P).

Perennial or annual herb, 40–160 cm high, (?flowering before leaves); stems erect, almost woody in larger plants, unbranched except for inflorescence, slightly arachnoid to glabrous, from a creeping rootstock. Stem leaves sessile, narrowly lanceolate to oblanceolate, 3–12 × 0.3–1 cm, attenuate to an exauriculate base, margins denticulate (but look entire), apex obtuse and apiculate to acute, glabrous except for sometimes a slightly arachnoid midrib, or occasionally puberulous. Capitula erect, radiate, many in a rather copious terminal corymbose panicles 10–40 cm across; peduncles of individual capitula 5–17 mm; involucre cylindrical, 5–7 × 3.5–4 mm, almost glabrous; calycular bracts 6–7, 1.5–2.5 mm long, glabrous, often red-tipped; phyllaries 10–13, 4.5–5.5 × 0.3–0.5 mm, glabrous except for apex which has a few hairs, tips often reddish. Ray florets 8–13, corollas yellow, tube 3–3.7 mm long, glabrous, ray limbs 3–6.5 × 1.2–3 mm, 4-veined. Disc florets 21–33, corollas yellow, narrowly infundibuliform, 4.5–6 mm long, tube gradually expanded above middle, glabrous, lobes 0.8–1 mm long. Achenes cylindrical, 1.2–2 mm long, ribbed, glabrous; pappus setae 4–6 mm long.

**Botswana**. N: Dube Island on Okavango R., 28.vi.1974, *Smith* 1062 (K, SRGH). **Zambia**. N: Kali dambo, 17.iii.1955, *Richards* 4985 (K). W: Kitwe, 23.iv.1970, *Fanshawe* 10790 (K,

NDO). C: Mumbwa, 25.iii.1964, *van Rensburg* 2871 (K). S: 20 miles N of Choma, 21.iv.1954, *Robinson* 714 (K). **Zimbabwe**. N: Trelawney Dist., 23.ii.1945, *Hopkins* GH 13266 (K, SRGH). W: Farm Besha Kobila, v.1957, *Miller* 4379 (K, SRGH). C: Ngezi River at Harare–Masvingo (Fort Victoria) road, 30.v.1972, *Gibbs Russell* 2010 (K, SRGH). E: Inyongombi River bank, below Rhodes Hiotel, 19.i.1948, *Chase* 596 (K, SRGH). S: Makoholi Experimental Station, 13.iii.1978, *Senderayi* 211 (K, SRGH). **Malawi**. N: South Vipya Plateau, Luwawa Dam, 8.v.1970, *Brummitt* 10507 (K, MAL).

Also known from D.R. Congo, Tanzania, Angola and Namibia. Marshy grassland, stream banks, wet dambos, sometimes in up to 45 cm deep water, often associated with sedges, *Phragmites, Miscanthidium*; may be locally common; 950–1700 m.

Conservation Status: With a wide distribution, a common habitat and a wide altitude range, this is most likely Least Concern.

45. **Senecio swynnertonii** S.Moore in J. Bot. **54**: 283 (1916). —Mapaura & Timberlake, Checkl. Zimb. Vasc. Pl.: 28 (2004). Type: Zimbabwe, Mt Chirinda, *Swynnerton* s.n. (BM holotype, K).

Coarse perennial herb 0.6–1.3 m high; root crown possibly woolly (specimens incomplete); stem apparently unbranched apart from inflorescence, slightly ribbed, slightly puberulous. Leaves thinly fleshy, of two kinds: basal leaves, long-stalked, elliptic to narrowly elliptic, (13)19–45 × (0.5)3–9 cm, base long-attenuate but very base sheathing, margins serrate-dentate, apex acute, glabrous or nearly so; cauline leaves sessile, narrowly elliptic, 6–21 × 1.5–5.5 cm, base narrowly sheathing to almost auriculate, margins remotely denticulate to regularly dentate, apex acute, glabrous or nearly so. Inflorescence a branched lax panicle, sometimes a bit corymbose, with few to several (3–20) capitula; peduncles of individual capitula 10–35 mm long; involucre campanulate; calycular bracts several, linear, to 6 mm long; phyllaries 22–26, 7.5–14 × 1.2–2 mm, base with some small hairs, margin pale and membranous, apex darker and with papillate tip. Ray florets 8–12, corollas yellow, tube 4.2–4.6 mm long, lamina 8–11(18) × 1.5–4.2 mm. Disc florets many (60+), corollas yellow, narrowly infundibuliform, 5.5–7.7(9.6) mm long of which lobes 0.8–1 mm. Achenes cylindrical, 1.4–5.4 mm long, glabrous; pappus setae white, 5.5–9 mm long.

**Zimbabwe**. E: Nyanga (Inyanga), Little Connemara, 2.i.1963, *West* 4467 (K, SRGH). **Malawi**. N: Mzuzu, 9.i.1965, *Hilliard* 4368 (K). S: Mt Mulanje, top of Litchenya path, 2180 m, 17.ii.1982, *Brummitt & Polhill* 15950 (K, MAL).

Also in Angola. Grassland, sometimes in dambo or on stream-bank; 1450–2500 m.

Conservation Status: While the habitat is a common one and the altitude range substantial, only sixteen specimens of this taxon are known: Data Deficient.

Most specimens were named *S. kacondensis* S.Moore, an Angolan species with fewer phyllaries per head, but otherwise looking much the same. *S. swynnertonii* might occur in Zambia, (Phiri 2005: 33) cites *S. kacondensis*) but I have not seen any specimens.

46. **Senecio syringifolius** O.Hoffm. in Bot. Jahrb. Syst. **20**(2): 236 (1894). —Jeffrey in Kew Bull. **41**(4): 885 (1986). —Lisowski, (Asterac. Fl. Afr. Cent. 2) Fragm. Flor. Geobot. **36** Suppl. 1: 261 (1991). —Mapaura & Timberlake, Checkl. Zimb. Vasc. Pl.: 28 (2004). — Jeffrey & Beentje in F.T.E.A., Compositae 3: 621 (2005). —Burrows & Willis, Pl. Nyika Plateau: 107 (2005). —Phiri, Checkl. Zamb. Vasc. Pl.: 33 (2005). —Gereau *et al.*, Lake Nyasa Climat. Reg. Florist. Checkl.: 31 (2012). Type: Tanzania, Usambara, Lutindi, Tewe, 14.vii.1893, *Holst* 3264 (B† holotype, HBG, M). FIGURE 6.4.**7**.

    *Senecio exsertiflorus* Baker in Bull. Misc. Inform. Kew **1898**(139): 154 (1898). Type: Malawi, Mt Zomba, alt. 4000–6000ʹ, xii.1898, *Whyte* s.n. (K000311516 holotype).

    *Senecio nyikensis* Baker in Bull. Misc. Inform. Kew **1898**(139): 154 (1898), nom. illeg., non Baker (1897: 271) (= *Kleinia abyssinica* (A.Rich.) A.Berger var. *abyssinica*). Type: Malawi, Nyika, *Whyte* 238 (K000311515 holotype).

    *Senecio subpetitianus* Baker in Bull. Misc. Inform. Kew **1898**(143): 303 (1898), nom. nov. pro *S. nyikensis* Baker (1898), non Baker (1897).

Fig. 6.4.**7**. SENECIO SYRINGIFOLIUS. 1, habit (× ⅔); 2, capitulum (× 3); 3, floret, (× 6); 4, achene with pappus setae (× 3). 1–3 from *Drummond & Hemsley* 2703; 4 from *Haarer* 1515. Drawn by Juliet Williamson. From Flora of Tropical East Africa.

Liana twining to 1.5–13 m; stems to 1.5 cm diameter, rather succulent, glabrous, developing a grey bark when old. Leaves slightly succulent, petioles glabrous or with a few hairs on margins near base, exauriculate, 1.4–5.6 cm long; lamina rhombic-triangular, oblong-triangular or hastate in outline, sometimes ovate, 3.8–10 × 1.5–9 cm, base rounded to truncate or weakly cordate and short-decurrent on petiole, margins entire or remotely dentate or slightly lobed in lower half to coarsely sinuate-dentate with apiculate teeth, apex obtuse to acuminate, glossy, glabrous to moderately densely pubescent, 5–7-veined from base (often only conspicuously 3-veined). Inflorescence dense, many-headed, of spreading compound corymbose panicles to 15 cm long, bracts at base of secondary branches reduced, leaf-like. Capitula discoid, sweet-scented; involucre cylindrical, 5.5–9 mm long; calyculus bracts 3–5, 1.5–5 mm long; phyllaries 8, 5–8.5 × 0.8-1.3 mm, green, glabrous except for ciliate apex margins and short-pubescent tip outside. Florets ± 20, corolla yellow, creamy yellow or greenish-white, glabrous, 6.5–9 mm long, tube narrowly infundibuliform, 5–8 mm long, lobes 1–2.4 mm long, often recurved; anthers yellow; style arms yellow. Achenes 3–4 mm long, few-ribbed, glabrous; pappus setae 5–9 mm long, barbellate, white, fragile and deciduous.

**Malawi**. N: Nyika Plateau, Zovochipolo, viii.1980, *Dowsett-Lemaire* 104 (K). S: Zomba Plateau, Mulumbe Peak, 29.vii.1986, *Nachamba & Usi* 359 (K, MAL).

Dowsett-Lemaire says this occurs also on the Zambian part of Nyika; I have not seen any specimens from there. Zimbabwean material identified as this species (*Dale* SKF 482 in SRGH, *Hopkins* in SRGH 8046) might be misidentified; this taxon seems to reach its southern extent in Malawi.

Also known from Kenya, Tanzania, Uganda and Rwanda. Evergreen forest (*Widdringtonia*, *Juniperus*, etc.) margins; 1550–2350 m.

Conservation Status: Most likely Least Concern due to a wide distribution, a common habitat and a wide altitude range.

47. **Senecio tabulicola** Baker in Bull. Misc. Inform. Kew **1898**(139): 155 (1898). —Jeffrey in
   Kew Bull. **41**(4): 895 (1986). —Jeffrey & Beentje in F.T.E.A., Compositae **3**: 637 (2005).
   Type: Malawi, Nyika plateau, *Whyte* 162 (K holotype).

Coarse short-lived perennial herb, 60–210 cm high; stems solitary, erect, unbranched but for inflorescence, at first slightly arachnoid, glabrescent, pale green with reddish longitudinal ridges; without hairy or fibrous root-crown. Basal leaves sessile, leathery to fleshy, narrowly lanceolate to elliptic, 19–70 × 2–10 cm (including 10–36 cm long petioloid base), gradually attenuate into a petioloid base, margins sinuate-dentate, apex shortly acuminate, green and glabrescent above, whitish to pale green and arachnoid but often glabrescent beneath; stem leaves smaller and becoming progressively smaller up stem, sessile, not or somewhat expanded-auriculate and semi-amplexicaul at base, sinuate-dentate to subentire. Capitula 3–35 in a lax to rather crowded terminal slightly corymbose panicle; peduncles of individual capitula 1–5 cm long; involucre cylindrical, 8–13 × 6–10 mm, ± woolly at base; calycular bracts 6–16, lanceolate, 2–7 mm long, ± arachnoid-tomentose; phyllaries 10–22, green with reddish or brownish tips, 7.5–12 × 0.8–1.2 mm, shortly lanate or glabrous except near base, apically papillate-bearded. Ray florets 8–14, corollas bright yellow, tube 3.5–4 mm long, rays 8–11 × 3–5 mm, 6–9-veined. Disc florets many (80+), corollas deep yellow to orange, narrowly infundibuliform, 6–10 mm long, of which lobes 0.8–1 mm long; anthers brown; style yellow. Achenes oblong (outer flattened?), 2–4 mm long, glabrous, many-ribbed; pappus setae 5–10 mm long.

**Zambia**. N: Kasama, 1.vii.1964, *Fanshawe* 8775 (K, NDO). **Malawi**. N: Mzuzu, Mazamba Forest, 27.iv.1967, *Salubeni* 689 (K, MAL, SRGH). C: Dedza Mt., 1.v.1989, *Pope et al.* 2254a (K, MAL). **Mozambique**. MS: 4 km from Vila Vasco da gama to Fingoe, 27.vi.1949, *Grandvaux Barbosa & Carvalho* 3352 (K, LSID).

Also known from Tanzania. *Brachystegia* or *Uapaca* woodland, grassland; twice called 'frequent', twice called sporadic or 'few'; (900)1250–2050 m.

Conservation Status: Most likely Least Concern due to a wide distribution, a common habitat and a wide altitude range.

*Chapama et al.* 407 from Malawi looks like *S. tabulicola* but has almost glabrous phyllaries and hairy achenes and so keys to *S. proprior*. These two spp. resemble each other; while Brenan (1954: 483) thought *S. tabulicola* was very close to *S. karaguensis*.

48. **Senecio tamoides** DC., Prodr. **6**: 403 (1838). —Hilliard, Compositae Natal: 493 (1977). —Da Silva *et al.*, Prelim. Checkl. Vasc. Pl. Mozamb.: 299 (1993) (as '*tanoides*'). —Da Silva *et al.*, Prelim. Checkl. Vasc. Pl. Mozamb.: 34 (2004). —Mapaura & Timberlake, Checkl. Zimb. Vasc. Pl.: 28 (2004). Type: South Africa, Umzimvubu (Omsamwoubo), 18.v.1832, *Drège* s.n. [5160] (G-DC holotype, E00239880, HAL, K, P0000651, P0000652, P0000653).

Herbaceous twiner to 10 m high in trees; stem slightly fleshy, glabrous. Leaves slightly fleshy, deltate to hastate in outline, 4–10 × 3.5–9.5 cm, base hastate to truncate, margin deltate-dentate to slightly lobed, apex acute; with several veins from base; sometimes hairy below; petiole 2–5 cm long. Capitula radiate, many in subumbellate or corymbose inflorescences at or near tip of leafy lateral branches, sweet-smelling; involucre cylindrical, 7–10 mm long; calycular bracts few and small, to 1.5 mm long; phyllaries 6–8, 6.5–9.2 × 0.5–1 mm, keeled, swollen at base, veins resinous, glabrous except very tip. Ray florets usually 5, corollas bright yellow, tube 5–7 mm long, blade 6.5–13 × 1.5–4 mm. Disc florets 10–18, tube very narrowly infundibuliform, 8–11.8 mm long, lobes 0.5–0.8 mm long with median resinous line, glabrous or with a few glands near apex. Achenes cylindric, 2.7–3.7 mm long, ribbed, glabrous; pappus setae white, 7–8.2 mm long.

**Zimbabwe**. E Chipinga, Chirinda Forest, vi.1962, *Goldsmith* 152/62 (K, SRGH). **Mozambique**. E: main Vumba road opposite Jevington road turn-off at 1722 m, 2005 (photo by B. Wursten but no specimen).

Also known from South Africa (Cape Province, Natal, Transvaal) and Eswatini. In forest margins, when in flower 'curtaining the trees with brilliant yellow'; 1100-1900 m.

Conservation Status: Probably Least Concern due to its distribution area, habitat and altitude range.

The capitula are prone to insect attack, resulting in hard top-shaped galls. Also cultivated as a garden ornamental, as in Harare in 1970 (*Biegel* 3238) and Maputo in 1971 (*Balsinhas* 1896).

To me this looks rather like a southern version of *S. syringifolius*, only differing in the presence of ray florets.

49. **Senecio teixeirae** Torre in Bol. Soc. Brot. sér.2, **44**: 289 (1970). Type: Angola, Bié, between Tarala and Nharea, 1650 m, 24.xi.1965, *Teixeira et al.* 9519 (LISC holotype, LUA).

Perennial herb, erect and to 70 cm high; stems slightly striate, glandular-pubescent. Basal leaves long-petiolate (petiole 3–9 cm), lamina in outline narrowly elliptic with attenuate base and acute apex, pinnatipartite to pinnatisect with lobes in their turn shallowly to deeply lobed, 7–12 × 3–5.4 cm, sparsely glandular-pubescent. Cauline leaves gradually becoming smaller and more sessile, lyrate-pinnatifid to pinnatilobed, base auriculate, apex obtuse to acute, sparsely glandular-pubescent. Capitula discoid, few to many in lax paniculate (sub-corymbose) inflorescences at or near tip of leafy lateral branches, sweet-smelling; calycular bracts few, lanceolate, 1–2.4 mm long, acute; phyllaries 10–13, lanceolate, 6–7.7 × 0.4–0.6 mm, acute, glandular-pubescent. Florets 20–30, corollas white or cream, infundibuliform, 4.8–6 mm long. Achenes 1.5–2.5 mm long, ribbed, puberulous; pappus setae 5–6 mm long.

**Zambia**. N: Mpika, 4.ii.1955, *Fanshawe* 1971 (K). W: SE of Dobeka Bridge, 16.xii.1937, *Milne-Redhead* 3701 (K).

Also known from Angola. Miombo; probably at 1400–1500 m (no altitude given on either specimen).

I have seen only two Zambian specimens, the cited ones. This is a new species for the FZ area; previously only known from the Bié area of central Angola.

Conservation Status: Data Deficient, as I have no idea of the status of this taxon in Angola.

50. **Senecio torticaulis** Merxm. in Proc. & Trans. Rhodesia Sci. Assoc. **43**: 69 (1951). —Mapaura & Timberlake, Checkl. Zimb. Vasc. Pl.: 28 (2004). Type: Zimbabwe, Marandellas, 10.viii.1942, *Dehn* 680 (M0105328 holotype, K).

Perennial herb with lowermost stem almost woody, but due to being burnt usually with multiple herbaceous stems from near ground level; stems 12–30 cm long, erect, glabrous or with minute arachnoid indumentum, mostly visible in leaf axils. Leaves linear, 1–4 × 0.05–0.15 cm, margins inrolled, apex obtuse, hairs on upper surface usually invisible due to inrolled margins. Capitula discoid, solitary at branch ends or in few-headed inflorescences, but whole plant resembling a large leafy inflorescence; stalk of individual capitula slender, glabrescent; involucre turbinate; calycular bracts 10–12, small, slightly below phyllaries; phyllaries 20–22, linear, (4.5)5–7.2 × 0.7–1 mm, apex acute to obtuse, green with paler margins, glabrous to minutely hairy along upper midrib, sometimes apex with a little tuft of hairs. Florets 60–80, corollas yellow, tube 4–6.3 mm long, lobes 0.4–0.7 mm long and glandular. Achenes cylindrical, 3.2–3.8 mm long, minutely puberulous; pappus setae 5.5–7 mm long, white.

**Zimbabwe**. C: Makoni, 3 miles NW of Headlands, 10.x.1965, *Chase* 8314 (K, SRGH). E: Umtali, Odanzi River, 1915, *Teague* 261 (BOL, K).

Grassland, burnt grassland, also on river banks and cultivated land (one record each); 1400–1600 m.

Conservation Status: While the habitat is a common one, specimens are few (eight) as well as old (all pre-1965, and most considerably so), and this must remain Data Deficient until more is known about habitat requirements and actual distribution.

This taxon resembles *S. randii* but lacks the ray florets; has longer phyllaries (with more hairs?) and inrolled leaves.

51. **Senecio triactinus** S.Moore in J. Linn. Soc. Bot. **40**(275): 119 (1911). —Mapaura & Timberlake, Checkl. Zimb. Vasc. Pl.: 28 (2004). Type: Mozambique, Jihu, Kurumadzi River, 17.xi.1906, *Swynnerton* 1878 (BM000798810 holotype, K000311505).

   *Senecio homoplasticus* S.Moore in J. Linn. Soc. Bot. **40**: 120 (1911). —Mapaura & Timberlake, Checkl. Zimb. Vasc. Pl.: 28 (2004). Type: Zimbabwe, Chirinda, x.1906, *Swynnerton* 288 (BM holotype, K), syn. nov.

   *Senecio homoplasticus* var. *tomentellus* S.Moore in J. Linn. Soc. Bot. **40**: 121 (1911). Type: Zimbabwe, Chirinda, 9.x.1906, *Swynnerton* 288a (BM holotype, K), syn. nov.

Perennial herb 0.9–1.8 m; stem apparently unbranched but for inflorescence (but I have not seen any specimen with a stem base), straight, light green, slightly angular when dry, minutely puberulous and becoming glabrous. Leaves ?all cauline, subsessile with petiole up to 0.7 cm long, elliptic or slightly ovate, 5–8.5 × 2–4.5 cm, base cuneate to obtuse, margins serrate-dentate, apex acute, young leaves with some tomentum but very soon glabrous except for slightly puberulous midrib. Capitula many, fairly small, in dense corymbose panicles to 25 cm across, lowermost branches subtended by leaves which gradually pass into hairy bracts; peduncles of individual capitula 4–15 mm long; involucre campanulate, calycular bracts subulate, small to 1.5 mm long; phyllaries 8–10, oblong, 2.8–4 × 0.3–0.6 mm, slightly puberulous, more so at base and apex. Ray florets 3–4(6), corollas white, cream or light yellow, tube 2–3 mm long, limbs 3.2–6 × 1.5–2.5 mm. Disc florets 12–33, narrowly infundibuliform, 3.7–4.5 mm long. Achenes cylindrical, 1.2–2.2 mm long, pilose-puberulous; pappus setae white, 3.5–5 mm long.

**Zimbabwe**. N: Makonde (Lomagundi) Dist., 24.i.1969, *Biegel* 2908 (SRGH, fide M. Hyde). C: Warwick Farm, NE of Lake McIlwaine, 10.i.1973, *Pope & Biegel* 1164 (K, SRGH). E: Umtali, S side of Murahwa's Hill commonage, 3700′, 6.ii.1963, *Chase* 7952 (K, SRGH). S: Masvingo (Victoria) Dist., Kyle National Park Game Reserve, 23.v.1971, *Grosvenor* 534 (SRGH, fide M. Hyde). **Mozambique**. MS: Mt Garuso, 1650 m, 18.i.1968, *Wild* 7674 (K, SRGH).

Endemic to our area. Habitat not very clear, possibly miombo and termitaria; 180–1650 m.

Conservation Status: Known from 13 specimens; this must remain Data Deficient until the habitat requirements are known better, as the determination of threats depends on that.

This is part of a group of closely related taxa; Moore (1911) himself compared *S. triactinus* to *S. crenatus* Thunb. (a species from the Cape), while Hilliard (1977) says it is close to three others: *S. homoplasticus* (from Zimbabwe), *S. pentactinus* Klatt (from NE South Africa, described in 1877) and *S. colensoensis* O.Hoffm. (from Natal, described in 1898). The type of *S. homoplasticus* looks very similar, and I place it into the synonymy of *S. triactinus* here; the only differences are the number of rays per head (variable with the larger number of specimens now available); the difference in phyllary length also links to *S. triactinus* by intermediates. *S. homoplasticus* var. *tomentellus* is only distinguished by the tomentellous (instead of araneose) stems and inflorescence axes; I believe it to be just a form, and also synonymous with the main taxon. The South African taxa are similar, but *S. colensoensis* is said to have glandular hairs, which are not present in our taxon. *S. pentactinus* seems to differ in its narrower leaves.

52. **Senecio urundensis** S.Moore in J. Linn. Soc. Bot. **35**(245): 355 (1902). —Jeffrey in Kew Bull. **41**(4): 896 (1986). —Lisowski, (Asterac. Fl. Afr. Cent. 2) Fragm. Flor. Geobot. **36** Suppl. 1: 288 (1991). —Jeffrey & Beentje in F.T.E.A., Compositae **3**: 642 (2005). —Gereau *et al.*, Lake Nyasa Climat. Reg. Florist. Checkl.: 31 (2012). Type: 'Urundi', 1893-1894, *Scott Elliot* 8181 (BM000924804 holotype, K000311383).

Herb with annual shoots from a perennial rootstock; crown of rootstock woolly; stems erect, leafy, with thick ridges, arachnoid-glabrescent, 20–120 cm high. Basal leaves sessile and prostrate (fide *Richards*), elliptic to slightly obovate, 14–39 × 4–11.5 cm (including 4–14 cm long petioloid base), long-attenuate into a petioloid base, margins sinuate-dentate, apex rounded; lower stem leaves with expanded membraneous exauriculate base; median and upper stem leaves sessile, broadly ovate to lanceolate, elliptic or oblanceolate-oblong, 3–12 × 1–5 cm, base tapered and exauriculate to expanded, auriculate and semi-amplexicaul, margins denticulate, apex shortly acuminate; all leaves thinly arachnoid, soon glabrescent except for a few hairs on midrib beneath. Capitula discoidor rarely radiate, many in copious congested to rather lax terminal corymbose panicles; stalk of individual capitula slender, thinly arachnoid, often glabrescent; involucre cylindrical, 4.5–7.5 × 3–4 mm, laxly arachnoid to almost glabrous at base; calycular bracts 2–6, lanceolate to linear, 2–7 mm long, usually thinly arachnoid and ciliate on margins; phyllaries 8–14, brown to sepia to pale green, 4.5–7 × 0.5–0.9 mm, thinly arachnoid, glabrescent, bearded at apex. Ray florets absent (rarely with 4 rays, see note). Disc florets 12–15, corollas yellow, 4.7–6.5 mm long, tube rather abruptly expanded in upper third, glabrous, lobes 0.8–1.2 mm long. Achenes 2–3 mm long, ribbed, hairy; pappus setae 5–7.5 mm long, white.

**Zambia**. N: Mbala, road to Kasulu Farm, 5500′, 4.i.1969, *Sanane* 388 (K, SRGH); W: Kasempa, 29.ix.1973, *Fanshawe* 12082 (K, NDO). **Malawi**. N: Vipya Link road, 15 miles in from southern end, 5500 ft., 11.i.1975, *Pawek* 8945 (K, MAL, MO, SRGH, UC).

Also known from D.R. Congo, Rwanda, Burundi and Tanzania. Grassland and sparsely wooded grassland, open bushland, especially where burnt; 1500–2400 m.

Conservation Status: With a large distribution area, common habitat requirements and a large altitude range, this must be Least Concern.

Two specimens are indistinguishable from the normal taxon except that they have four rays per head: ray limbs 6–11 × 2 mm. **Malawi**, Misuku Hills, alt. 1500 m, 10.i.1959, *Richards* 10597 (K); Misuku Hills, alt. 1410 m, 19.xii.1972, *Pawek* 6226 (K, MAL, UC).

53. **Senecio vulgaris** L., Sp. Pl. **2**: 867 (1753). —Oliver & Hiern, F.T.A. **3**: 411 (1877). — Jeffrey in Kew Bull. **41**(4): 904 (1986). —Mapaura & Timberlake, Checkl. Zimb. Vasc. Pl.: 28 (2004). —Jeffrey & Beentje in F.T.E.A., Compositae **3**: 669 (2005). Type: 'Habitat in Europae cultis, ruderalis, succulentis', Herb. Clifford: 406, *Senecio* 1, sheet A (BM000647048 lectotype), lectotypified by Jeffrey in Jarvis *et al.*, Regnum Veg. **127**: 87 (1993).

Annual herb 14–37 cm high; stems at first white-pubescent, glabrescent. Basal leaves sessile, oblanceolate or oblanceolate-elliptic in outline, tapered into an expanded exauriculate petioloid base, pinnately toothed or lobed, rounded, soon disappearing; stem leaves sessile, oblong to broadly oblong in outline, 2.3–6 × 0.15–0.6 cm (1–1.7 cm across lobes), base auriculate and semi-amplexicaul, margins shallowly to deeply pinnately lobulate with shortly oblong to narrowly oblong-linear, apiculate-dentate lobes; all leaves with scattered long white hairs above on midrib and beneath especially on midrib.

Capitula discoid, in rather congested terminal corymbs; stalk of individual heads with long white pubescence, glabrescent; involucre cylindrical, 7–8 × 2.5 mm; calycular bracts about 8, imbricate, lanceolate, glabrous, up to 2 mm long; phyllaries about 19, glabrous, 6.5–7.5 × 0.2–0.4 mm. Florets many (several tens), corollas yellow, 5 mm long, tube glabrous, expanded in upper ⅓, narrowly tubular below, lobes 0.5 mm long. Achenes 2–2.5 mm long, ribbed, hairy in grooves; pappus setae 5–5.5 mm long.

**Zimbabwe**. C: Harare, weed in landscape nurseries, 12.xi.1970, *Wood* SRGH 207461 (K, SRGH).

Widely distributed in temperate Eurasia and N Africa, and widely naturalized in many areas. A weed of cultivation.

Conservation Status: Least Concern.

### 54. **Senecio** sp. aff. **serra**

Herb 90–120 cm high with erect and branched stems, glabrous. Leaves (all cauline, I presume, but basal parts of plant are missing) narrowly elliptic, 3–6 × 0.5–0.8 cm (but missing lower leaves might be larger!), base cuneate and with two narrow 1–6 mm long lobers on each side of very base, margin serrate, apex acute; glabrous. Capitula radiate, many in corymbose panicles, with bracts or ever-decreasing leaves; peduncles of individual capitula 4–18 mm long; phyllaries 12–14, 3.1–4.2 × 0.5–0.8 mm, glabrous but for (?reddish) glandular-papillate apex, and with narrow membranous margins. Ray florets 11–15, corollas yellow, ray limbs 3–4 × 1.5–2 mm. Disc florets ± 33, 4.5 mm long. Achenes 1 mm long, glabrous; pappus setae white, 4–4.3 mm long.

**Zimbabwe**. E: Nyanga, banks of Inyongambi River below Rhodes Hotel, 19.i.1948, *Chase* 596 (K, SRGH); E of Troutbeck Hotel, 11.ii.1965, *Chase* 8264 (K, SRGH).

? Not known elsewhere. Streambanks, about 1860 m.

If the leaf base lobes are disregarded, this would key to *S. strictifolius*, but is not that species. It seems closer to the South African *S. serra* Sond., which looks very similar but has auriculate leaf bases. A collection including the lower part of the plant would help sort out these matters!

### 55. **Senecio sp. A** (= *Nuvunga et al.* 329)

Erect herb, '40 cm high' (K duplicate is just over 20 cm), sparsely branched; glabrous. Leaves ovate in outline, 2–8 × 0.7–3 cm, of which petiole 1–3 cm long, a pair of subopposite proximal lobes to 1.3 × 0.7 cm (and slightly lobed in turn), a bare axis of up to 1 cm, and a distal blade part that is ± deltoid, to 3.5 × 3 cm, base cordate, margins deeply dentate to slighty lobed, apex acute; glabrous. Capitula radiate, in few-flowered panicles (damaged in Kew duplicate); calycular bracts very few, minute; phyllaries 11–14, 3.8–4.3 × 0.4–0.6 mm. Ray florets number unclear, corollas yellow, limbs 3.2 × 0.8 mm. Disc florets ?27, 3–3.5 mm long. Achenes 1.7 mm long, glabrous; pappus setae white, 3.5 mm long.

**Mozambique**. GI: Chipenhe, Chirindzene, 19.ix.1980, *Nuvunga et al.* 329 (K, LMU).
Edge of forest, maybe around 50 m.

The only other species in our area with such a divided leaf lamina is *S. auriculatissimus*, but that has a completely different blade dissection. I have not found any other taxon looking remotely like this, either from our area or from South Africa. As the specimen is rather poor, I feel I have to leave this as a '*sp.*' More material is likely to show this is an as yet undescribed species.

*Species for which material has not been traced*

**Senecio sp. 1** of Macnae & Kalk, Nat. Hist. Inhaca Isl., revis. ed.: 155 (1969); *Mogg* 27698.
  'Deciduous climbing shrubby herb of NE forest (maritime) with small leaves; corymbs of small heads, March'.

**Senecio sp. 2** of Macnae & Kalk, Nat. Hist. Inhaca Isl., revis. ed.: 155 (1969); *Mogg* 27625. 'Evergreen climber with large lobed leaves found chiefly in SE (Ponta Torres) forest; heads not seen'.

*No specimens seen, but likely to occur in our area*

**Senecio conrathii** N.E.Br. in Bull. Misc. Inform. Kew **1914**(2): 79 (1914). —Hilliard, Compositae Natal: 451 (1977). —Mapaura & Timberlake, Checkl. Zimb. Vasc. Pl.: 28 (2004). Type: South Africa, Transvaal, near Modderfontein, xii.1896, *Conrath* 1320 (K holotype).

    *Senecio pachythelis* E.Phillips & C.A.Sm., Rep. Director Veterin. Serv. Anim. Industr. S. Afr. **17**(2, 6): 649 (1931). Type: South Africa, Mount Currie/Kokstad District, Westlands farm on Kopjie, 29.xii.1930, *Phillips* 3488 (PRE0223811-0 holotype, PRE0587162-0).

Herbaceous perennial to 70 cm high; stems 5–7 mm diameter, ribbed, arachnoid pubescent (cottony) or glabrous. Cauline leaves lanceolate, 3.5–27 × 0.6–6 cm, base narrowing and semiamplexicaul, midrib prominent below, margins callose-dentate, apices obtuse to acute. Inflorescences of corymbose panicles, ultimate branches with 2 or 3 capitula; bracts 6–40 × 3–5 mm. Capitula discoid; calycular bracts few; phyllaries 15, 7–8 × 1.5–2.5 mm, glabrous, margins membranous, glandular and dark-coloured towards apex, apices obtuse. Florets many, corollas yellow, infundibuliform, 7.7–7.8 mm long. Achenes 2.5 mm long, terete, striate, beaked, glabrous; pappus setae 7 mm long, barbellate, deciduous.

I have seen no specimens, but this might occur in Zimbabwe (and appears on Mapaura & Timberlake 2004).

Known from South Africa (Cape Province, Natal, Transvaal) and Eswatini. Grassland, often in large colonies.

**Senecio gregatus** Hilliard in Hilliard & Burtt, Notes Roy. Bot. Gard. Edinburgh **43**(3): 352 (1986), as nom. et stat. nov. pro *S. serratuloides* DC. var. *gracilis* Harvey. Type: South Africa, Natal, *Grant* s.n. (K holotype).

    *Senecio serratuloides* var. β *gracilis* Harvey in Harvey & Sonder Fl. Cap. **3**: 382 (1865). —Hilliard, Compositae Natal: 410 (1977).

    *Senecio serratuloides* var. δ *rehmannii* Thell. in Vierteljahrsschr. Nat. Ges. Zürich **66**: 246 (1921). Type: South Africa, Transvaal, Hogge Veld, Pages Hotel, 1875-80, *Rehmann* 6853 (Z000003984, Z000061802 syntypes).

    *Senecio serratuloides* var. ε *dieterlenii* Thell. in Vierteljahrsschr. Nat. Ges. Zürich **66**: 246 (1921), as 'dieterleni'. Types: Lesotho (Basuto-Land), Léribé, 1911, *Dieterlen* 778 (K, NH, P00122371, PRE, Z000061795, Z000061796, Z000061797, Z000061803 syntypes); Transvaal, 1892, *Fehr* 57 (Z000003981 syntype).

    *Senecio serratuloides* var. ζ *holubii* Thell. in Vierteljahrsschr. Nat. Ges. Zürich **66**: 246 (1921). Types: South Africa, Transvaal, Hogge Veld, between Porter and Trigardsfontein, 1875–80, *Rehmann* 6627 (Z000003982, Z000003983 syntypes); Linokana, iv.1887, *Holub* 3773/4 (Z000061799 syntype); Phoberg, *Holub* 4786 (Z000076870 syntype); v.1887, *Holub* 4788 (Z000061800 syntype); Matebequellen, v.1887, *Holub* s.n. (Z000061798 syntype); Natal, Howick, 1890, *Junod* 16 (?syntype).

Stems simple or branching low down, branches stiffly erect, glabrous, closely leafy. Leaves to 12 × 0.2–1.5 cm, flat, not folded, apical lobe linear to lanceolate, tapering at both ends; basal lobes sometimes reduced to one pair, margins thickened, serrate or serrulate, tips of teeth not incurved, both surfaces glabrous or nearly so.

Reported (Hilliard & Burtt 1986) to occur in Zimbabwe. I have not seen any specimens.

Known from South Africa (Cape Province, Natal, Orange Free State, Transvaal), Lesotho, Eswatini, and Angola. Streambanks, often in water.

Hilliard & Burtt (1986) noted that the leaves of *S. gregatus* were narrower than *S. serratuloides*, have straight teeth (not falcate) which are smaller and blunter and the basal lobes fewer and smaller.

**Senecio inaequidens** DC., Prodr. **6**: 401 (1838). —Macnae & Kalk, Nat. Hist. Inhaca Isl.: 155 (1969). —Hilliard, Compositae Natal: 405 (1977). —Da Silva *et al.*, Prelim. Checkl. Vasc. Pl. Mozamb.: 34 (2004). Type: ?South Africa, no further location, 1835, *Drège* s.n. [5879] right specimen (G-DC lectotype, P), lectotypified by Jeanmonod *et al.*, Complément. Prodr. Flore Corse, Asterac. **2**: 47 (2004).

Perennial herb to 1 m (Macnae & Kalk say annual), branching vigorously from base, branches branching again, woody, glabrous or very sparsely hairy, often with axillary leaf tufts. Leaves linear to narrowly lanceolate, to 10 × 1 cm in moist shady conditions but usually much narrower, base half-clasping and often hastately eared, margin ± revolute and denticulate to dentate or even pinnately lobed, apex acute. Capitula radiate, few to many in corymbose panicles; phyllaries ± 20, 4–7 mm long, keeled, resinous; calycular bracts few, often dark-tipped. Ray florets 7–13, corollas canary yellow, often rolled. Disc floret lobes with short median resinous line. Achenes cylindric, 2–2.5 mm long, hispid between ribs; pappus setae size unknown.

Not yet found in our area, but from the South African distribution this could occur in Mozambique and Botswana.

Known from South Africa (Cape Province, Natal, Orange Free State and Transvaal) and Namibia. Widely naturalized in Europe (Belgium, France, Italy) as far north as Great Britain (where it has been recorded as a wool shoddy alien).

Rock outcrops, grassy slopes, along watercourses, disturbed areas, sometimes as a roadside weed; 1400–2850 m in Natal.

Toxic to goats, fide Macnae & Kalk (1969).

*Uncertain species and uncertain identifications*

**Senecio forbesii** Oliv. & Hiern, F.T.A. **3**: 420 (1877). Type: Mozambique, *Forbes s.n.* (holotype, not found).

Scandent with elongated wiry glabrous stem and branches. Leaves elliptical, narrowed at both ends, coarsely dentate, puberulous, 3.7–7.5 cm long; petioles ranging up to 1.3 cm long, not auriculate at base. Capitula discoid, campanulate, 16 mm long, on rather short slender and puberulous peduncles in few-headed pedunculate cymes. Calyculus consisting of unequal narrow and linear appressed bracts. Phyllaries c. 12, narrow and linear, acute, glabrous, nearly equalling florets. Florets 'numerous'. Achenes glabrous.

**Mozambique**. locality not given; specimen not found.

This is a problem: the protologue does not give a flower colour, and I have not been able to trace the type – even though Forbes' types are supposed to be at Kew. From the rather scanty description this does not resemble any other *Senecio* climber. It will remain a mystery until the type is found; I have been unable to include it in the key.

**Senecio serratuloides** DC. Prodr. **6**: 395 (1838). —Hilliard, Compositae Natal: 409 (1977). —Da Silva *et al.*, Prelim. Checkl. Vasc. Pl. Mozamb.: 34 (2004). —Mapaura & Timberlake, Checkl. Zimb. Vasc. Pl.: 28 (2004). Type: South Africa, Port Natal, *Drège* s.n. [5146] (?G-DC holotype, HAL0113137, P).

Perennial herb from a creeping woody rootstock, stem solitary, to 100 cm high, densely leafy in upper part, glandular-pilose to glabrous. Leaves lyrate-pinnatifid, to 12 × 2–2.5 cm, becoming smaller upwards, apical lobe very large and often folded lengthwise, lanceolate to narrowly eliptic, margins with 2–4 pairs of lower lobes and double-serrate, apex acute; lower lobes 1–4 on each side of petiole-like base, lanceolate, very small, entire or few-toothed; lower surface thinly cobwebby with a few long gland-tipped hairs on veins. Capitula radiate, many in teminal corymbose panicles; phyllaries 10–12, 3–3.5 mm long, keeled, with 2–3 resinous veins, glabrous; calycular bracts similar, descending on stalk. Ray florets 8–13, corollas yellow, short, soon recurved. Disc florets 30–40, lobes with a median orange resinous line. Achenes cylindric, 1.7 mm long, ribbed, glabrous; pappus 3.5–5 mm long, barbellate, deciduous.

Known from South Africa, plants of grassland and open woodland. Despite the reported occurrence in Mozambique (fide Da Silva *et al.* 2004) and Zimbabwe (fide Mapaura & Timberlake 2004), I have not seen any specimens from our area.

**Senecio subsessilis** Oliv. & Hiern, F.T.A. **3**: 415 (1877). —Burrows & Willis, Pl. Nyika Plateau: 107 (2005). Type: Ethiopia, Gaffat, 27.xi.1863, *Schimper* 1532 (K000311441, K000311442, BM000797687, BM000924651, E syntypes).

    *Senecio denticulatus* Engl. in Abh. Königl. Akad. Wiss. **1891**: 442 (1892), nom. illeg., non DC. (1838). —Brenan in T.T.C.L.: 158 (1949). Type: Tanzania, Kilimanjaro, *Meyer* 72 (B† holotype).

    *Senecio thomsianus* Muschl. in Bot. Jahrb. Syst. **43**(1–2): 59 (1909). Type: Tanzania, E Usambara Mts, *Keil* 180 (B† holotype).

    *Senecio trichopterygius* Muschl. in Wiss. Ergebn. Deut. Zentr.-Afr. Exped. (1907-1908), Bot. **2**: 392 (1911). —Andrews, Fl. Pl. Sudan **3**: 49 (1956). Type: D.R. Congo, Rugege, Rukarara, ix.1907, *Mildbraed* 905 (B† holotype).

    *Senecio dernburgianus* Muschl. in Wiss. Ergebn. Deut. Zentr.-Afr. Exped. (1907-1908), Bot. **2**: 394 (1911). Type: D.R. Congo, Karisimbi, *Mildbraed* 1613 (B† holotype, BR).

    *Senecio fistulosus* De Wild., Pl. Bequaert. **5**: 105 (1929). Type: D.R. Congo, Ruwenzori, 19.iv.1914, *Bequaert* 3830 (BR0008879561, BR0008879899, BR8879233 syntypes).

    *Senecio fistulosus* var. *mukuleensis* De Wild., Pl. Bequaert. **5**: 106 (1929). Type: D.R. Congo, Ruwenzori, 27.ix.1914, *Bequaert* 5925 (BR000887924, BR000888009 syntypes).

    *Senecio late-alatopetiolatus* De Wild., Pl. Bequaert. **5**: 109 (1929). Type: D.R. Congo, Lanuri, 29.v.1914, *Bequaert* 4538 (BR0008879523, BR0008879196, BR0008879851 syntypes).

Perennial herb, woody herb or weakly woody shrub 45–300 cm high; stems densely to thinly arachnoid or sparsely pubescent to glabrescent; roots thick. Basal leaves petiolate, upper leaves sessile; leaves lanceolate, ovate, ovate-triangular to broadly elliptic, 11–50 × 3–13 cm or up to 27 cm wide in basal leaves, base cordate, subtruncate, rounded or cuneate, shortly decurrent onto petiole, margins sinuate-dentate to coarsely bidentate, apex obtuse or rounded and minutely apiculate, subacute or acuminate; petiole unwinged, exauriculate at base in basal leaves, broadly auriculate and semi-amplexicaul to exauriculate in stem leaves, 11–67 cm long; all leaves coriaceous to chartaceous, thinly arachnoid or floccose-tomentose and glabrescent above, tomentose or arachnoid or pubescent beneath. Capitula radiate, erect, usually many in a copious congested to lax sometimes divaricately-branched terminal corymbiform cyme; peduncles of individual capitula arachnoid-tomentose or shortly pubescent; involucre cylindrical, 6–11 × 3–7 mm, arachnoid or woolly at least at base or glabrescent; calycular bracts (2)6–12, lanceolate or linear-lanceolate, 1.5–10 mm long, thinly arachnoid to glabrous and sometimes bearded at apex; phyllaries 8–20, green, 5.5–10 mm long, thinly arachnoid or pubescent to glabrous, bearded and shortly tapered or sometimes slightly expanded at apex. Ray florets 5–8(12), corollas pale to bright yellow, tube 4.2–5 mm long, glabrous or sometimes shortly glandular-hairy or with a few long straight multicellular eglandular hairs in lower part, ray limbs 6–18 × 2–5 mm, 4–10-veined. Disc floret corollas yellow, 6.5–9 mm long, tube gradually expanded above middle or from just below middle, glabrous or sometimes with long multicellular hairs near base, lobes 1–1.5 mm long. Achenes 2–5 mm long, glabrous; pappus setae 5–9 mm long, white.

Reported to occur in Malawi (Nyika Plateau) (Burrows & Willis 2005, which record is based on '*Mill* 1979, *Patel* 1999, without localities or collecting details'). I have not seen any specimens.

Also in South Sudan, Ethiopa, D.R. Congo, Rwanda, Burundi, Uganda, Kenya, Tanzania. Montane bushland, *Hagenia* woodland, often in clearings.

*Excluded taxa*

**Senecio abruptus** Thunb., Prodr. Pl. Cap. **2**: 159 (1800). —Setshogo, Prelim. Checkl. Pl. Botsw.: 38 (2005). Type: none cited in protologue; [South Africa:] Swartland, *Thunberg* s.n. (UPS-V-132269, lectotype) as cited by Thunb., Fl. Cap. (Thunb.): 685 (1823): 'Crescit in arenosis Swatlandiae. Floret Septembri', here lectotypified.

*Kleinia acaulis* (L.f.) DC., Prodr. **6**: 339 (1838).
   *Senecio heteroclinius* DC., Prodr. **6**: 380 (1838). Type: 'in Africa Capensi ad Paarl legit', 25.ix.1833, *Drège* 5885 (G-DC holotype, HAL, K, P).

Known from South Africa (Cape Province). Not mentioned [and that includes synonyms] in Hilliard (1977). The distribution of the main population makes it very unlikely that this taxon occurs in our area; I have not seen any specimens.

**Senecio hadiensis** Forssk., Fl. Aegypt.-Arab.: 149 (1775); Icones Rerum Nat.: tab. XIX. (1776). —Jeffrey in Kew Bull. **41**(4): 898 (1986). —Mapaura & Timberlake, Checkl. Zimb. Vasc. Pl.: 28 (2004). —Jeffrey & Beentje in F.T.E.A., Compositae **3**: 646 (2005). Type: Yemen, Bolghose,'in montibus', 1763, *Forsskål* 1714 (C holotype, BM – an unnumbered Forsskål collection, ex herb. Banks).
   *S. subcrassifolius* De Wild., Pl. Bequaert. **5**: 120 (1929). Type: D.R. Congo, Bukumbo, 12.viii.1914, *Bequaert* 5266 (BR8877710, BR8878984 syntypes).

Reported to occur in Zimbabwe: I have not seen any specimens. From the distribution elsewhere it seems doubtful this taxon would occur in our area.
   Known from D.R. Congo, Rwanda, Burundi, Ethiopia, Somalia, Uganda, Kenya, Tanzania, Madagascar, Yemen, Saudi Arabia.

**Senecio kacondensis** S.Moore in J. Bot. **56**: 229 (1918). —Phiri, Checkl. Zamb. Vasc. Pl.: 33 (2005). Type: Angola, 'along the river near Kaconda', 24.ii.1907, *Gossweiler* 4239 (BM holotype, COI).

Several FZ specimens have been misinterpreted (which is how the taxon wound up in the Zambia checklist) and are really *Senecio swynnertonii* S.Moore.

**Senecio karaguensis** O.Hoffm. in Pflanzenw. Ost-Afrikas C: 417 (1895). —Jeffrey in Kew Bull. **41**(4): 895 (1986). —Lisowski, (Asterac. Fl. Afr. Cent. 2) Fragm. Flor. Geobot. **36** Suppl. 1: 284 (1991). —Da Silva *et al.*, Prelim. checkl. Vasc. Pl. Mozamb.: 34 (2004). — Jeffrey & Beentje in F.T.E.A., Compositae **3**: 638 (2005). Type: Tanzania, Lake Province, *Stuhlmann* 1679 (B† syntype); 1978 (B† syntype); 2122 (B†, K syntypes).

This species is reported to occur in Zambia (Lisowski 1991) and in Mozambique (Da Silva *et al.* 2004). This species is known from Rwanda, Burundi, Uganda and northern Tanzania. I consider it unlikely that it occurs in the FZ area.
   The specimens cited as *S. karaguensis* ('new for Malawi') by Brenan *et al.* (1954: 483) were later named *S. tabulicola*, a closely related species that does occur in our area.

**Senecio linifolius** L., Syst. Nat., ed. 10, **2**: 1215 (1759), as '*linifolia*', non (L.) L. (1763), nom. illeg. —Setshogo, Prelim. Checkl. Pl. Botsw.: 38 (2005). Type: 'Habitat ad Cap. b. spei. Burmannus' Herb. Burmanm (G00360056 lectotype), lectotypified by Wijnands, Bot. Commelins: 82 (1983).

I have not seen any specimens; the confirmed population is far away from the FZ area!

**Senecio littoreus** Thunb., Prodr. Pl. Cap. **2**: 158 (1800). —Da Silva *et al.*, Prelim. Checkl. Vasc. Pl. Mozamb.: 34 (2004). Type: South Africa, 'crescit in cultis prope Cap, in et extra hortos', *Thunberg* s.n. (not found).

I have not seen any specimens. From what Hilliard writes in Compositae Natal (1977) it is very unlikely this taxon occurs in our area.

**Senecio luembensis** De Wild. & Muschl. in Bull. Soc. Roy. Bot. Belgique **49**: 233 (1913).
—Lisowski, (Asterac. Fl. Afr. Cent. 2) Fragm. Flor. Geobot. **36** Suppl. 1: 279 (1991).
—Phiri, Checkl. Zamb. Vasc. Pl.: 33 (2005). Type: D.R. Congo, Petite Luembe valley,
ii.1910, *Hock* s.n. (BR887920 holotype).

Reported to occur in Zambia (Phiri 2004) but I have not seen any specimens that could be
this taxon. Known from D.R. Congo and Burundi.


**Senecio lydenburgensis** Hutch. & Burtt Davy in Bull. Misc. Inform. Kew **1936**(1): 82
(1936). —Hilliard, Compositae Natal: 455 (1977). —Da Silva *et al.*, Prelim. Checkl. Vasc.
Pl. Mozamb.: 34 (2004). —Mapaura & Timberlake, Checkl. Zimb. Vasc. Pl.: 28 (2004).
Type: South Africa, Lydenburg, xi.1884, *Wilms* 809 (K holotype).
   *Hertia kuntzei* O.Hoffm. in Kuntze, Revis. Gen. **3**(3): 157 (1898). Type: South Africa, Transvaal,
   Pretoria, 17.ii.1894, *Kuntze* s.n. (NY holotype).
   *Senecio verdoorniae* R.A.Dyer in Bothalia **4**: 186 (1941), nom. nov. pro *Hertia kuntzei*, non *Senecio kuntzei*
   O.Hoffm. [= *S. glaberrimus* DC.] —Da Silva *et al.*, Prelim. Checkl. Vasc. Pl. Mozamb.: 299 (1993).

Reported to occur in Zimbabwe (Mapaura & Timberlake 2004) and in Mozambique (Da
Silva *et al.* 2004); but I have not seen any specimens. Possibly a case of mistaken identity with
*S. proprior* specimens?
Known from South Africa (Transvaal, Natal) and endemic to South Africa (Hilliard 1977).


**Senecio macroglossus** DC., Prodr. **6**: 404 (1838). —Hilliard & Burtt in Notes Roy. Bot.
Gard. Edinb. **32**(3): 382 (1973). —Hilliard, Compositae Natal: 494 (1977). —Jeffrey in
Kew Bull. **41**(4): 933 (1986). —Lisowski, (Asterac. Fl. Afr. Cent. 2) Fragm. Flor. Geobot.
**36** Suppl. 1: 262 (1991). —Da Silva *et al.*, Prelim. Checkl. Vasc. Pl. Mozamb.: 34 (2004).
—Mapaura & Timberlake, Checkl. Zimb. Vasc. Pl.: 28 (2004). Types: South Africa,
Cape, Albany, *Drège* 5123 (G-DC lectotype), lectotypified by Hilliard & Burtt (1973: 382);
Cape of Good Hope, Mt Tabularis, *Zeyher* s.n. (G? syntype); ad Zwarte Omsamcaba et
Omsamcubo, *Drège* s.n. (G-DC, HAL syntypes); ad Albany, *Drège* s.n. (G-DC syntype).

Herbaceous twiner with slightly succulent stem and deltate, lobed leaves, sometimes
cultivated as an ornamental (e.g. in Nairobi, Kenya, and in Kasai, D.R. Congo).
Reported to occur in Zimbabwe and Mozambique; *B.S. Fischer* 1545 (SRGH) collected
in 1948 in the Vumba Mts at 1680 m is possibly a naturalised plant (Drummond; Hyde,
pers. comm.).
Known from South Africa (Cape Province, Natal).


**Senecio pellucidus** DC., Prodr. **6**: 380 (1838). Type: South Africa, Zuurbergen
[Zeeurebergen], *Drège* s.n. (G-DC holotype, E, HAL, HBG, P).

This taxon was listed in Da Silva *et al.* (2004). This plant is a species of the southern and
western Cape and part of the *Senecio madagascariensis* complex – and clearly confused with it.


**Senecio quinquelobus** (Thunb.) DC., Prodr. **6**: 404 (1838). —Macnae & Kalk, Nat.
Hist. Inhaca Isl.: 155 (1969). —Hilliard, Compositae Natal: 496 (1977). —Da Silva *et
al.*, Prelim. Checkl. Vasc. Pl. Mozamb.: 34 (2004). Type: not stated in protologue, but
interpreted as South Africa: Cape, *Thunberg* s.n. (Herb. Thunb. 20904) (UPS holotype).
   *Cacalia quinqueloba* Thunb., Prodr. Pl. Cap. **2**: 142 (1800).

The specimens on which the southern Mozambique 'records' were based, *Borle* 127 and
*Mogg* 31471 are really *S. helminthioides*. *S. quinquelobus* occurs in South Africa (Cape Province
and as far North as Natal); Hilliard (1977) says it is often confused with that taxon.

**Senecio retrorsus** DC., Prodr. **6**: 387 (1838). —Hilliard, Compositae Natal: 487 (1977). —Mapaura & Timberlake, Checkl. Zimb. Vasc. Pl.: 28 (2004). Types: South Africa, Zuurbergen [Zneeurobergen], *Drège* s.n. [2088] (G-DC lectotype), lectotypified by Hilliard (1977: 487); Uitenhagen, *Ecklon* s.n. (G-DC syntype).

    *Senecio retrorsus* var. *subedentulus* DC., Prodr. **6**: 387 (1838). Type: South Africa, between Storm and Wittebergen [Willbergen], *Drège* s.n. [2088] (G-DC syntype); ad Tambukiland, *Ecklon* 1322 (G-DC syntype).

    *Senecio barbellatus* DC., Prodr. **6**: 388 (1838). Type: South Africa, Cafferland, *Ecklon* 700 (G-DC syntype); Albany, *Ecklon* 1423 (G-DC syntype).

    *Senecio latifolius* var. *retrorsus* (DC.) Harv. in Harvey & Sonder, Fl. Cap. **3**: 377 (1865).

    *Senecio latifolius* var. *barbellatus* (DC.) Harv. in Harvey & Sonder, Fl. Cap. **3**: 377 (1865).

    *Senecio graminifolius* C.A.Sm. in J. S. African Bot. **9**: 120, pl. 3, 3 (1943), as '*graminicola*', nom. illeg. non. Sch.Bip. (1844). Type: South Africa, Richmond, Blackwood, *Smith* 7000 (PRE holotype).

Reported to occur in Zimbabwe by Mapaura & Timberlake (2004), but I have not seen any specimens. Known from South Africa (Cape Province, Natal).

## 100. **SOLANECIO** (Sch.Bip.) Walp.[23]

**Solanecio** (Sch.Bip.) Walp., Repert. Bot. Syst. **6**: 273 (1846). —Jeffrey in Kew Bull. **41**(4): 920–923 (1986). —Lisowski, (Asterac. Fl. Afr. Centr. 2) Fragm. Flor. Geobot. **36** Suppl. 1: 411–420 (1991). —Mesfin Tadesse in Kew Bull **49**(1): 137–141 (1993). —Bremer, Asterac. Cladist. Classif.: 509 (1994). —Jeffrey & Beentje, F.T.E.A., Compositae **3**: 670–680 (2005). —Nordenstam in Kubitzki, Fam. Gen. Vasc. Pl. **8**: 231 [2006](2007).

    *Senecio* L. subgen. *Solanecio* Sch.Bip. in Flora **25**(2): 441 (1842).[24]

    *Senecio* L. sect. *Crassuli* Muschl. in Bot. Jahrb. Syst. **43**(1): 38 (1909).

    *Senecio* L. sect. *Tuberosi* Muschl. in Bot. Jahrb. Syst. **43**(1): 38 (1909).

Perennial herbs, often scandent, rarely suffrutices, shrubs or small trees, or sometimes epiphytes. Rootstock of terrestrial herbs sometimes tuberous; stem and leaves often somewhat succulent, pith solid or lamellate when seen in longitudinal section, leaf scars sometimes conspicuous in arborescent species. Leaves alternate, rarely in terminal rosettes, simple, entire to pinnatilobed, often somewhat succulent, glabrous or sparsely to densely variously pubescent, usually tomentose or arachnoid, venation pinnate, usually petiolate, auriculate or exauriculate at base. Inflorescences paniculate-corymbose, sometimes in dense terminal corymbs on inflorescence branches. Capitula homogamous discoid (very rarely reported with some material obviously radiate, or disciform),[25] florets few to many; involucre usually cylindrical, calyculate, calycular bracts few, usually lanceolate, glabrous or pubescent, rarely floccose; phyllaries uniseriate; receptacle epaleaceous, glabrous. Florets hermaphrodite, corollas yellow to orange, sometimes exserted from phyllaries, corolla tube usually dilated at base, glabrous, corolla lobes 5, triangular; basal anther appendages obtuse to sagittate; style arms bifid, apices truncate, sometimes long-penicillate, papillae usually few. Achenes cylindrical, ribbed, glabrous or setuliferous; pappus setae capillary, persistent, barbellate, white or off-white.

---

[23] Jeffrey (1986: 920) cited *Senecio tuberosus* Sch.Bip. ex A.Rich. (= *Solanecio tuberosus* (Sch.Bip. ex A.Rich.) C.Jeffrey) as the 'holotype' of the genus, something repeated by Lisowski (1991: 412). However, Schultz Bipontinus's minimal diagnosis of the subgenus indicated that there were two (unnamed) species he considered belonged in the genus, so there would be no holotype. *Index Nominum Genericorum* states that the type has not been designated.

[24] Lisowski (1991: 411) considered this as nom. inval., and that publication of the name was validated by Walpers; Schultz Bipontinus's minimal prose diagnosis was perfectly adequate to validate the name, with Walpers 'description' merely a Latin translation of the German.

[25] *Solanecio kangae* Q.Luke & Beentje, from Tanzania, was described with heterogamous and disciform capitula, a novelty in the genus which the authors did not comment on, apparently they were more concerned with corolla colour and changing aromas! Jeffrey & Beentje (2005: 676) also reported that one collection of *S. angulatus* had been seen with a few ray florets, noting it was 'extremely unusual'.

17 spp. in tropical Africa, Yemen and Madagascar; three spp. are recorded from the Flora area.

1. Leaves auriculate at base, lamina mostly lyrate-pinnately lobed, lobes spreading or recurved, glabrous; achenes setuliferous . . . . . . . . . . . . . . . . . . . . . . . . . . . . . . .**2.** *angulatus*
   - Leaves exauriculate, lamina not lyrate-pinnately lobed and entire or serrate, tomentose beneath; achenes glabrous. . . . . . . . . . . . . . . . . . . . . . . . . . . . . . . . . . . . . . . . . . . . 2
2. Florets 20–40 per capitulum; capitula 11–15 mm tall; plants erect or scrambling shrubby herbs . . . . . . . . . . . . . . . . . . . . . . . . . . . . . . . . . . . . . . . .**1.** *cydoniifolius*
   - Florets 6–12 per capitulum; capitula 6–8(9) mm tall; plants erect, thick-stemmed shrubs or small trees . . . . . . . . . . . . . . . . . . . . . . . . . . . . . . . . . . . . . . . . . . . . . **3.** *mannii*

1. **Solanecio cydoniifolius** C.Jeffrey in Kew Bull. **41**(4): 921, fig. 6e (1986). —Lisowski, (Asterac. Fl. Afr. Centr. 2) Fragm. Flor. Geobot. **36** Suppl. 1: 414 (1991). —Agnew & Agnew, Upland Kenya Wild Fl., ed. 2: 220 (1994).

   *Senecio cydoniifolius* O.Hoffm. in Physik. Abh. Preuss. Akad. Wiss. (in Engler's Glieder. Fl. Usambaras), **1894**(Abh. I): 19 (20 & 42, list) (1894), as '*cydoniaefolius*', nom. illeg., non Steud. (1841: 560).[26] ?Type/s: [Tanzania:] 'II. Formationen der Creekzone (auf recentem Kalk). ... e Dürres Creekbuschgehölz. ... (II*f*, V*e*)[27] ... [p.20] II*c*, V*c*). ... V. Formationen des Buschsteppenvorlandes. ... [p. 42] e. In Sümpfen des Buschvorlandes ... II*e*, II*f*).'[28] *Holst* 2106 (B†, HBG505249, K00037774 syntypes).[29]

   *Senecio stuhlmannii* Klatt in Bot. Beibl. Leopoldina (Neue Afrik. Compos.) **1895**: 2 (1895). — Brenan in T.T.C.L.: 159 (1949).[30] —Verdcourt & Trump, Common Poison. Pl. E. Afr.: 153 (1969). Type: [Tanzania:] 'Statio: Pangani, leg. *Dr. Stuhlmann*, d. 10. Dec. 1889. (Herb. Mus. bot. Hamburg.)' (HBG505292 – '36', the material det. F. W. Klatt, and the label stamped 'HOLOTYPUS', holotype).[31]

---

[26] A nom. nud. = *Senecio yegua* (Colla) Cabrera.

[27] '*f* = Uferwald

[28] Based on the minimal label information the following would suffice as a transcription 'Tanga[nyika] [Lushoto District, Usambara Mts], Bachüfer. II. 1893. *Holst* [2106]'.

[29] Although Jeffrey (1986: 921), Lisowski (1991: 414), and Jeffrey & Beentje (2005: 674) stated '(holotype B†; isotype K!)', it is unclear if the three mentions of the species by Engler, under the three different habitats, were supported by herbarium collections, presumably all made by *Holst*. Although both extant duplicates were determined by Hoffmann, the HBG material is still in relatively tight bud, and that of the K material has a few more-mature florets on the longer-stiped inflorescence. Since neither Jeffrey (1986: 921), Lisowski (1991: 414), nor Jeffrey & Beentje (2005: 674) selected a lectotype, it would be sensible to select K000377741 as the lectotype.

[30] Brenan (1949: 159–160), had clearly seen a copy of the Botanische Beiblatt, as he mentioned *Senecio stuhlmannii*. However, it is interesting to note that he cited several specimens: '... *Stuhlm.* s.n., *Holst* 2106!, 3035, 3490 (T. Nos.) *Haarer* 2491!!.' (where 'T. Nos.' = Type numbers, in this case representing syntypes of *Senecio cydoniifolius*, of which he had clearly only seen one, and the non-type *Haarer* collection at K). It must be assumed that the *Holst* collections are the syntypes implied under Hoffmann's treatment of *Senecio cydoniifolius*, although Brenan did not mention that name; no *Holst* collections were mentioned by Klatt.

[31] Cited incorrectly in Jeffrey (1986: 921) and Jeffrey & Beentje (2005: 675). as 'Tanzania, Pangani, Stuhlmann s.n., (B†, holo).' The publication in which this name was published is always cited incorrectly (e.g. in FTEA, WFO, POWO, etc., etc.), as simply 'Leopoldina'; the Botanisches Beiblatt were bibliographically independent, separately paginated, additions to that journal and are exceedingly rare. Examination of one copy (from MO) indicated Klatt had clearly designated a holotype collection; this is still extant.

*Senecio dewevrei* O.Hoffm. in Bull. Soc. Roy. Bot. Belgique **39**(2: Comptes-Rendus Roy. Soc. Bot. Belgique): 35 (1900).[32] Type: [D.R. Congo:] 'Congo [(Kinshasa), near Lake Tanganyika,] (*Alfr. Dewèvre*, n. 935).' (BR8878410 – a leafy flowering shoot apex, holotype, ?BR8879912 – a sterile leafy shoot, most probably also used to prepare the species' description).[33]

*Senecio elskensii* De Wild., Pl. Bequaert. **5**: 103 (1929). Type: [Burundi:] 'Brousse de la colline de Kitete (Urundi), 8 novembre 1922 (*Elskens*, n. 111. – Plante assez grasse; d'environ 1 m. 50 de hauteur; feuilles grises sur la face supérieure, argentées sur la face inférieure. Fleur jaune. – Nom. ind.: Ilalile). [D.R. Congo:] Entre Beni et Kasindi, 7 août 1914 (*J. Bequaert*, n. 5177. Village dans le savane; arbuste de environ 1 m. 50 à fleurs jaunes).' *Elskens* 111 (BR8880116 lectotype), lectotypified by Jeffrey (1986: 922); *Bequaert* 5177 (BR9861404 – 4 leaves, BR9861732 – 1 leafy portion of stem, 1 leafy flowering stem apex and a capsule with 2 inflorescences, syntypes).

*Senecio mearnsii* De Wild., Pl. Bequaert. **5**: 114 (1929). Type: 'Vicinity of Kisingo, on the trail from Entebbe, Victoria Nyanza, to Butiaba, Albert Nyanza, Uganda, altitude 650-1110 m., décembre 1909 (Smiths. Afric. Exped. dir Col. Th. Roosevelt, leg. *Dr Edg. Mearns*, n. 2593).' (BR8879813 holotype).

*Senecio giorgii* De Wild. in Contr. Fl. Katanga Suppl. **3**: 154 (1930). Type: [D.R. Congo:] 'Plante herbacée, Albertville [= Kalemie], août 1922 (*De Giorgi*, n. 13).' (BR8879585 holotype).

*Senecio simulans* Chiov., Racc. Bot. Miss. Consol. Kenya: 70 (1935). Type: [Kenya:] 'Mt. Kenya NE, Meru sui monti nella steppa, 9-VIII-1911 (*Balbo* n. 31).' (FT003893 holotype).[34]

Erect to scrambling semi-succulent shrubby herb, 0.6–3.6(7) m high or long. Stems succulent (becoming ridged when dry, and described as pithy), densely white-tomentose, ± glabrescent. Leaves alternate, petiolate, petiole 1.5–8 cm long (sometimes very short and leaves appearing sessile), densely tomentose, exauriculate, leaf lamina ± succulent (described as plicate when young), silvery-white or grey-green to light pale green, simple, ovate, lanceolate, ovate-rhombic or oblong, 4–25 × 2.3–16 cm, base rounded to cuneate and short-decurrent on petiole, lamina densely tomentose to thinly floccose above, densely tomentose beneath, margins appearing almost entire or remotely sinuate-denticulate or conspicuously sinuate-serrate (sometimes drying as conspicuously revolute in some material), apex obtuse to rounded, apiculate. Inflorescence of numerous capitula in large terminal stalked lax- to contracted-thyrses of stalked congested subumbelliform corymbs, branches spreading to ascending, 1–9 cm long; pedicels 4–13 mm long, arachnoid-tomentose. Capitula homogamous and discoid; involucre cylindrical, 6.5–14 mm long, ± 2.5 mm diam.; calycular bracts 2–6, narrowly lanceolate, 1.5–6 mm long, arachnoid at least at base and margins; phyllaries 8–13, green, 6–14.2 × c. 2 mm, glabrous or sparsely hairy and glabrescent. Florets 20–40, corollas yellow, 6–12 mm long, tube gradually expanded from below middle, glabrous or sparsely pubescent upwards, lobes 1–1.7 mm long. Achenes 3–3.7 mm long, glabrous or shortly sparsely setuliferous; pappus setae 6–12 mm long.

**Zambia**. N: Ifuna R., 17.viii.1958, *Fanshawe* F4695 (K, NDO).

Also in D.R. Congo, Rwanda, Burundi, Uganda, Kenya, Tanzania. Lakeside mushitu, forest margins, cultivation edges, thicket, coral rocks near the sea; 0–2100 m, flowering August–December.

Conservation Status: Widespread and common throughout its distribution although apparently restricted, or very rarely collected, in the Flora area; it is best recorded as LC (Least Concern) throughout its range.

Uses: Minor medicinal, bark used like cotton thread; possibly poisonous

---

[32] Hoffmann is clearly the sole author as declared in the Septième Fascicule introduction and for the new species of Compositae!

[33] There is a large glued white paper label on the flowering holotype collection which reads: 'Herbier A. Dewevre No. 935./ ZAIRE/ Lololo (Tanganika) douéamboutou (ikwangoula) matosa- (Tanganika) appliquent les feuilles entières sur les poutas. Tige et partie inférieure couverte de poils blancs.' On the sheet of the vegetative specimen the large brown paper label reads 'Motosa (Ka__)/ 935/ Composée/ Kar appliquent feuilles/ sur fistules.'

[34] There are two labels on the bottom of the type sheet. The original handwritten one bears Chiovenda's published name; the typed label bears the unpublished name '*Senecio fallax* Chiov.' However, it is probable that Chiovenda realized that the name pre-existed, Mattfeld having used the name over a decade earlier.

2. **Solanecio angulatus** (Vahl) C.Jeffrey, Kew Bull. **41**(4): 922 (1986). —Lisowski, (Asterac. Fl. Afr. Centr. 2) Fragm. Flor. Geobot. **36** Suppl. 1: 415, t. 88 (1991). —Beentje, Kenya Trees Shrubs Lianas: 562, map. (1994). —Agnew & Agnew, Upland Kenya Wild Fl., ed. 2: 220, t. 92 (1994). Type: [Yemen:] 'Cacalia sonchifolia. Forssk. Catal. pl. Arab. no. 485. Habitat in Arabia felici.'[35] *Forsskål* s.n. (C10001884 – a sterile leafy shoot, '1026', C10001885 – a flowering shoot, '1028', C10001886 – a flowering stem and a leafy sterile shoot, '1027', C10001887 – a sterile shoot, '1029' syntypes).[36] FIGURE 6.4.**8**.

 *Cacalia sonchifolia* sensu Forssk., Fl. Aegypt.-Arab. CXIX, 485 (1775), non L. (1753: 835)[= *Emilia sonchifolia* (L.) DC.]

 *Cacalia angulata* Vahl in Symb. Bot. **3**: 92 (1794).

 *Kleinia angulata* (Vahl) Willd. in Sp. Pl., ed. 4, **3**(3): 1739 (1803).

 *Emilia? angulata* (Vahl) DC., Prodr. **6**: 303 (1838).

 *Senecio bojeri* DC., Prodr. **6**: 376 (1838). —Brenan in T.T.C.L.: 159 (1949). Type: 'in ins. Madagascar circà urbem Tananarivon legit cl. *Bojer*. Sen. leontodontifolius Bojer! in litt. ... (v.s.)' (G-DC-G00468811 holotype,[37] P00558761 – s.n. the label is printed, with only 'Madagascar. m. Bojer' on it).

 *Senecio leontodontifolius* Bojer ex DC., Prodr. **6**: 376 (1838), nom. nud. pro syn.[38]

 *Senecio subscandens* A.Rich. in Tent. Fl. Abyss. **1**: 434 (1848). —Oliver in Oliver & Hiern, F.T.A. **3**: 421 (1877). Types: 'Crescit in provincia Choa (*Ant. Petit*), et in Abyssinia, sine speciali indicatione (*Schimper*).' Choa, *Quartin-Dillon & Petit* s.n. (P0001248 syntype);[39] *Schimper* Abyss. Sect. III, 1926 (MO391700, MPU011872, P0001249, P0001256, P0001257, S08-7005 syntypes).

 *Senecio gabonicus* Oliv. & Hiern in F.T.A. **3**: 421 (1877). Type: 'Upper Guinea. By River Gaboon, Mann!' *Mann* 995 (K000377700, K000377701,[40] lectotypes, [41] P0001226 – fertile shoots, P0001227 – sterile stems),[42] first-step lectotypification by Jeffrey (1986: 922).

 ?*Gynura fischeri* O.Hoffm. in Pflanzenw. Ost-Afrikas C: 416 (1895). Type: [Tanzania: North Mara District:] '17 (Ukira. – *Fischer* n. 362).' (B† holotype).[43]

 *Crassocephalum subscandens* (A.Rich.) S.Moore in J. Bot. **50**(595): 211 (1912)

 *Crassocephalum bojeri* (DC.) Robyns in Fl. Spermatophyt. Parc Nat. Albert **2**: 544 (1947). —Lind & Tallantire, Comm. Fl. Pl. Uganda, ed. 1: 182 (1962).

Scrambling or rambling perennial succulent herb (often described as a climber), occasionally prostrate or trailing, sometimes rooting when touching soil, 0.6–3(10) m long or high. Stems rather succulent, pale green, often variegated with pale yellow green and purple, slightly angular, glabrous, narrowing markedly into inflorescence. Leaves rather succulent, pinnatisect or lyrately-lobed or compound, very rarely simple (and then beneath inflorescence rachis), 3.7–21.5 × 1–14 cm, with 1–4 pairs of spreading or ± recurving ovate to oblong lobes and a larger ± triangular, sometimes basally deeply 2-lobulate, terminal lobe, margins (sometimes very) coarsely serrate, rachis narrowly or broadly winged, leaf base with conspicuous

---

[35] There are 4 sheets in C that have been determined as syntypes by C. Jeffrey. However, only the sheet annotated as '*Forsskål* 1027', C10001886, is also annotated as '*Cacalia forskolei/sonchifolia* Forssk. CXIX. no. 485', the only specific material cited in the protologue.

[36] All the syntype sheets are databased as '[Yemen] Al Hadiyah'.

[37] There is a long, French description on the label, the plant originally being named 'Senecio leontodifolius Boj. herb. Nro 6.', with the note 'j'ai le recuilly sur les arbres autour de la Capitale Tananarivon, Madagascar/ Mr Bojer 1835'.

[38] Based on the herbarium name on the *Bojer 6* material in G-DC.

[39] Only localized to 'Choa', but in a different hand, the collectors 'Q. Dillon et Petit' are added at the bottom of the label.

[40] Both collections ex herb. Hookerianum, the first with Mann's handwritten label on bluish paper.

[41] Jeffrey (1986: 922), and Jeffrey & Beentje (2005: 676), merely noted 'holotype K!' without specifying which of the two sheets. This can be considered as first-step lectotypification, but taking this further is unnecessary.

[42] Both P collections were duplicates from K.

[43] '17' = Seengebiet. Jeffrey (1986: 930), although listing the name under 'Species insufficiently known', suggested that, from Hoffmann's description, this was most probably a synonym.

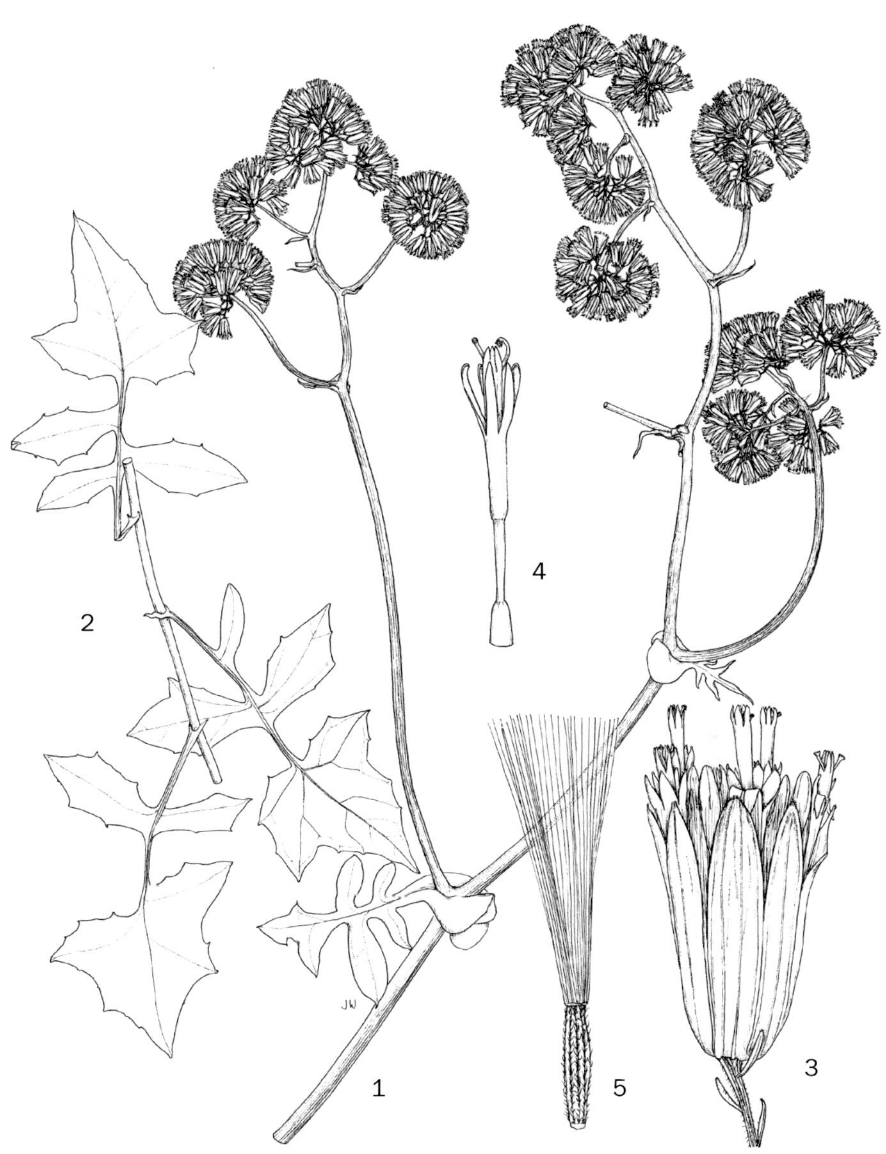

Fig. 6.4.**8**. SOLANECIO ANGULATUS. 1, flowering stem (× ½); 2, portion of leafy stem (× ½); 3, capitulum (× 4); 4, corolla (× 6); 5, achene (× 6). 1 from *Bally* 41630; 2 from *Archbold* 3015; 3, from *Whittall* 128; 4 from *Archbold* 2700; 5 from *Bally* 11583. Drawn by Juliet Williamson. From Flora of Tropical East Africa.

basal semi-amplexicaul rounded auricles (to 20 × 15 mm), lamina discolorous, glabrous or almost so (rarely slightly puberulous), often rather glaucous, especially beneath, margins coarsely sinuate-dentate or sinuate-laciniate to sinuate-lobulate, apex acute, often attenuate. Inflorescences often pendent, terminal and simple or branched, or terminal and axillary, of numerous capitula, branches sometimes spreading and distally decumbent, sometimes in a loose thyrse of congested subumbelliform corymbs, capitula pedicellate, pedicels 4–8 mm long, glabrous or pubescent, braceolate or ebracteolate, bracteoles scale-like or narrow-lanceolate. Capitula strongly scented; involucre cylindrical, 6.5–10.5 × 2–3 mm, calyculate, calycular bracts 2–4, lanceolate, 1.2–2 mm long, glabrous or shortly pubescent on margins and apex; phyllaries 5–8, 6–10 mm long, pale green, often dull purple tinged at apex and base. Florets 10–11, corollas dull golden yellow, 5.5–9.5 mm long, tube glabrous, gradually slightly expanded from below middle, lobes 1.3–2 mm long. Achenes 3–3.5 mm long, ribbed, sinuses setuliferous, often densely so, setulae mostly adpressed to achene, apices acute and apical cells adnate; pappus setae 5.5–8 mm long.

**Zambia**. N: Abercorn, 1615 m, 11.v.1951, *Bullock* 3868 (K, LCS). W: Ndola, 5.v.1954, *Fanshawe* F1161 (K, NDO). S: Mapanza West, R. Munyeke, 1067 m, 14.v.1953, *Robinson* 221 (K). **Zimbabwe**. N: Near Chinhoyi (Sinoia) by Hinyani River, 22.iv.1948, *Rodin* 4379 (K, OBI). W: Hwange (Eankie) Dist., Victoria Falls, 880 m, 7.vi.1979, *Msshasha* 202 (K, SRGH). C: Salisbury Dist., Mukware, 1372 m, 13.v.1946, *Wild* 1089 (K, SRGH). E: Melsetter, Right on confluence of Haroni R./ Chisengu R., 500 m, vii.1973, *Goldsmith* 31/73 (K, SRGH). **Malawi**. N: Njakwa Gorge, S. Rukuru river bridge, 1067 m, 13.vi.1971, *Pawek* 4901 (K, MAL). S: In savannah near Nkopola Lodge, Mangochi, Lake Malawi, 475 m, 11.v.1980, *Townsend* 2158 (K). **Mozambique**. Z: Metolola, 9.ix.1949, *Barbosa & Carvalho* 4003 (K, LMJ). GI: 5 miles south of Maimelane (Maxixe-Mambone Road), ± 152 m, 7.x.1963, *Leach & Bayliss* 11860 (K, SRGH). M: Inhaca Island, 23 mls E of Lourenco Marques, Ponte Torres, 2.x.1957, *Mogg* 27625 (K).

Also known from Angola, Cameroon, Equatorial Guinea, D.R. Congo, Rwanda, Burundi, Sudan, Ethiopia, Somalia, Uganda, Kenya, Tanzania, South Africa, Comoro Is., Madagascar and Yemen.

Rainforest, or forest especially in riverine or lakeshore localities, dambo margin woodland, *Brachystegia* woodland, bushland, scrubby grassland, thickets, ravines, over rocks, sandy soils on sandstone rocks; (250)600–1800(2500) m; usually flowering between April–October.

Conservation Status: Widespread and common throughout its range and in the Flora area; considered LC (Least Concern).

Vernacular name: 'Moleza' (Chinyanja) (*Brass* 17848).

3. **Solanecio mannii** (Hook.f.) C.Jeffrey in Kew Bull. **41**(4): 922, fig. 6f (1986). —Lisowski, (Asterac. Fl. Afr. Centr. 2) Fragm. Flor. Geobot. **36** Suppl. 1: 418 (1911). — Beentje, Kenya Trees Shrubs Lianas: 562, fig., map (1994). —Agnew & Agnew, Upland Kenya Wild Fl., ed. 2: 220 (1994). Type: [Equatorial Guinea:] 'Fernando Po, alt. 6000 ped. (fl. April.) [*Mann* 282]' (K000377702 – ex Herb. Hookerianum, with Mann's original blue paper label – lectotype, K000377703 – ex Herb. Hookerianum, with a transcribed label, 'rec'd 1861', GH00549503), first-step lectotypification by Jeffrey & Beentje (2005:78), second-step lectotypification designated here.[44]

> *Senecio mannii* Hook.f. in J. Proc. Linn. Soc., Bot. **6**: 14 (1861). —Oliver & Hiern, F.T.A. **3**: 418 (1877).
>
> *Senecio multicorymbosus* Klatt in Annal. Naturh. Hofmus. Wien **7**(1): 103 (1892). —Brenan in T.T.C.L.: 159 (1949). Type: 'Angola, Malange, leg. Aug. 1879, [*Mechow*] No. 187.' (?B† holotype, BR8879592).[45]

---

[44] Jeffrey & Beentje (2005: 678) merely indicated the 'holotype' was in K, without specifying which of the sheets was considered the holotype, nor annotating the material.

[45] Although Klatt titled the paper *Compositae Mechowianae*, he did not cite where the *Mechow* collections were located; they appear not to be in W where he apparently wrote the paper.

*Senecio congolensis* De Wild. in Ann. Mus. Congo Belge, Bot. sér. 5, **1**(Fasc.1):[46] 86 (1904). Type: [D.R. Congo:] 'Kisantu, 1900 (*J. Gillet*, n. 1356)' (BR8879165 – a leafy shoot apex with 2 longitudinal ½ stems, BR8879493 – an apical inflorescence mounted alongside a long portion of stem; BR8879615 – a flowering, leafy stem apex, with an original 'jeweller's tag attached to the stem – syntypes).[47]

*Senecio morrumbalaensis* De Wild. in Pl. Nov. Minus Cogn. Herb. Hort. Thenensis **2**(3): 135, t. 101 (1910). Type: [Mozambique:] 'Hab. – Montagnes du Morrumbal. 18 août 1900 (*Éd. Luja*, n. 321 [sic! = '32']).' (BR8879608 holotype).[48]

*Senecio acervatus* S.Moore in J. Linn. Soc., Bot. **40**(275): 121 (1911). Type: [Zimbabwe:] 'Chirinda Forest, 3700–4000 ft.; in fl. Oct.; [*Swynnerton*] n. 665.' (BM000924605 holotype, K000377739 – with a detailed pencil dissection of the style arm apices drawn by Peter Taylor on the sheet above the label).

*Crassocephalum multicorymbosum* (Klatt) S.Moore in J. Bot. **50**(595): 211 (1912).

*Senecio bogoroensis* De Wild., Pl. Bequaert. **5**: 99 (1929). Type: [D.R. Congo:] 'Bogoro, 9 juillet 1914 (*J. Bequaert*, n. 4966. Dans un village; semi-ligneux, buissonnant, de env. 1 m. 50 de haut, fleurs jaune-pâle).' (BR8879820 holotype, BR8880024).[49]

*Senecio mannii* var. *kikuyensis* Chiov., Racc. Bot. Miss. Consol. Kenya: 71 (1935). Type: [Kenya: Fort Hal District:] 'Mt. Aberdare E, Tusu nella brughiera, 15-II-1911 (*Balbo* n. 97).' (FT003894 holotype).

*Crassocephalum mannii* (Hook.f.) Milne-Redh. in Kew Bull. **5**(3): 377 (1951). —Andrews, Fl. Pl. Sudan **3**: 17 (1956). —Blundell, Wild Fl. E. Afr.: fig. 360 (1987).

Erect shrub (sometimes described as a subshrub, or 'arborescent shrub') or small tree 1–8(10) m tall. Stems thick and succulent, green, with prominent persistent leaf-scars, branching dichotomously forming a rounded crown; trunk up to 15 cm diam. with a thin, smooth greyish or greenish bark. Leaves crowded towards branch apices, sometimes appearing rosulate, slightly succulent (frequently described as fleshy), obovate, oblanceolate, or elliptic, 10–55 × 2–16 cm, attenuate into pseudopetiole/petiole, 1–4.5 cm long, exauriculate, lamina, discolorous, often shiny, glabrous or finely pubescent to thinly floccose especially on main veins above, midrib drying paler than lamina above, paler and glabrous or sparsely shortly pubescent to thinly floccose or quite densely tomentose beneath, sometimes lower midrib purplish, prominent, margins shallowly to sinuate-serrate or sinuate-laciniate, or prominently serrate, rarely subentire, apex acute, acuminate-apiculate. Inflorescences of numerous capitula in large thyrses of ± congested subumbelliform corymbs; inflorescence branches 5–30 cm long; pedicels glabrous or shortly pubescent, 4–13 mm long. Capitula homogamous and discoid, scented; involucre cylindrical, 4–7.5 ×1.5–2 mm; calycular bracts 3–4, lanceolate or oblong-lanceolate, pubescent at least on margins, 1–3 mm long; phyllaries 5–8, 3.5–7 mm long, green, apices sometimes purple, abaxially glabrous or sparsely pubescent. Florets 6–12, all hermaphrodite, corollas yellow, sometimes orange or white, 4.5–8.5 mm long, tube expanded in upper half and glabrous or pubescent, lobes 0.7–2 mm long. Achenes 2–3.5 mm long, ribbed, sinuses setuliferous; pappus setae 3.5–8.5 mm long.

**Zambia**. N: Abercorn Dist., Ndundu, 1740 m, 3.x.1959, *Richards* 11484 (K). W: Mufulira, cultivated shrub/tree, 13.ix.1973, *Fanshawe* F12072 (K, NDO). **Zimbabwe**. C: Makoni, 22.iii.1952, *Dehn* R85/53 (K, SRGH). E: Chipinge Dist., Manicaland, Chirundu Forest, Mt Selinda, 1200 m, 20.414°S, 32.708°E, 11.x.1994, *van Wyk* BSA2650 (K, PRU). S: Zimbabwe Dist., on 'Acropolis', 1067 m, 1.x.1949, *Wild* 3014 (K, SRGH). **Malawi**. N: Nkhata Bay Dist., ½ ml E Chikangawa, 1737 m, 12.xi.1977, *Phillips* 3053 (K, MO). C: Dedza mountain, 14°21'37.08"S, 34°19'47.52"E, 1742 m, 30.x.2015, *Chapama et al.* MW/MSBP 1154 (K, MAL). S: Mangochi Dist., Uzuzu Mt., 1250 m, 10.vi.1978, *Iwarsson & Ryding* 762 (K). **Mozambique**. Z: Serra de Milange, 11.ix.1949, *Barbosa & Carvalho* 4033 (K, LMJ). MS: Vumba Mountain, region of Manica, 1200 m, 25.viii.1962, *Gomes e Souza* 4790 (K – of 2 sheets).

---

[46] Several references, and databases, cite this name as published in Vol. 1, Fasc. 2 – which starts on p. 89!

[47] Some accounts merely state the holotype is in BR, although it is clear there are three sheets, all labelled 'TYPE'; BR8879615 could be considered the holotype, or selected as the lectotype, although unnecessary.

[48] The collecting number on Luja's label is clearly only '32', not '321' as in the protologue.

[49] Both of these duplicates appear identical, in terms of being flowering stems, although only the former has a complete label; they may have represented type material available to De Wildeman when described but have subsequently been mounted as two specimens.

Also in Nigeria, Bioko, Cameroon, D.R. Congo, Rwanda, Burundi, Sudan, Ethiopia, Uganda, Kenya, Tanzania, Angola. Forest or forest glades or margins, muchitu, scrub, grassland on rocky slopes, amongst rocks, near cultivated areas; 80–2700 m; flowering August–November, but frequently flowering sporadically throughout the year.

Conservation Status: Widespread and common, and best recorded as LC (Least Concern).

Uses: Planted as a hedge around homesteads, or in crops to keep root-eating rodents away; stems used as light rafters in huts (see Jeffrey & Beentje, 2005: 678 for cited material); minor medicinal.

## 101. **KLEINIA** Mill.

**Kleinia** Mill., Gard. Dict. Abr., ed. 4: 729 (28.i.1754), non Guett. (1754: 377), nec Jacq. (1760: 28) [= *Porophyllum* Guett. –Heliantheae], nec Crantz (1766: 488) [= *Quiscalis* L. – Combretaceae], nec Juss. (1803: 423) [= *Jaumea* Pers. –Heliantheae].[50] —Dyer, Gen. S. Afr. Fl. Pl. **1**: 714 (1975). —Jeffrey, Kew Bull. **41**(4): 923–929 (1986). —Halliday in Hooker's Icon. Pl. **39**(4): 1–135 (1988). —Nordenstam in Kubitzki, Fam. Gen. Vasc. Pl. **8**: 231–232 [2006](2007).
*Notonia* DC. in Arch. Bot. **2**: 518 (1833).
*Senecio* subgen. *Kleinia* (Mill.) O.Hoffm., Nat. Pflanzenfam. **4**, 5(74): 301 (1892). —Jeffrey, Kew Bull. **41**(4): 923–929 (1986). —Halliday, Hooker's Icon. Pl. **39**(4): 1–135 (1988).
*Senecio* subgen. *Notonia* (DC.) O.Hoffm., Nat. Pflanzenfam. **4**, 5(74): 301 (1892).
*Senecio* sect. *Kleinia* (Mill.) Benth. & Hook.f., Gen. Pl. **2**(1): 449–450 (1873).
*Notoniopsis* B.Nord., Opera Bot. **44**: 69 (1978).
*Kleinia* subgen. *Notonia* (DC.) C.Jeffrey, Kew Bull. **41**(4): 926 (1986).

Perennial herbs or subshrubs, erect or prostrate, rarely clamberer, often with tuberous roots. Stem succulent, cylindrical or ± angled. Leaves cauline, alternate, simple, fleshy, flat, sometimes reduced or rudimentary, entire or serrate-lacerate, rarely pinnatilobed, glabrous. Inflorescences of solitary terminal capitula or of few capitula in corymbs, panicles or dense glomerules. Capitula homogamous and discoid (rarely heterogamous and disciform), all florets hermaphrodite (rarely marginal florets functionally female); involucre campanulate or cylindrical, usually calyculate; phyllaries uniseriate; receptacle flat or slightly convex, sometimes alveolate or fimbrillate, glabrous. Corollas tubular, 5-dentate, tube widened at base, white, yellow, orange or red; filament collar cylindrical, subbalusterform or balusterform, anthers obtuse at base, appendaged at apex; style arms compressed, with long or short, papillose, conical to elongated terminal appendages. Achenes cylindrical, fusiform or oblong, ribbed, glabrous or pubescent; carpopodium obscure; pappus multiseriate, many fine bristles, frequently accrescent, barbellate, white.

About 50 spp., mostly in tropical Africa but a few in N Africa, the Canary Is., South Africa, Madagascar, Arabia, India, Sri Lanka, and Thailand.

Many species may flower when leafless.

1. Leaves at stem apices or restricted to branchlets, if scattered then not reducing in size upwards. . . . . . . . . . . . . . . . . . . . . . . . . . . . . . . . . . . . . . . . . . . . . . . . . . . 2
– Leaves crowded towards base of stem, if scattered then reducing upwards . . . . . . . . . 5
2. Phyllaries 5–6. . . . . . . . . . . . . . . . . . . . . . . . . . . . . . . . . . . . . . . . . . . . . . . . . . . . 3
– Phyllaries 9–16. . . . . . . . . . . . . . . . . . . . . . . . . . . . . . . . . . . . . . . . . . . . . . . . . . . . 4
3. Scandent, clambering shrubs; leaves restricted to branchlets; achenes glabrous. . . . . . . . .
. . . . . . . . . . . . . . . . . . . . . . . . . . . . . . . . . . . . . . . . . . . . . . . . . . . . . **1.** *viminalis*

---

[50] Hill, in the *Supplement* of *Mr Chambers's Cyclopedia* (vol. 2), published under the editorship of Scott in 1753, has an entry for *Kleinia*, based upon the same root as Linnaeus's *Cacalia* (from *Genera Plantarum* ed. 4). However, the work is *opera utique oppressa* and is not added to the homonymy. The plant Guettard's referred to is *Kleinia neriifolia* Haw., which was Miller's *Kleinia* of 'The first Sort' and *Cacalia kleinia* L.

- Erect shrubs with erect to spreading stems; leaves set on slight shoulder along young stems; achenes setose between ribs . . . . . . . . . . . . . . . . . . . . . . . . . . . .**2.** *longiflora*
4. Leaves petiolate, margin entire; capitula ecalyculate; phyllaries 9–13 . . . . . . . .**3.** *mweroensis*
- Leaves sessile, margin coarsely and irregularly sinuate-serrate; capitula calyculate; phyllaries 13–16. . . . . . . . . . . . . . . . . . . . . . . . . . . . . . . . . . . . . . **4.** *schweinfurthii*
5. Leaves with midrib abaxially not enlarged towards base; capitula ecalyculate or calyculus obscure . . . . . . . . . . . . . . . . . . . . . . . . . . . . . . . . . . . . . . . . . . . . . 6
- Leaves with midrib abaxially enlarged towards base; capitula calyculate . . . . . . . . . . . 7
6. Shrub up to 90 cm; leaf apex acute to obtuse, margin entire to pinnatilobed, frequently toothed, lateral veins usually indistinct abaxially. . . . . . . . . . . . . . . . . . . . . . . . .**5.** *fulgens*
- Robust shrub 1–2.4 m; leaf apex shortly acuminate and apiculate, margins almost always entire, higher order venation and veinlets visible . . . . . . . . . . . **6.** *abyssinica* var. *abyssinica*
7. Leaves crowded towards base of stem, higher order venation and veinlets plainly visible; corollas yellow; achenes glabrous. . . . . . . . . . . . . . . . . . . . . . . . . . **7.** *chimanimaniensis*
- Leaves scattered, reducing in size upwards, lateral veins usually indistinct; corollas red to orange, achenes densely pubescent . . . . . . . . . . . . . . . . . . . . . . . . . . . .**8.** *mashavensis*

1. **Kleinia viminalis** (Bremek.) Loeuille & D.J.N.Hind, comb. nov. Type: [South Africa:] 'Loc. Waterpoort (Zoutpansberg), 22.I.31.' *Bremekamp & Schweickerdt* 308 (PRE0228182-0 holotype, PRE0587156-0).

> *Senecio viminalis* Bremek. in Ann. Transvaal Mus. **15**: 263 (1933). —Hilliard, Compositae Natal: 498 (1977).
> *Senecio garcianus* Schltr., nom. non rite publ. —Rowley, Succul. Compositae: 115 (1994). — Rowley in Eggli, Illustrated Handb. Succul. Pl.: dicotyledons: 36 (2002).
> *Kleinia garciana* (Schltr.) Torre, nom. non rite publ.

Scandent, clamberer, perennial, succulent shrub, moderately- to densely-branched. Stems cylindrical, mature stems c. 1 cm thick, younger stems c. 5 mm thick, light green with purplish ribs, ± glaucous, farinose, sulcate, glabrous, only branchlets leafy, leaf-scars ± prominent, rounded, pale grey. Leaves sessile or with petiole 0.4–0.6 cm long, deciduous; lamina broadly spatulate, wide elliptic or suborbicular, sometimes narrow elliptic to narrow oblanceolate, 0.9–3 × 0.3–1.2 cm, apex acuminate or mucronate, base attenuate to cuneate, margins entire, glabrous, green, midrib impressed above, slightly prominent beneath, lateral veins usually distinct. Inflorescence of (8)10–20(27) capitula, terminal, congested corymbose panicle; peduncles 1–2.5 cm long, green, leafy bracts, 2.3–5 mm long, few, deciduous, filiform to linear, base truncate, margins entire, frequently densely puberulent adaxially. Capitula homogamous, erect; involucre cylindrical, 0.9–1.5 × 0.35–0.5 cm, calycular bracts 2–3, narrowly lanceolate to linear, 1–4 mm long, deciduous; phyllaries 5, linear or narrowly oblanceolate, 9–15 × 1.3–2.3 mm, glabrous, at least 3 veins, margins scarious, stramineous, apex acute, slightly puberulent. Florets 4–5, corollas white or pale yellow, 1.4–1.9 cm long, tube glabrous, expanded in upper third, lobes 1.5–1.8 mm long, erect, glandular, apex thickened, ± cucullate, mammillate; filament collar subbalusterform. Achenes cylindrical, 3.4–5.5 mm long, glabrous; pappus setae 1.4–1.7 cm long (flowering), 2.4–3.1 cm long (fruiting).

**Zimbabwe**. N: Mutoko (Mtoko) Dist., Nyatsene, 31.xii.1961, *Wild* 5583 (K, SRGH). C: Harare (Salisbury), 30.xii.1929, *Eyles* 6891 (K). E: Mutare (Umtali), 3200', 12.ii.1954, *Chase* 5493 (BM, K, SRGH). S: Chiredzi Dist., 'Chitsa' stone area, Pombadzi river area, lower Lundi on gorge side, 5.ii.1971, *Sherry* 130/71 (K, SRGH). **Mozambique**. N: Antonio Enes, Dist. Mocambique, 13°14'S, 39°54'E, 20.x.1965, *Mogg* 32441 (LISC). GI: Mocucuni Island, Inhambane Bay, 23°51–53'S, 35°22–24'E, 26.xi.1958, *Mogg* 29280 (LISC). M: Maputo (Lourenço Marques), próximo de Porto Henrique, 1.vii.1961, *Balsinhas* 512 (K, LMA).

Also known from Eswatini and South Africa (KwaZulu-Natal, Limpopo, Mpumalanga). *Brachystegia* woodland, *Combretum/Acacia* vlei, granite hills, dune vegetation and as a weed in cultivated fields; 0–975 m; flowering between December and February.

Conservation Status: LC.

Specimens from Zimbabwe differ by their narrow elliptic to narrow oblanceolate leaves (vs. broadly spatulate, wide elliptic or suborbicular) but they share the same scandent habit and identical inflorescence and capitula, so they are treated here as *K. viminalis*.

2. **Kleinia longiflora** DC., Prodr. **6**: 337 (1838). —Halliday in Hooker's Icon. Pl. **39**(4): 129–131, Tab. 3900 (1988). Types: [South Africa:] 'ad Cap. Bonae Spei in regione Trans-Gariepinà inter Klaar-Water et Nu-Gariep legit cl. *Burchell* et mecum sub n. 1718 cat. geogr. comm. et ad Nieuweweld cl. *Drege*! *Cineraria angulosa* E. Mey.! in *Drege* coll. ... (v.s.)' [Between Spingslang and Vaal river', 25. Oct. 1811. (Griqualand West, Herbert Div.)] *Burchell* 1718 (G-DC00471377 lectotype, K000377711, K000377712), lectotypified by Hallyday (1988: 129); [13/10[?]26. ... den Nieuweveldsbergen, bei Beaufort, 4000′ (II, d)] *Drège* 748[51] (G-DC00471443, HAL0113442, P0001497, P0001498, TUB005688 syntypes).

    *Senecio longiflorus* (DC.) Sch.Bip. in Flora **28**: 499 (1845). —Oliver & Hiern, F.T.A. **3**: 421 (1877), p.p.

Erect perennial succulent shrub, (30)60–150(180) cm high, densely branched, with tuberous root. Stems erect to spreading, stiff, cylindrical, ± angled later, c. 1 cm thick, sometimes furrowed, longitudinally marked with 3 darker purplish-green parallel stripes, grey-green or stained purple to mauve, occasionally glaucous, glabrous, leaf-scars conspicuous, prominent, rounded to triangular. Leaves sessile or subsessile, deciduous, on slight shoulders along young stems apices; lamina oblong to elliptic, sometimes oblanceolate, 1.9–6.5 × 0.6–1.1 cm, apex rounded, mucronate, base attenuate, margins entire, glabrous or rarely glabrescent, dark green to grey-green, midrib impressed above, inconspicuous beneath. Inflorescence of 1–4(8) capitula, terminal and lateral, sometimes branched, umbelliform or corymbose; peduncles 0.3–1.3 cm long, pale green, rarely slightly purple, sometimes glaucous, glabrous, bracts sessile, 1.2–4 mm long, linear to lanceolate, at first fleshy, later scale-like, deciduous, dark green, often glaucous, frequently densely puberulent adaxially. Capitula homogamous, erect, appearing before leaves; involucre narrowly cylindrical, 1.3–1.9 × 0.3–0.8 cm; calycular bracts 2–3, lanceolate, 1.4–3.5(7) mm long, glabrous; phyllaries 5–6, somewhat contorted, linear, 1.3–1.9 cm × 1.4–2.1(2.9) mm, green, purplish to mauve, sometimes glaucous, glabrous, striate with at least 3 veins, margins scarious stramineous, apex acute, usually darker, slightly puberulent. Florets 5–12, corollas white or yellow, 1.8–2.1 cm long, tube glabrous, narrowly tubular, very slightly expanded in upper quarter, scarcely swollen at base, lobes 0.9–1.2 mm long, erect or patent, glandular, apex mammillate; filament collar subbalusterform. Achenes cylindrical to fusiform, 5.5–6.5 mm long, setose between ribs; pappus setae 1.6–2.9 cm long (flowering), 3.5–4.5 cm long (fruiting).

**Botswana**. N: Road from Maun-Nokaneng on termite hill near fence, vii.1967, *Lambrecht* 230 (K, SRGH). SW: Ghanzi, 21°42'S, 21°39'E, on limestone gravel in shrubby, overgrazed and trampled area in village, 2.viii.1977, *Skarpe* S-189 (K). SE: Gaborone, university campus, 3250′, 21.viii.1974, *Mott* 526 (K, UCBG). **Zambia**. N: near junction of Munde & Nagupande rivers. Binga, Sebungwe, 866 m, 15.ix.1958, *Lovemore* 548 (K, SRGH). **Zimbabwe**. W: Hwange (Wankie) Dist., Wankie Reserve Forest, 2500′, viii.1960, *Davies* 2803 (K, SRGH). E: Nyanga (Inyanga) Dist., St. Swithin's Tribal Trust Land, 17.i.1967, *Biegel* 1780 (K, SRGH). S: Beitbridge Dist., road from Masere to Tuli, bridge station, 30.viii.1959, *West* 4044 (K, SRGH). **Malawi**. S: Chikala Hills, Liwonde F/Res., Grid Ref. 36LYU5525, c. 950 m., 26.xi.1981, *Chapman & Tawakali* 6013 (K, MAL).

Also known from Angola, South Africa (Cape Province, Transvaal) and Namibia. Mopane woodland, *Acacia-Combretum* scrub, pan edges, among rocks over sandstone or limestone gravel and termite mounds; 450–1200 m; flowering mostly between July and September.

Conservation Status: LC. Sometimes cultivated: one record from Mozambique, *Cedro* 3916 (LMA).

---

[51] The number '743' on P0001498 is a later addition based on a misreading of the label on P0001497.

3. **Kleinia mweroensis** (Baker) C.Jeffrey, Kew Bull. **41**(4): 927 (1986). —Halliday in Hooker's Icon. Pl. **39**(4): 63–64, Tab. 3886 (1988). Type: [Zambia:] 'Habitat. – Kalougwizi river, Mwero, west of Lake Tanganyika, *Carson*, 15 of 1894 collection' *Carson* 15 (K000377725 holotype).

*Senecio mweroensis* Baker, Bull. Misc. Inform. Kew **1895**(107): 290 (1895).

Erect, perennial succulent herb, becoming sprawling or prostrate, tuberous-rooted. Stems semi-prostrate, cylindrical, 8–20(90) cm long, c. 2 cm thick, tuberculate, longitudinally marked with darker purplish-green stripes radiating from prominent stramineous rounded leaf scars, at first glaucous, later greenish, glabrous. Leaves several at top of stem on new growth, fugitive, with petiole 1.1–1.8 cm long, slender, frequently persisting as a straw-coloured, hardened stump; lamina narrowly elliptic, elliptic or ovate-lanceolate, 4–11× 2–3 cm, apex acute to acuminate, base attenuate, margins entire, glabrous, midrib slightly impressed above, prominent beneath, thin. Inflorescence of 1–3 capitula, terminal, cyme; peduncles 6.5–21(25) cm long, often pink-tinged, glabrous or glabrescent, bracts few, 4–8 mm long, linear. Capitula homogamous, usually succeeding leaves, sometimes produced simultaneously; involucre campanulate, 15–23 × 11–20 mm; calycular bracts absent or obscure, linear, 3–4 mm long, glabrous; phyllaries 9–13, oblong or oblanceolate, 15–23 × 1.7–4 mm, glabrous but puberulous at apex, margins scarious whitish or unfrequently purplish, apex acute to acuminate. Florets 30–80, corollas orange-red, 1.7–3.5 cm long, tube glabrous, slightly expanded in upper two-thirds, lobes 3–4 mm long, patent, glandular, mammillate; filament collar balusterform. Achenes somewhat curved, ± fusiform, angled, 3–7 mm long, glabrous; pappus setae 1.3–2.5 cm long, somewhat coherent at base.

**Zambia**. N: Mweru Wantipa (Mweru-wa-Ntipa), Bulaya, 15.ix1958, *Fanshawe* 4841 (K).
Also known from Tanzania, along the shores of Lake Tanganyika and Lake Mweru Wantipa or among rocks, often near waterfalls; 760–1900 m; flowering from July to October.
Conservation status: known only by seven specimens and restricted to a small area in Tanzania and Zambia, VU.

4. **Kleinia schweinfurthii** (Oliv. & Hiern) A.Berger in Monatsschr. Kakteenk. **15**: 11 (1905). —Jeffrey in Kew Bull. **41**(4): 926 (1986). —Jeffrey & Beentje in F.T.E.A., Compositae **3**: 687 (2005). Type: [Sudan:] 'Nile Land. Dar Fertit, by Damuri on the Pango', 10.i.1871, *Schweinfurth* 28 (K000377734 holotype).

*Notonia schweinfurthii* Oliv. & Hiern, F.T.A. **3**: 407 (1877). —Agnew & Agnew, Upland Kenya Wild Fl., ed. 2: 225 (1994).

*Notonia dalzielii* Hutch. in F.W.T.A. **2**: 149 (1931). Type: 'Northern Nigeria, Gimi, Katagum District', v.1908, *Dalziel* 351 (K000377736 holotype).

*Notonia incisifolia* P.R.O.Bally in Journ. E. Afr. Nat. Hist. Soc. **18**: 127 (1946), nom. non rite publ.

*Senecio ballyi* G.D.Rowley in Nat. Cact. & Succ. J **10**: 31 (1955), nom. non rite publ.

*Notoniopsis schweinfurthii* (Oliv. & Hiern) B.Nord., Opera Bot. **44**: 73 (1978).

Erect, perennial succulent herb, tuberous-rooted. Stems erect, cylindrical, 3–8 cm high, leafy, brownish, glabrous; flowering stems precocious, arising before leaves, 15.5–31 cm long, c. 1 cm thick, bearing scattered small scale-like leaves. Leaves sessile, lamina elliptic, 3–12.5 × 1.2–6 cm, apex obtuse to acute, minutely apiculate, base attenuate, margins coarsely and irregularly sinuate-serrate, glabrous, midrib impressed above, slightly prominent or impressed beneath. Inflorescence of 1–3 capitula, terminal, cyme; peduncles 5–19.5 cm long, greenish, glabrous, bracts 0.9–2.5 cm long, elliptic to linear, scale-like, reddish-brown. Capitula homogamous; involucre campanulate, 1.7–2.9 × 1.8–2.6 cm; calycular bracts (1)2–4, lanceolate, 4–6 mm long, glabrous; phyllaries 13–16, oblong or oblanceolate, 1.7–2.9 cm × 1.9–3.5 mm, glabrous, faintly striate, margins scarious whitish, apex acuminate or triangular. Florets ± 50, corollas vivid scarlet, 1.6–2.1 cm long, tube glabrous, slightly expanded in upper half, swollen at base, lobes 1–1.6 mm long, patent, glandular, slightly mammillate; filament collar balusterform. Achenes cylindrical, 7–7.5 mm long, glabrous; pappus setae 1–1.7 cm long (flowering), 2.8–3 cm long (fruiting).

**Malawi**. S: Mangochi Dist., Namingundi River, upper reaches, 13.x.1954, *Jackson* 1374 (K). Also known from Sudan, Nigeria, Kenya, Tanzania. Dry bushland, Grassland?; 200–900 m; flowering between September and January.

Conservation Status: Two specimens from the FZ area; presumably LC because of the wide distribution.

5. **Kleinia fulgens** Hook.f. in Bot.Mag. **92**: t. 5590 (1866). —Halliday in Hooker's Icon. Pl. **39**(4): 71–73, Tab. 3888 (1988). —Lisowski, (Asterac. Fl. Afr. Cent. 2) Fragm. Flor. Geobot. **36** Suppl. 1: 420–421 (1991). Type: 'Amongst these is the subject of the accompanying Plate, which was sent from Port Natal, by Mr. Plant, to our indefatigable horticulturist W. W. Saunders, F.R.S., and in whose succulent-house it flowered in May of the present year' (K, clastotype –a 'mixed' sheet, with *Cooper* 2537 mounted with the voucher material of *K. fulgens* – simply 2 leaves provided, and with profuse notes, by N.E. Brown from Saunders original cultivated material grown at Kew);[52] the original watercolour illustration of *Kleinia fulgens*, used to produce t. 5590 of the protologue (K lectotype), here lectotypified;[53] [Angola:] 'HUILLA. –A very beautiful plant. In dry thickets about Lopollo; fl. and fr. April [1860]', *Welwitsch* 3572 [see Hiern, Cat. Afr. Pl. **1**: 597 (1898)] (LISU218535 –lh specimen, from the April collection –'Hab. pulcherrima planta in dunetis siccis circa Lopolo. c. flor. Apr. 1860 *CW*' epitype; BM000924657, K000377724), here epitypified.[54]

 *Senecio fulgens* (Hook.f.) G.Nicholson, Ill. Dict. Gard. **3**: 420 (1887), non Rydb. (1900: 177) [= *Packera streptanthifolia* (Greene) W.A.Weber & Á.Löve]. —Hilliard, Compositae Natal: 500 (1977).

 *Senecio welwitschii* O.Hoffm. in Bol. Soc. Brot. **13**: 33 (1896). Types: 'Angola (*Welwitsch*, n.º 3582 [*sic* –see Hiern's comment (1898: 597) = 3572] «*Podachaenium*»). –Huilla (*Antunes*)' *Antunes* 29 (COI syntype); *Welwitsch* 3572 (BM000924657, COI, K000377724 –the label annotated with '*Podachaenium* Welw.', LISU21853555 –at the lh bottom is a typical blue-green paper label dated 'c. flor. Apr. 1860 *CW*', at the rh bottom of the sheet is a handwritten label –bearing a Latin description of the plant – clearly marked as 'Distr. Huilla 20/1 60. No. 3572' with a pencilled

---

[52] The colour plate of *Kleinia fulgens* was drawn and lithographed by Walter Hood Fitch, presumably the painting based on the cultivated material in the 'succulent-house' of the horticulturist W.W. Saunders. It is also possible cultivated material eventually came to Kew but that was long after Fitch painted the original; Brown's note on the sheet suggests it came to Kew from Saunders in 1899. Fitch's original plate (from which the lithograph used in Curtis's Bot. Mag. originated) is in K, and is now housed in the Illustrations Collections. It is unclear if 'Mr. [Robert William] Plant' (1818-1858) only sent live material to Saunders [and Stevens], or prepared herbarium material as well (and sent that to George Bentham). Accounts of his travels (Hooker's J. Bot. Misc. **4**: 222–223, 257–265) suggest he sent live material to both Mr Samuel Stevens (24, Bloomsbury Square, London) as well as W. Wilson Saunders (East Hill, Wandsworth, London).

[53] A detailed search of the collections has only revealed the material mentioned above, which provides a conundrum in the typification of the name –see also note 54, pertaining to *Welwitsch* 3572. The journal's extended title at this period was '… comprising the plants of the Royal Gardens of Kew, and of other botanical etablishments in Great Britain, with suitable descriptions; by Joseph Dalton Hooker …', and the specimen found, of only two leaves, would clearly be insufficient to confirm the diagnosis and description of the plant in Curtis's. It is considered that the leaves are syntype material, along with the original illustration of the species. The herbarium specimen purporting to be [clasto]type material by N.E. Brown (K), is considered ambiguous in identifying the species.

[54] It is unclear if the other flowering duplicate in LISU (LISU218536) is the January 1860 collection (see rh specimen on LISU218535), or from the April flowering epitype.

[55] LISU218535 possesses a det. slip by A. R. Torre, oddly determining the material as '*Kleinia fulgens* (Nichols.) Torre', from November 1973, patently unaware of Hooker's earlier description of the species.

'Podachenium gen. nov.', at the top of the sheet is a small set of pencilled dissections on a pinned sheet, LISU218536, LISU218537 –sterile).[56]

*Notonia welwitschii* (O.Hoffm.) Hiern, Cat. Afr. Pl. **1**(3): 596 (1898).

*Notonia fulgens* (Hook.f.) Guillaumin in Bull. Mus. Natl. Hist. Nat., sér. 2, **11**(1): 155 (1939).

*Senecio hookerianus* H.Jacobsen in Sukkulentenk. **4**: 90 (1951), nom. illeg. superfl., as nom. nov. pro *Kleinia fulgens* Hook.f.

*Notoniopsis fulgens* (Hook.f.) B.Nord. in Opera Bot. **44**: 70 (1978).

Erect, perennial, succulent subshrub, up to 90 cm tall, poorly branched near base, with rhizomatous rootstock, supplemented by slender, fleshy, adventitious roots. Stems ascending or erect, frequently with fleshy swollen (potato-like) base, cylindrical, 30–75 cm long, mature stems c. 1 cm thick, glaucous, green, glabrous, ± woody, leaf-scars prominent, lenticular, pale grey, younger stems c. 7 mm thick, blue-green with heavy detersile bloom. Leaves crowded at base, often forming a pseudorosette, sessile or subsessile, lamina spatulate, oblanceolate, unfrequently oblong or obovate, 5.9–13.5 × 2.4–4.2 cm, apex acute to obtuse, base long attenuate, frequently decurrent, margins entire to pinnatilobed, usually coarsely, unevenly, ± remotely toothed, at least in upper half, glabrous, glaucous, bluish-green, often purple beneath, midrib deeply impressed above, prominent beneath, lateral veins usually indistinct. Inflorescence of 1–5(10) capitula, terminal, usually branched, lax cyme, frequently subcorymbose; peduncles 10–40 cm long, glaucous-green, glabrous, ± swollen directly beneath capitulum, leafy bracts, 0.6–2.3 cm long, linear to lanceolate, reducing in size upwards and becoming bractiform, base cuneate or truncate, margins entire. Capitula discoid and homogamous, sometimes disciform and heterogamous, erect; involucre campanulate, rarely cylindrical, 1.5–2.7 × 1–2.9 cm, ecalyculate; phyllaries 7–9, linear, lanceolate or oblong, 1.5–2.7 × 2–4.3(6.4) mm, glabrous, at least 3 veins, margins scarious, stramineous, apex acute, minutely pubescent or puberulent. Florets ± 60, marginal florets sometimes functionally female, corollas vermillion deepening to crimson as they age, 1.8–2.1 cm long, tube glabrous, expanded in upper half, swollen at base, lobes 2.5–3 mm long, patent, glandular, apex thickened, ± cucullate; filament collar subbalusterform. Achenes cylindrical, sometimes ± curved, 2.3–6 mm long, glabrous; pappus setae 0.8–1.6 cm long.

**Zambia**. B: Machili, 8.vi.1963, *Fanshawe* 7828 (K, NDO). C: Mumbwa, 4.vi.1961, *Fanshawe* 6648 (K, NDO). **Zimbabwe**. N: E side of Umvukwe Mountains near Dawson, 28–29.iv.1948, *Rodin* 4472 (K). C: 'Salisbury Distr.', Twentydales, 4500ʹ, 1.iii.1947, *Wild* 1879 (GH 16225) (K). E: 'Odzani River Valley, Dist Manica, Div Umtali', 1914, *Teague* 241 (K). S: Chibi Dist., 17 miles N. of Lundi River on Salisbury-Beit Bridge Road, 24.iv.1961, *Leach* 10803 (K, SRGH). **Mozambique**. GI: Gaza, Massingir, piste vers l'ancien poste, 27.vi.1980, *Schafer* 7175 (K, LMU). M: Lourenço Marques, about 10 miles N. of Moamba, 28.vi.1964, *Leach* 12262 (K, SRGH).

Also known from Angola, Eswatini, D.R. Congo, South Africa (KwaZulu-Natal, Transvaal). Mopane and miombo woodlands, usually among rocks on slopes, over dolorite outcrops and anthills, rarely in *Hyphaene* clumps on dambo grassland, at altitudes of between 30–1400 m; flowering mostly between March and August.

Conservation Status: LC. Sometimes cultivated.

A very polymorphic species regarding the leaf morphology, ranging from entire to pinnatilobed and coarsely toothed margin (sometimes in a single specimen). The two specimens from Zambia represent an extreme of that variation with smaller and nearly entire leaves, strongly resembling *K. grantii* Hook.f. But they do not have the longer and narrower petioles of that species. *K. grantii* has also a different distribution (Mali, Guinea, D.R. Congo, Ethiopia, Somalia, Kenya and Tanzania).

---

[56] Hiern (1898: 597) correctly pointed out Hoffmann's error in citing the *Welwitsch* collection as '3582' –all collections, including those annotated with labels reading '*Podachaenium*', are numbered '3572'. The numbered *Antunes* collection is cited by Mendonça, Contrib. Conhec. Fl. Angola **1** (Compositae): 115 (1943). Hiern's catalogue gave the collection locality as: [Angola:] 'HUILLA. –A very beautiful plant. In rocky elevated parts of Morro de Lopollo, very sparingly fl. and fr. Jan. 1860. In dry thickets about Lopollo; fl. and fr. April [*Welwitsch*] No. 3572. Flowers large, cinnabar-red. Lopollo, fl. and fr. May 1860. COLL. CARP. 670.' It is highly probably that Hooker's mention of 'a very similar plant in cultivation, brought from Angola by Dr. Welwitsch, …' is the same collection as *Welwitsch*'s 3572.

6. **Kleinia abyssinica** (A.Rich.) A.Berger in Monatsschr. Kakteenk. **15**: 11 (1905). —Jeffrey in Kew Bull. **41**(4): 928 (1986). —Halliday in Hooker's Icon. Pl. **39**(4): 77, Tab. 3889 (1988). —Lisowski, (Asterac. Fl. Afr. Cent. 2) Fragm. Flor. Geobot. **36** Suppl. 1.: 423, t. 89 (1991). —Jeffrey & Beentje in F.T.E.A., Compositae **3**: 694–696 (2005). Type: [Ethiopia:] 'Crescit in locis humidis prope *Kouiaetha*, non procul a convalle fluvii *Mareub* (Quartin Dillon)' *Quartin Dillon & Petit* s.n. (P0001418 holotype).

   *Notonia abyssinica* A.Rich., Tent. Fl. Abyss. **1**: 444, t. 59 (1848). —Oliver & Hiern, F.T.A. **3**: 407 (1877).

   *Senecio (Kleinia) nyikensis* Baker in Bull. Misc. Inform. Kew **1897**(128–129): 271 (1897), non Baker (1898), nom. illeg. (=*Senecio syringifolius* O.Hoffm.). Type: [Malawi:] 'BRITISH CENTRAL AFRICA. Nyika Plateau, alt. 6000–7000 ft. [July 1896] *Whyte*' (K000377723 holotype).

   *Notonia opima* S.Moore in J. Bot. **45**: 329 (1907). Type: [Uganda:] 'Hab. Semliki Valley, Toro; [2800', 26.x.1906,] *Bagshawe*, 1276' (BM000924659 holotype).

   *Senecio superbus* De Wild. & Muschl. in Bull. Soc. Roy. Bot. Belg. **49**: 236 (1913). Type: [D.R. Congo:] 'Tanganika (Kassner, n. 3025a. –30–V–1908)' (BR8875747, E00239788, HBG504770, K000377735, P0001415 syntypes).

   *Notoniopsis abyssinica* (A.Rich.) B.Nord. in Opera Bot. **44**: 70 (1978).


## Var. **abyssinica**

Erect perennial succulent herb, 1–2.4 m high, unbranched or sparsely branched, with fleshy potato-like tuberous or rhizomatous rootstock. Stems erect, cylindrical, c. 2.5 cm thick near base, at first white-glaucous, purplish, later green, glabrous, leaf-scars conspicuous, wedge or butterfly-shaped. Leaves succulent, sessile, persistent, crowded towards base of stem, less so above; lamina broadly ovate to elliptic, obovate to oblanceolate, 6–26(30) × 2.5–8.4(13.2) cm, apex acute to rounded, shortly acuminate-apiculate, base cuneate to attenuate or frequently semi-amplexicaul and ± cordate, margins almost always entire, occasionally with few ± coarse teeth, glabrous, pale green, usually variously variegated with red or purple and/or with whitish veins, midrib impressed above, prominent beneath, higher order venation and veinlets visible. Inflorescence of 1–10 capitula, terminal, lax bracteate cyme, frequently subcorymbose; peduncles 4–60 cm long, white-glaucous or greenish, glabrous, swollen directly beneath capitulum, bracts subsessile, 0.8–2.2 cm long, ovate, elliptic to lanceolate, leaf-like, deciduous, pale green, often glaucous. Capitula homogamous, nodding in bud, erect in flower; involucre widely campanulate, 1.8–3.3 × 2–4.9 cm; calycular bracts obscure or absent; phyllaries (8)13(14), oblong, lanceolate, sometimes linear, 1.8–3.3 cm × 2–6.7 mm, dark blue-green to pale green, often glaucous, glabrous, faintly striate, margins scarious stramineous or pinkish, apex acute, slightly puberulent. Florets 70–80, buds yellow, corollas pink, bright red or orange-red, 2–2.3 cm long, tube glabrous, slightly expanded in upper two-thirds and lower third narrowly tubular, swollen at base, lobes 3–5 mm long, patent, glandular, apex thickened, slightly mammillate; filament collar balusterform. Achenes subcylindrical, somewhat angled, 3–5.5 mm long, glabrous, with minute but distinct apical rim; pappus setae 1.3–2.3(3) cm long.

**Zambia**. N: Abercorn Dist., woodland near Kilema, 1500 m, 12.v.1962, *Richards* 16461 (K). S: 9 m E. of Choma, dry bush, 1400 m, 29.v.1955, *Robinson* 1282 (K). **Malawi**. N: Chitipa Dist., Mugesse For. Res., 9°40'S, 33°32'E, 24.v.1989, *Pope et al.* 2341 (BR, K).

Also known from Central African Republic, D.R. Congo, Burundi, Sudan, Ethiopia, Uganda, Kenya and Tanzania. Grassland with scattered trees or bushes, thicket edges, wooded grassland, woodland, often in rocky sites; 500–2250 m; flowering in April and May. Conservation Status: LC.

*Richards* 22383 differs by its corolla shape: base of tube gradually swollen and upper third abruptly expanded in upper third (vs. slightly expanded in upper two-thirds and lower third narrowly tubular, swollen at base), scape bracts are also more numerous. Vegetative parts are lacking. Flowering in October! (vs. April-May).

*Kleinia abyssinica* var. *hildebrandtii* (Vatke) C.Jeffrey has not been recorded in the flora area, it differs from the typical variety by its habit (often prostrate or arching), lower size (15–50 cm high), leaves with narrow base and smaller and narrower inflorescence bracts.

7. **Kleinia chimanimaniensis** van Jaarsv. in Bradleyia **33**: 122 (2015). Type: 'Zimbabwe: 1932-(Chimanimani): Outward Bound, Chimanimani Mountains, on sheer W. facing cliffs (Mutare region), (-DD), 17 June 2014. *Van Jaardsveld* 25444' (SRGH holotype).

Erect, perennial, succulent subshrub, up to 60 cm tall, moderately branched, stoloniferous from base, roots fibrous. Stems ascending to decumbent, slightly flattened or cylindrical, 5–30 cm long, 1–4 cm thick, glaucous, later green and sometimes pale brown and purplish, glabrous, leaf-scars slightly prominent, lenticular, pale grey. Leaves crowded at base, sometimes forming a pseudorosette, sessile, lamina narrowly oblanceolate, oblanceolate or unfrequently oblong, 5–17.6 × 1.6–6.4 cm, apex obtuse to rounded, shortly acuminate, base attenuate to decurrent, margins entire, glabrous, glaucous, later dark green, midrib enlarged towards base, impressed above, prominent, carinate beneath, higher order venation and veinlets plainly visible. Inflorescence of (1)2–6(12) capitula, terminal, compound cyme or panicle; peduncles (5)9–30(41.5) cm long, glaucous, pale green, sometimes purplish, glabrous, leafy bracts, 1.1–5.6 cm long, lanceolate to ovate, base usually cordate, frequently subamplexicaul. Capitula homogamous, nodding in bud and flower, erect in fruit; involucre campanulate, 1.9–2.3 × 1.4–3 cm; calycular bracts 2–4, ovate or elliptic, base obtuse or frequently cordate, 0.6–1.5 cm long, glabrous; phyllaries 8, oblong or lanceolate (1–3 linear), 19–23 × 2.1–7 mm, glabrous, inconspicuously striate, margins scarious, stramineous or unfrequently purplish, apex acute, slightly puberulent. Florets 30–80, corollas yellow, 1.4–1.5 cm long, tube glabrous, slightly expanded in upper half, swollen at base, lobes 0.9–1.6 mm long, erect or patent, glandular, scarcely mammillate, filament collar cylindrical. Achenes cylindrical, ± angled, 2.9–7 mm long, glabrous; pappus setae 0.9–1.3 cm long.

**Zimbabwe**. E: Chimanimani (Melsetter), 4800′, 29.v.1959, *Leach* 9058 (K, SRGH). **Mozambique**. MS: Manica Province, South of Manica (Vila Pery), Zembe Mountain, vi.1961, *Leach* 11097 (K, SRGH).

Endemic from the Chimanimani-Nyanga Highlands. Montane grassland, savannah, among rocks on sheer cliffs and exposed hill tops; 450–2000 m; flowering in May and June.

Conservation Status: NT? because its distribution area is rather small and its habitat restricted. Cultivated.


8. **Kleinia mashavensis** Loeuille & D.J.N.Hind, sp. nov.[57] Type: Zimbabwe, 'S. Rhodesia, Mashuba, Flws. bright red with orange styles, Flrd. Prinshof May 1958', Mashaba Dist., 23.viii.1957, *Leach* 7176 (K000374796 holotype, PRE).[58] FIGURE 6.4.**9**.

Erect perennial succulent herb, unbranched, with fleshy potato-like tuberous. Stems erect, cylindrical, c. 1 cm thick, at first glaucous, later green with purplish lines, glabrous, leaf-scars conspicuous, lenticular. Leaves succulent, sessile, persistent, reducing in size upwards, lamina spatulate or oblanceolate, 6–18 × 1.9–6 cm, apex acute to rounded, shortly acuminate-apiculate, base attenuate, decurrent, margins almost always entire, occasionally with few ± coarse teeth, glabrous, pale green,

---

[57] **Kleinia mashavensis** Loeuille & D.J.N.Hind sp. nova, *K. fulgenti* Hook.f. foliis spatulatis vel oblanceolatis, nervis lateralis indistinctis similis, sed foliis dispersis (non aggregatis basin versus), foliorum nervis mediis infra amplificatis basin versus (non sine amplificatione), capitulis calyculum habentia (non sine calyculo) et achaeniis dense pubescentis (non glabris) differt. Type: Zimbabwe, 'S. Rhodesia, Mashuba, Flws. bright red with orange styles, Flrd. Prinshof May 1958', Mashaba Distr., 23.viii.1957, *Leach* 7176 (K holotype, PRE).

[58] Despite having the same collector number, the two specimens of *Leach* 7176 at K cannot be considered as a single gathering. K000374795 consists of two separate capitula and two single leaves; at the bottom of the label it has been added later 'Flowering at Greendale, Salisbury. 2/4/1958'; I consider this specimen as a collection from the wild since the label has a more precise locality information 'sandveld near Mashaba' and earlier date 'April, 1956'. K000374796 consists of a much better-preserved flowering stem; the label indicating 'Flrd. Prinshof May 1958', suggesting that this was a plant cultivated at the University, and the label typed long after the material was collected. Moreover, the collecting dates do not match between the two specimens: K000374795 is from 'April 1956' and K000374796 indicates '23.8.57' (one will be a collecting date, the other will be a potential preparation date).

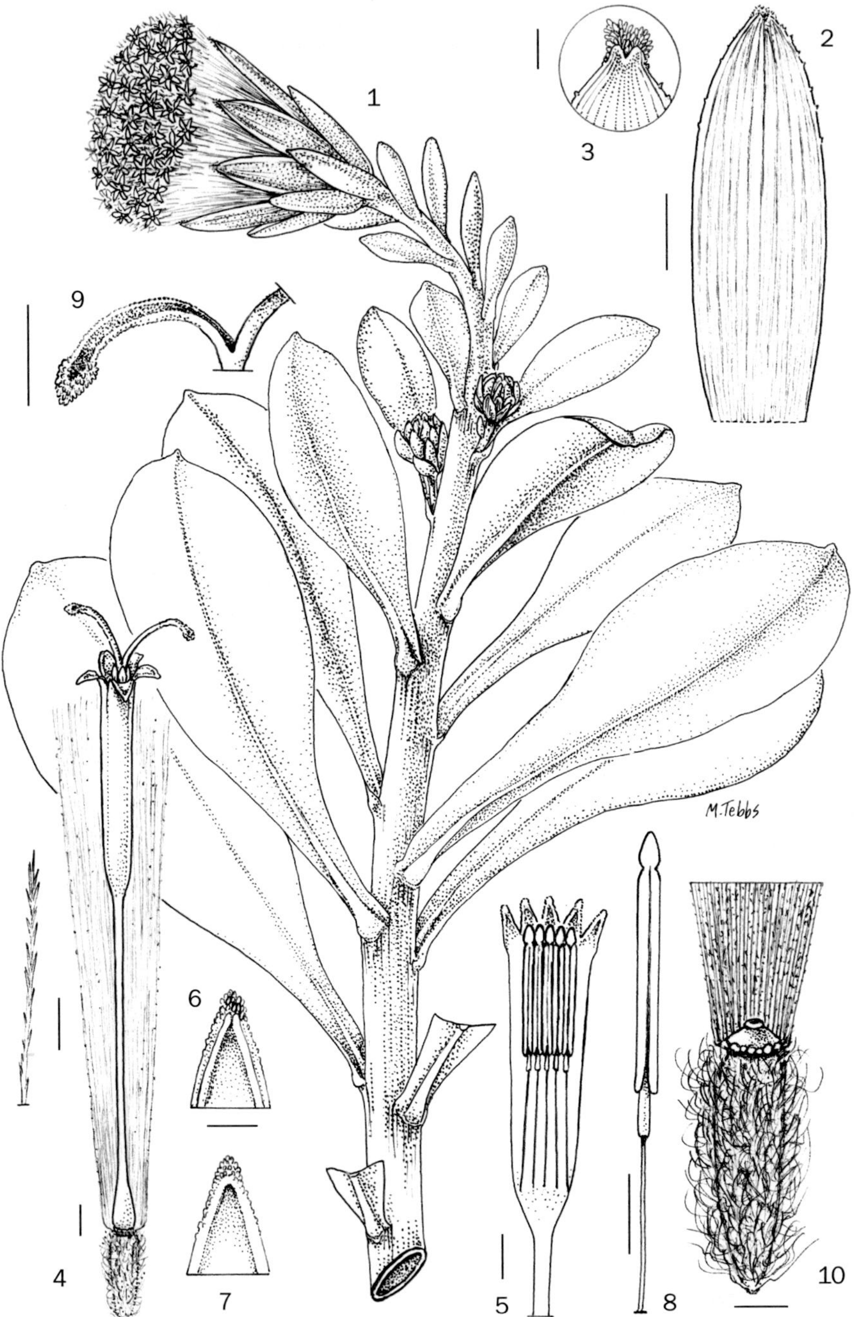

Fig. 6.4.**9**. KLEINIA MASHAVENSIS. 1, flowering stem; 2, phyllary; 3, detail of phyllary apex; 4, floret; 5, corolla opened out to show attachment point of filaments; 6, detail of apex of corolla lobe, adaxial view; 7, detail of apex of corolla lobe, abaxial view; 8, stamen; 9, detail of style arm; 10, achene and base of pappus setae; 11, detail of apex of pappus seta. All from *Leach* 7176. Scale bars: 1 = 10 mm; 2 = 5 mm; 3–5, 8–10 = 1 mm; 6, 7, 11 = 0.5 mm. Drawn by Margaret Tebbs.

midrib impressed above, prominent and slightly carinate beneath, midrib enlarged towards base, lateral veins ± indistinct. Inflorescence of 1–2 capitula, terminal; peduncles 6–13 cm long, greenish, glabrous, swollen directly beneath capitulum, bracts sessile, numerous, 1.6–4.9 cm long, oblanceolate or oblong, leaf-like, base cuneate or truncate, green. Capitula homogamous, nodding in flower and fruits; involucre campanulate, 3.4–4 × 2.3–3.5 cm; calycular bracts 2–6, lanceolate or narrowly elliptic, base truncate, 1.6–2.6 cm long, glabrous; phyllaries 7–8, oblong (sometimes 1–2 linear), 3.4–4 cm × 4.5–11.7 mm, green, ± glaucous, glabrous, faintly striate, margins scarious stramineous, apex shortly acuminate, slightly puberulent. Florets 60–80, corollas bright red or deep orange, 2.5–3.1 cm long, tube glabrous, slightly expanded in upper third and lower two-thirds narrowly tubular, swollen at base, lobes 1.1–1.8 mm long, glandular, apex and margins thickened; filament collar cylindrical. Achenes cylindrical, 3–4 mm long, densely pubescent, with distinct darker apical rim; pappus setae 1.9–3.2 cm long, somewhat coherent at base.

**Zimbabwe**. S: Masvingo (Victoria) Dist., 4 miles W of Mashava (Mashaba), iv.1956, *Leach* 7165 (K, SRGH); sandveld near Mashaba, iv.1956, *Leach* 7176 (K000374795, SRGH).

Sandveld in partial shade; flowering in May (in cultivation).

Conservation status: probably CE. Only known by two collections from the 1950's, apparently of very restricted distribution and thus potentially vulnerable particularly by asbestos mining activities in the area. Cultivated.

The species is unique in the combination of large capitulum with conspicuous calyculus and densely pubescent achenes. Closely related to *K. fulgens* by its spatulate to oblanceolate leaves with lateral veins ± indistinct but the new species does not have its leaves crowded towards the base, abaxially the midrib is enlarged towards the base and *K. fulgens* has ecalyculate heads and glabrous achenes. Also similar to *K. abyssinica* by its robust habit and large heads but that species has an obscure calyculus, frequently absent, and glabrous achenes

*Uncertain identification*

**Kleinia grantii** (Oliv. & Hiern) Hook.f. in Bot. Mag. **125**: t.7691 (1899). —Jeffrey in Kew Bull. **41**(4): 928 (1986). —Halliday in Hooker's Icon. Pl. **39**(4): 87, Tab. 3891 (1988). —Lisowski, (Asterac. Fl. Afr. Cent. 2) Fragm. Flor. Geobot. **36** Suppl. 1: 421 (1991). —Jeffrey & Beentje in F.T.E.A., Compositae **3**: 692–694 (2005). Type: [Tanzania:] 'Mozamb. Distr. M'bǔmi, 6°56'30" S. Lat. [, 1860,] *Col. Grant!*' (K000377726 holotype).

   *Notonia grantii* Oliv. & Hiern, F.T.A. **3**: 407 (1877).

   *Notonia coccinea* Oliv. & Hiern, F.T.A. **3**: 407 (1877). Type: [Ethiopia:] 'Nile Land. Abyssinia, [Somak Efat, iv.1842,] *Roth!* [*sic*]'[59] (K000377730 holotype).

   *Senecio longipes* Baker in Bull. Misc. Inform. Kew **1895**(105): 217 (1895), nom. illeg., non Hook.f. (1846). Type: [Somalia:] 'Somali-land: Golis range. *Miss Edith Cole.*' (K000377729 holotype).

   *Kleinia coccinea* (Oliv. & Hiern) A.Berger in Monatsschr. Kakteen. **15**: 11 (1905).

   *Senecio coccineus* (Oliv. & Hiern) Muschl. in Bot. Jahrb. Syst. **43**(1): 55 (1909), comb. illeg., non Klatt (1886).

   *Senecio phellorrhizus* Muschl. in Bot. Jahrb. Syst. **43**(1): 70 (1909). Type: [Tanzania:] 'Kilimandscharozone: [Masai District, Lamuniane Mts,] In humusreichem Boden (JAEGER n. 365ª)' (B† holotype).

   *Notonia bequaertii* De Wild., Pl. Bequaert. **5**: 444 (1929). Type: [D.R. Congo:] 'Kabare, 19 août 1914 (J. Bequaert, n. 5357. – Steppe herbeuse des bords du lac. – Fleurs rouges-orangées).' (BR8875433 lectotype, BR8875785), here lectotypified.[60]

---

[59] Although the collector is cited by Oliver & Hiern as Roth it is clear on the label that the name reads Rohr. In addition to the original number the label has '365' pencilled below the name together with 'Flowers of the brightest scarlet' in the original handwriting.

[60] Both specimens in BR are considered to be parts of the original material. However, the one considered as isolectotype does not bear the detailed habitat information of the lectotype, nor an indication that it is sheet 2 of 2.

*Senecio coccineus* (Oliv. & Hiern) H.Jacobsen, Handb. Sukkulent. Pfl. **2**: 1022 (1954), comb. illeg., non Klatt (1886).

*Senecio coccineiflorus* G.D.Rowley in Nat. Cact. & Succ. J. **10**: 31 (1955), nom. nov. pro *S. coccineus* (Oliv. & Hiern.) Muschl. Type as for *Notonia coccinea*.

*Notoniopsis coccinea* (Oliv. & Hiern) B.Nord. in Opera Bot. **44**: 70 (1978).

*Notoniopsis grantii* (Oliv. & Hiern) B.Nord. in Opera Bot. **44**: 70 (1978).

Known from Guinea, Mali, D.R. Congo, Rwanda, Burundi, Ethiopia, Somalia, Kenya and Tanzania, plants of grassland, woodland and open bushland, often in stony sites. Despite the reported occurrence in Zambia (Phiri, Checkl. Zamb. Vasc. Pl., 2005), I have not seen any specimens from our area.


# 102. **GYNURA** Cass.

**Gynura** Cass. in Dict. Sci. Nat. **34**: 391 (1825), nom. cons. —Davies in Kew Bull. **33**(2): 335–342 (1978). —Jeffrey in Kew Bull. **41**(4): 929–930 (1986). —Lisowski, (Asterac. Fl. Afr. Cent. 2) Fragm. Flor. Geobot. **36** Suppl. 1: 425–434 (1991). —Jeffrey & Beentje in F.T.E.A. Compositae **3**: 697–702 (2005). —Nordenstam in Kubitzki, Fam. Gen. Vasc. Pl. **8**: 232 [2006](2007). —Vanijajiva & Kadereit in J. Syst. Evol. **49**(4): 285–314 (2011). —Chen & Nordenstam in Fl. China **20-21**: 538–542 (2011).

*Gynaecura* Hassk., Cat. Hort. Bot. Bogor.: 103 (1844), nom. inval., orth. var.[61]

*Senecio* L. sect. *Gynura* (Cass.) Baill., Hist. Pl. **8**: 260 (1882).

Perennial herbs, erect, scandent or climbing; roots fibrous or tuberous. Stems often fleshy, sometimes woody. Leaves sessile or petiolate, petiole base auriculate or exauriculate, alternate, mostly cauline, sometimes basally rosettiform, lamina linear-lanceolate, oblong, ovate to deltoid, pale to dark green, sometimes purplish beneath, simple, dentate or lobed, glabrous or variously pubescent, base cuneate, truncate or rounded, rarely unequal, margins entire, crenate, minutely denticulate to coarsely dentate or lyrate to pinnately-lobed, apex apiculate to acuminate. Inflorescence of terminal corymbose cymes, and/or of axillary cymose panicles, rarely of solitary capitula, capitula pedicellate. Capitula homogamous and discoid; involucre cylindrical or narrow-campanulate, calyculate, calycular bracts linear or subulate, glabrous or variously pubescent; phyllaries uniseriate, free, herbaceous, margins narrow- or broad-scarious; receptacle epaleaceous, glabrous, flat. Florets all hermaphrodite, numerous, corollas orange to yellow, sometimes red or purple, upper part campanulate or infundibuliform, corolla lobes 5, oblong-lanceolate, apices acute; filament collar balusterform; basal anther appendages obtuse or slightly sagittate, apical anther appendages ovate or lanceolate; style arms prominent, long, exserted, gradually tapered, sometimes coloured, penicillate. Achenes oblong to cylindrical, ribbed, glabrous or setuliferous, brown; carpopodium annular or cylindrical; pappus setae many, capillary, finely barbellate, persistent, white, off-white or yellowish.

A genus of about 40 spp., mostly in SE Asia but also in mainland Asia, Africa, and Madagascar; three spp. are known from the Flora area as natives, contrary to the suggestions of Vanijajiva & Kadereit (2011). *Gynura aurantiaca* (Blume) DC., a native of India, China and Indonesia, widely grown in horticulture, is known from cultivation in Malawi.

1. Climbing or scrambling herbs, 1.5–12 m tall/long . . . . . . . . . . . . . . . . . . . . . . . **1.** *scandens*
   – Erect or decumbent herb . . . . . . . . . . . . . . . . . . . . . . . . . . . . . . . . . . . . . . . . . . . . 2
2. Capitula 2–3; leaves glandular . . . . . . . . . . . . . . . . . . . . . . . . . . . . . . . . . . **2.** *pseudochina*
   – Capitula 5 or more; leaves eglandular . . . . . . . . . . . . . . . . . . . . . . . . . . . . . . **3.** *valeriana*

---

[61] Hasskarl cited the genus as '*Gynaecura* Css. Endl. Gen. 2792', but provided a breakdown and translation of the name he provided, even though Endlicher's account only used the generic name *Gynura* Cass.

1. **Gynura scandens** O.Hoffm. in Pflanzenw. Ost-Afrikas C: 416 (1895). —Agnew, Upland Kenya Wild Fl., ed. 1: 471 (1974). —Davies in Kew Bull. **33**(2): 337 (1978). —Jeffrey in Kew Bull. **41**(4): 929 (1986). —Blundell, Wild Fl. E. Afr.: fig.425 (1987). —Halliday & Hind in Kew Mag. **7**(3): 109–113 (1990). —Lisowski, (Asterac. Fl. Afr. Cent. 2) Fragm. Flor. Geobot. **36** Suppl. 1: 427, t. 90 (1991). —Agnew & Agnew, Upland Kenya Wild Fl., ed. 2: 225, t. 95 (1994). Types: [Tanzania: Usambara, Lutindi] '13. (Usb., Lutndi, Bemgarru, 1500 m. – *Holst.* n. 3315) 17 (Buddu und Bu. – *Stuhlm[ann]*. n. 1029, 1182, 4015). – Im Gebirgsbuschwald kletternd.' [VII. 1893] *Holst* 3315 (K000377685 lectotype, BM000924540, M105362, P00101984, W1894-0006531), lectotypified by Davies (1978: 337); *Stuhlmann* 1029 (B† syntype); *Stuhlmann* 1182 (B† syntype); *Stuhlmann* 4015 (B† syntype).

> *Crassocephalum scandens* (O.Hoffm.) Hiern, Cat. Afr. Pl. **1**(3): 595 (1898).
>
> *Crassocephalum ruwenzoriensis* S.Moore in J. Linn. Soc., Bot. **35**(245): 352 (1902). Type: [Uganda:] 'Hab. Ruwenzori Mountain, 7–8000 feet; *G. F. Scott Elliot* no. 7777.' (BM000924542 holotype, K000377686).
>
> *Crassocephalum auriforme* S.Moore in J. Linn. Soc., Bot. **37**(259): 171 (1904). Type: [Uganda:] 'Hab. Island of Buvúma, Lake Victoria Nyanza. Fl. March [27.1904 *Bagshawe*] 657.' (BM000924536 holotype).
>
> *Gynura taylorii* S.Moore in J. Bot. **44**(517): 23 (1906), as '*taylori*'. Type: [Tanzania:] 'Rabai, [31] December [1885 *Taylor* s.n.]' (BM924539 holotype).
>
> *Senecio seretii* De Wild. in Ann. Mus. Congo Belge, Bot. sér. 5, **3**(2): 315 & tab. XXXII fig. 2 (1910). —Brenan in T.T.C.L.: 159 (1949). Type: [D.R. Congo:] 'Entre Arebi et Gumbari, 20 décembre 1906 (*F. Seret*, n. 707)' (BR8679277 – with Seret's original label – holotype, BR8679284).
>
> *Senecio agathionanthes* Muschl. in Wiss. Ergebn. Deut. Zentr.-Afr. Exped. (1907–1908), Bot. **2**: 385 (1911). Type: [Rwanda:] 'Rugege-Wald: Rukarara, ca. 1900 m ü. m. Prächtige, groß, klimmende Staude mit dunkelgoldgelben Blüten, nicht selten (blühend und fruchtend Mitte August 1907 – [*Mildbraed*] n. 901).' (B† holotype).
>
> *Gynura auriformis* (S.Moore) S.Moore in J. Bot. **50**(595): 212 (1912).
>
> *Gynura ruwenzoriensis* (S.Moore) S.Moore in J. Bot. **50**(595): 213 (1912).
>
> *Gynura brownii* S.Moore in J. Bot. **54**(646): 281 (1916). Type: 'Uganda, Umbendi road 100 m NW of Kampala; [savannah, 4200 ft., June 1915] *E. Brown* (*Dümmer* 2723)' (BM924537[62] syntype).
>
> *Senecio rutshuruensis* De Wild., Pl. Bequaert. **5**: 130 (1929). Type: [D.R. Congo:] 'Rutshuru, 19 octobre 1914 (*J. Bequaert*, n. 6054. –Steppe à Andropogon; fleurs rouge-orangé' (BR8679949 holotype).
>
> *Senecio seretii* var. *divergentiramus* De Wild., Pl. Bequaert. **5**: 132 (1929). Type: [D.R. Congo:] 'Fendula (Nyakolonge), 1928 dans la brousse cuisante du Congo (*Scaetta*, n. 666)' (BR8874870 holotype).
>
> *Senecio subalatipetiolatus* De Wild., Pl. Bequaert. **5**: 134 (1929). Type: [D.R. Congo:] 'Blukwa, [8.] 1921 (*J. Claessens*, n. 1488)' (BR8679611 holotype).
>
> *Senecio variostipellatus* De Wild., Pl. Bequaert. **5**: 135 (1929). Type: [D.R. Congo:] 'Entre Mboga et Lesse [savane herbeuse, ± 1300 m] 18 Mars 1914 (*J. Bequaert*, n.2980).' (BR8679604 holotype, BR8679932).

Weak perennial herb, climbing to scandent or scrambling, 2–6(12) m tall, sometimes subfleshy with an unpleasant smell when bruised or crushed; rootstock of fibrous roots. Stems freely branched, leafy towards top, older stems often clothed with remains of withered leaves (at least in nature) and with a thin papery 'bark', younger stems often purple-tinged. Leaves petiolate, petioles 20–60(95) mm long, base often auriculate, lamina ovate or triangular, 60–90(140) × 50–80(130) mm, green, puberulent, 3–5(6)-veined, apex obtuse or acute, base cuneate, truncate or cordate, margins finely toothed to shallowly lobed, or rarely lyrate. Inflorescences terminal, corymbose. Capitula usually few (3)5–8(30), c. 14 mm × c. 7 mm, 30–50(80)-flowered; involucre campanulate, calyculate, calycular bracts linear-lanceolate, 3–7 mm long, ciliate; phyllaries 10–13, 8–14 × 2–3 mm, broadening below apex, sparsely pubescent; receptacle naked, glabrous, surface alveolate between florets. Corollas orange-yellow, 12–16 mm long, limb c. 5 mm long, corolla lobes 1.2–2 mm long, tips cuculate, epapillose; filament collar

---

[62] The division of the sheet suggests that this sheet has 2 duplicates mounted on it, the lh material has the R. Dümmer original label, with R. Dümmer crossed out with E. Brown written above, the rh label suggest that material was 'ex Hb. Dümmer'.

distinctly enlarged and balusterform; style arms acuminate, papillate. Achenes 3.5–4.5 ×1 mm, 7–9 ribbed, setuliferous in sinuses, setulae of twin-hairs; carpopodium annular; pappus setae 8–13 mm long, persistent, barbellate, white.

**Zambia**. N: Penza Village, 19.x.1968, *Sanane* 334 (K). **Malawi**. N: Nkhata Bay, Chikangwa, 21.viii.1978, *Phillips* 3807B (K, MO).

Also in Angola, Burundi, D.R. Congo, Kenya, Rwanda, Uganda, Tanzania. Moist forest, especially in forest margins and clearings, also in secondary vegetation, riverine forest, *Acacia xanthophloea* woodland; 1–2550 m; flowering sporadically throughout the year.

Conservation Status: LC (Least Concern) – a not uncommon widespread species.

Jeffrey & Beentje (2005: 699) noted that *G. scandens* looks quite like *Solanecio nandensis* (S.Moore) C.Jeffrey, providing distinctions between the two, although the form of the inflorescence alone should be enough of a distinction, and I doubt the two can be confused that easily.

2. **Gynura pseudochina** (L.) DC., Prodr. **6**: 299 (1838). —Davies in Kew Bull. **33**(4): 638 (1979). —Jeffrey in Kew Bull. **41**(4): 930 (1986). —Blundell, Wild Fl. E. Afr.: fig. 366 (1987). —Lisowski, (Asterac. Fl. Afr. Cent. 2) Fragm. Flor. Geobot. **36** Suppl. 1: 432 (1991). —Agnew & Agnew, Upland Kenya Wild Fl., ed. 2: 225 (1994). Type: 'Habitat in India.' Type not designated.[63]

    *Senecio pseudochina* L., Sp. Pl. **2**: 867 (1753).

    *Gynura miniata* Welw., Apont.: 586 (1859). —Oliver & Hiern in F.T.A. **3**: 403 (1877). —Bally in J. East Afr. Nat. Hist. Soc. **18**: 124, t. 20/4 (1944). —Adams in F.W.T.A., ed., 2 **2**: 243 (1963). —Agnew, Upland Kenya Wild Fl., ed. 1: 471 (1974). —Davies in Kew Bull. **33**(2): 339 (1978). Type: [Angola: Pungo Andongo, Caghuy:] 'No. 29. Regio III. Gynura miniata Welw. Herb. Angol. ... In dumetis District. Pungo Andongo 1857 leg. *W[elwitsch]*.' *Welwitsch* 3595 (BM000924532 holotype, BR8678980, K000377683 – the sheet has a pencilled dissection of the style arms by Peter Taylor), LISC219549 – 'Am Wasser fall in Calundo bei Caghuy. Dcbr 856. W.', LISC219550, P00101957).

    *Gynura miniata* var. *orientalis* O.Hoffm. in Pflanzenw. Ost-Afrikas C: 416 (1895). —Chiovenda, Fl. Somala **2**: 270, fig. 155 (1932). Type: [Tanzania:] '11 [= Sansibar] (Usaramo. – *Stuhlm[ann]*. n. 7776).' (B† holotype).

    *Gynura rusisiensis* R.E.Fr., Wiss. Ergebn. Schwed. Rhodesia–Kongo Exped. **1**: 342 (1911). Type: [Burundi:] 'Deutsch-Ostafrika: Rusisi-Tal, zwischen Mpanda und Mecherenge auf schattigen Standorten in Gebüschen (blühend 10. Dez. – [*R. E. Fries*] n. 1435).' (UPS046496 holotype).

    *Senecio somalensis* Chiov., Res. Sci. Somalia Ital. **1** [Coll. Bot.]: 106 (1916). Type: [Somalia:] 'Boscaglia di Baidoa 5 [*sic*], XI, 1913 (*Paoli* n. 1110).' (FT003805 – 'Paoli No. 1110. Boascalia di Baidoa, 4 Novbre 1913.', holotype).[64]

    *Gynura eximia* S.Moore in J. Bot. **56**(668): 225 (1918). Type: 'Angola, along rivulets in moist shrubby situations at Kaconda; *Gossweiler*, 3638. ... To be referred here is *Gossweiler* No. 4315 from marshes of the Seculu river near Kaconda, with somewhat smaller leaves.'[65] (BM000924534 holotype,[66] COI00077163 – oddly databased as 's.n.', but is quite clearly numbered, K000377682, LISC014901); *Gossweiler* 4315 (BM000924533, COI00033764 'paratypes').

---

[63] Davies (1979: 638; in Kew Bull. **35**(2): 364, 1980) cited both *Royen* 164 & fig. 335, t. 258 in Dillenius, Hort. Eltham: 345 [as 'Senecio Madraspatanus rapifolio Dill.'] (1732) as syntypes, later followed by Jeffrey (1986: 960), but none of these statements effects a formal typification. The *van Royen* material in L is Herb. Lugd. Bat. No. 900288-368 [= L0053096], but much of the single capitulum has been destroyed by historical insect damage.

[64] The sheet with a TYPUS label, further annotated as 'HOLO').

[65] Most authors have considered *Gossweiler* 3638 as the type collections and *Gossweiler* 4315 as the paratype collection; arguments can certainly be made that they are syntypes, however, I am following the majority.

[66] 'An annual total height 8 ft. stems erect; leaves fleshy green: capitula scales atropurple; florets orange: along the rivulets in moist shrubby situations at Kaconda 21 – 2 – 07.', databased as 'both isotype and co-type.'

?*Gynura variifolia* De Wild., Pl. Bequaert. **5**: 93 (1929). Type: [D.R. Congo:] 'Rutshuru, 10 septembre 1914 (*J. Bequaert*, n. 5627. – Steppe à Andropogon; fleurs rouge-vermillon).' (BR8679598 holotype, BR8679260).[67]

    *Gynura amplexicaulis* sensu Hutch. & Dalziel in F.W.T.A. **2**: 148 (1931), non Oliv. & Hiern

    *Gynura somalensis* (Chiov.) Cufod. in Nuovo Giorn. Bot. Ital., ser. 2, **50**: 112 (1943).

Perennial herb 0.4–1 m high, with unpleasant musky smell; rootstock tuberous, woody, tubers to 10 × 2–5 cm. Stems ± fleshy, leaves rosettiform or stems leafy to half their length, or with an apical rosette of leaves, ± pubescent. Leaves slightly fleshy, green or purplish, pubescent or glabrous, glandular, basal leaves ovate or spatulate, 4–22 × 2.5–11 cm, attenuate into a petioloid base, margins ± entire, apex obtuse, middle and upper leaves narrower, elliptic, obovate or narrowly obovate, lobed to pinnatisect, 6–25 × 2–7 cm, base clasping stem in uppermost leaves, margins 1–6-lobed, lobes toothed or lobed, subopposite. Inflorescences scapose or subscapose, terminal, capitula 1–3(6) together, campanulate, erect. Involucre campanulate, 8–13 × 13–15 mm; calycular bracts 1–3 mm long, margins ciliate; phyllaries 10–16, (6)8–13 mm long, pubescent and sometimes glandular. Florets 45–100, corollas 9–12.5 mm long orange or yellow, expanded in limb, corolla lobes 0.9–1.7 mm long, resin canals present, apices broad, papillose. Achenes 3 mm long, setuliferous or glabrous; pappus setae 7–11 mm long.

**Zambia**. N: Mprokoso Dist., Kabwe Plain, Mweru-Wantipa, 1000 m, 15.xii.1960, *Richards* 13710 (K). **Malawi**. S: Chiradzulu Mt., 5000′ [1524 m], 24.ix.1944, *Benson* 470 (K).

Also in Sierra Leone to D.R. Congo, Burundi, Ethiopia, Somalia, Kenya, Tanzania, Angola, India, Sri Lanka, Bhutan, China, Burma, Thailand, Vietnam. Boggy grassland, black cotton soil, often with *Acacia drepanolobium*, less often in bushy grassland or secondary woodland, in evergreen woodland, sometimes growing on ant hills; 0–1800 m; flowering mainly April to November, but sporadically throughout the year.

Conservation Status: LC (Least Concern). A very widespread species, but not a commonly collected species in the Flora area.

Clearly marked on the distribution map for this species as occurring in eastern Africa (Kenya, Uganda and Tanzania) in Vanijajiva & Kadereit (2011), the authors mistakenly stated the species was only known from Madagascar and Sri Lanka, even with the type from Tanzania and supporting specimens listed for Kenya and Tanzania.

3. **Gynura valeriana** Oliv. in Hooker's Icon. Pl. **16**(1): t. 1507 (1886); —in Trans. Linn. Soc., ser. 2, Bot. **2**: 339 (1887). —Agnew, Upland Kenya Wild Fl.: 471 (1974). —Davies in Kew Bull. **33**(2): 338 (1978); —Jeffrey in Kew Bull. **41**(4): 929 (1986). —Lisowski, (Asterac. Fl. Afr. Cent. 2) Fragm. Flor. Geobot. **36** Suppl. 1: 433 (1991). — Agnew & Agnew, Upland Kenya Wild Fl., ed. 2: 225 (1994). Type: [Tanzania:] 'Hab. Kilimanjaro, 5,500 ft., *H. H. Johnston* [103]' (K000377684 holotype).[68] FIGURE 6.4.**10**.

    *Senecio valeriana* Oliv. in H. Johnston, The Kilim. Exped. Append.: 342 (1886), nom. illeg., nom. nud. in list.

Perennial herb 0.6–1.8(3) m high, with unpleasant smell (smelling of valerian); stem fleshy, to c. 10 mm diam. at base, erect or procumbent, leafy throughout or only in lower part, usually drying with conspicuous ribs, indumentum of scattered hairs. Leaves petiolate, petiole up 10 cm long, lamina thin and fleshy, dark green or grey-green, purplish beneath, ovate to elliptic, 15–50 (including petiolate base) × 8–20 cm, base decurrent and auriculate, auricles reniform, lamina margins deeply lobed to pinnatisect with 1–10 pairs of oblong lobes, lobe margins denticulate or dentate, apex acute, glabrous or with scattered hairs. Inflorescences large, axillary and terminal, capitula in crowded groups of 5–20 capitula, pedicellate, pedicels 1–3 cm long. Capitula homogamous and discoid; involucre obconical, 6–12 mm long, calyculate, calycular bracts few (3–5), 1–6 mm long; phyllaries 10–15, 6–12 × 1 mm, apices darker, glabrous or with

---

[67] Lisowski (1991: 432) indicated the name is in synonymy 'sensu auct. non De Wild.; Davies (1978: 340)', although Davies stated she had not seen type material; otherwise this name is provided by Lisowski as a synonym of *Gynura amplexicaulis* Oliv. & Hiern., it keying out by virtue of the glabrous achenes.

[68] A pencilled dissection drawing of the style arms was drawn on the sheet by Peter Taylor.

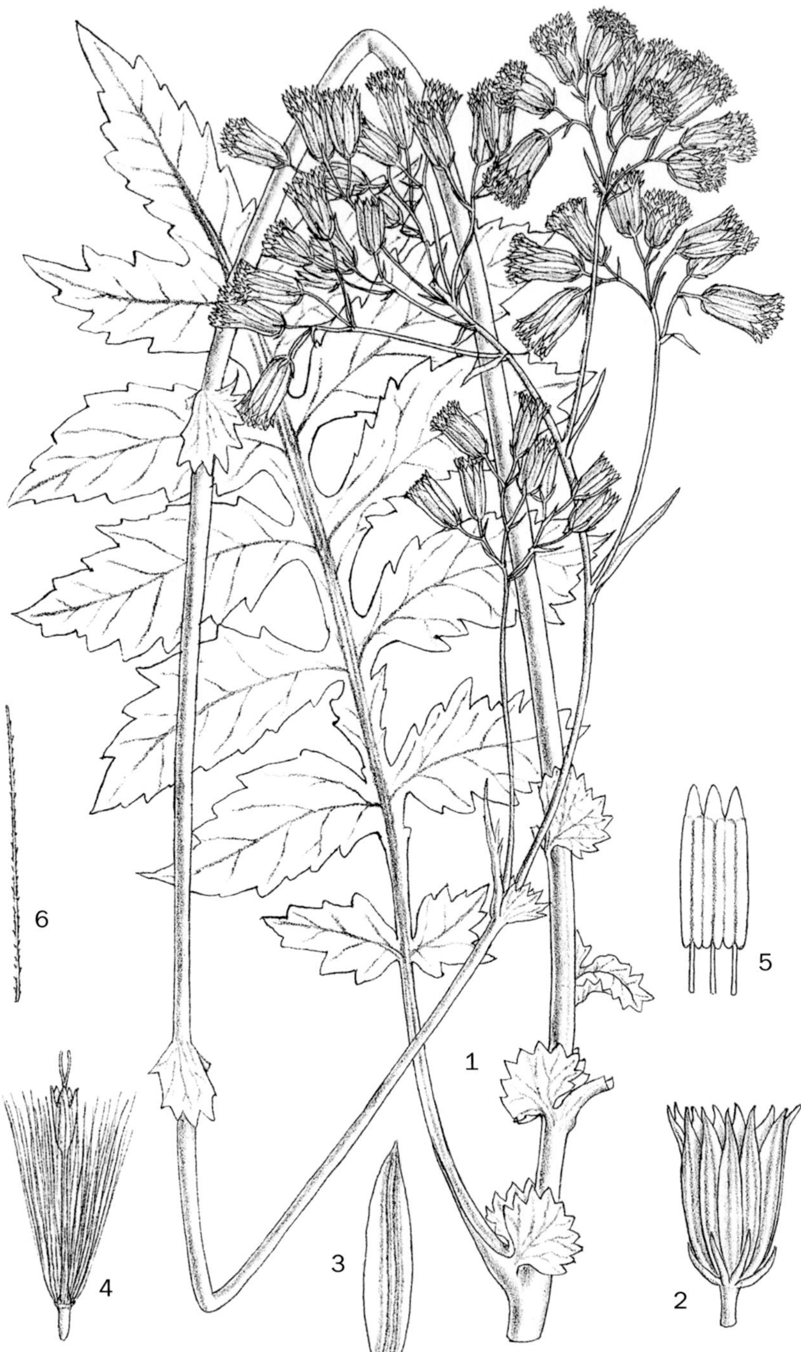

Fig. 6.4.**10**. GYNURA VALERIANA. 1, stem leaf and terminal inflorescence (× 1); 2, capitulum (× 2); 3, phyllary (× 2.5); 4, floret (× 2.5); 5, three stamens (greatly magnified); 6, apex of pappus seta (greatly magnified). All from *Johnston* 103. Drawn by Matilda Smith, from Hooker's Icones Plantarum, ser. 3, **6**(1): pl. 1507 (1886).

scattered hairs. Florets 35–55, corollas yellow or orange, less often orange-red, night-scented, corolla tube 10–12 mm long, expanded in upper third, limb 3–3.5 mm long, corolla lobes 1–1.2 mm long, with median resin canal; anthers conspicuously yellow. Achenes 2–5 mm long, brown, glabrous; pappus setae 7–10 mm long, white or off-white.

**Malawi**. S: Mlange Dist., Mlanje Mountain, half way up Little Ruo path, 1160 m, 29.vii.1970, *Brummitt* 12344 (K, MAL).

Also in D.R. Congo, Kenya, Tanzania. Moist sites in forest, on streamside rocks, and sometimes persisting after forest clearing, rarely as epiphyte in trees; (500)800–2150 m; flowering sporadically throughout the year.

Conservation Status: Two Kenyan specimens, more than twenty Tanzanian specimens. Presumably LC (Least Concern).

Whilst clearly listing material (including the type), and mapping its distribution, from Kenya and Tanzania, Vanijajiva & Kadereit (2011) stated that the species was 'only known from Madagascar and Sri Lanka (Ceylon)'! I suspect this was in error as they also noted it grows in 'tropical Africa', and did not map the species in either Madagascar or Sri Lanka. *Gynura zeylanica* Trimen is the only species they stated was from 'Madagascar and Sri Lanka', but only mapped from Sri Lanka (based on cited material).

## 103. **CRASSOCEPHALUM** Moench.

**Crassocephalum** Moench, Methodus: 516 (1794). —Muschler in Bot. Jahrb. Syst. **43**(1): 1–74 (1909). —Moore in J. Bot. **50**(590): 209–213 (1912). —Muschler in Repert. Spec. Nov. Regni Veg. **11**(274–278): 113–119 (1912).[69] —Belcher in Kew Bull. **10**(3): 455–465 (1955). —Humbert, Fl. Madagasc. Composées **189**: 829–843 (1963). —Dyer, Gen. S. Afr. Fl. Pl. **1**: 713 (1975). —Gibbs Russell in Kirkia **10**(2): 502 (1977). —Jeffrey in Kew Bull. **41**(4): 904 –908 (1986). —Belcher in Kew Bull. **44**(3): 533–542 (1989). —Jeffrey & Beentje in F.T.E.A., Compositae **3**: 598–611 (2005). —Nordenstam in Kubitzki, Fam. Gen. Vasc. Pl. **8**: 232 [2006](2007). —Vanijajiva & Kadereit in Syst. Biodiv. **7**(3): 269–276 (2009). —Beentje, Fl. Gabon **56**: 42–47 (2021).
*Cremocephalum* Cass. in Dict. Sci. Nat., ed. 2, **34**: 390 (1825), nom. illeg. superfl. pro *Crassocephalum* Moench.
*Senecio* subgen. *Gynuropsis* Muschl. in Bot. Jahrb. Syst. **43**(1): 38 (1909).

Annual or perennial herbs, rarely shrubs. Stems erect or scrambling. Leaves alternate, cauline, sessile or petiolate, sometimes amplexicaul, entire, serrate, lobed or pinnatifid. Capitula solitary or few to many and laxly to densely corymbose. Capitula homogamous and discoid or rarely heterogamous and radiate; involucre cylindrical to campanulate, calyculate; phyllaries narrow; receptacle epaleaceous. Ray florets female, fertile, ray limb deep pink or pinkish-mauve, tube very narrow. Disc floret corollas or all corollas with narrow tube, tapering towards apex, 5-lobed, creamy white, yellow, orange, pink, red, mauve, purple or blue; filament collar subalusterform, basal anther appendages obtuse or slightly sagittate, anthers ecaudate; style arms with short to long penicillate tuft. Achenes cylindrical to oblong, ribbed, sometimes angled, glabrous or variously short-setuliferous, setulae of thickened, spiralled 'twin-hairs'; carpopodium indistinct; pappus setae subequal, numerous, capillary, whitish.

A genus of c. 25 spp. of tropical Africa, Madagascar, Mascarenes and Yemen. Two spp. are weedy and introduced in many tropical countries.

---

[69] In what was clearly an academic disagreement between Spencer Le Marchant Moore and Reno Muschler, Muschler (1909) having newly described *Senecio* subg. *Gynuropsis* was clearly animated when Moore (July 1912) commented that the diagnostic characters were already well known in the genus *Crassocephalum* Moench. However, taking exception to this, Muschler (July! 1912) promptly transferred all of Moore's names to either *Senecio* L. or *Gynura* Cass., presumably in a fit of pique. Belcher (1955) has admirably provided a synopsis of this disagreement, and provided a more than adequate explanation of what should be accepted – with *Crassocephalum* Moench being maintained

1.  Capitula radiate . . . . . . . . . . . . . . . . . . . . . . . . . . . . . . . . . . . . . . . . . . . . . . . . . **1.** *radiatum*
 −   Capitula discoid . . . . . . . . . . . . . . . . . . . . . . . . . . . . . . . . . . . . . . . . . . . . . . . . . . . . . 2
2.  Corollas yellow or orange . . . . . . . . . . . . . . . . . . . . . . . . . . . . . . . . . . . . . . . . . . . . . 3
 −   Corollas purple, blue, mauve, pink, white or brick red . . . . . . . . . . . . . . . . . . . . . 9
3.  Capitula few to many in ± congested corymbs; not swamp plants . . . . . . . . . . . . . 4
 −   Capitula 1–10, lax (or if congested then swamp plants, leaves narrow) . . . . . . . . . . . 5
4.  Corolla lobes often dull orange-brown to reddish; style-arm appendages distinct;
     involucre 9–13.5 mm long . . . . . . . . . . . . . . . . . . . . . . . . . . . . . . . . **2.** *crepidioides*
 −   Corolla lobes yellow; style-arm appendages very short; involucre 4.5–10.5 mm long . .
     . . . . . . . . . . . . . . . . . . . . . . . . . . . . . . . . . . . . . . . . . . . . . . . . . . **3.** *montuosum*
5.  Subshrubs . . . . . . . . . . . . . . . . . . . . . . . . . . . . . . . . . . . . . . . . . . . . . . **4.** *torreanum*
 −   Annual or perennial herbs . . . . . . . . . . . . . . . . . . . . . . . . . . . . . . . . . . . . . . . . . . . . 6
6.  Plant subscapose; achenes glabrous . . . . . . . . . . . . . . . . . . . . . . . . . . . . . . **5.** *uvens*
 −   Plants with leafy stems; achenes setuliferous . . . . . . . . . . . . . . . . . . . . . . . . . . . . . . 7
7.  Involucre 1.5–2.5 × as long as wide; leaves ovate to linear-lanceolate, cuneate to
     attenuate at base . . . . . . . . . . . . . . . . . . . . . . . . . . . . . . . . . . . . . . . . . . . . . . . . . . . 8
 −   Involucre 1–1.5 × as long as wide; leaves broadly ovate to elliptic, cordate to subtruncate
     to cuneate at base . . . . . . . . . . . . . . . . . . . . . . . . . . . . . . . . . . . . . . . **6.** *vitellinum*
8.  Capitula narrow, involucre 2.5–5 mm diam., ± 2.5 × as long as wide; leaves elliptic to
     linear-lanceolate; phyllaries green throughout . . . . . . . . . . . . . . . . . . . . . . **7.** *paludum*
 −   Capitula wide, involucre 4–8 mm diam., less than 2.5 × as long as wide; leaves ovate to
     lanceolate to narrowly elliptic; phyllaries green with purplish or brown tips . . **8.** *picridifolium*
9.  Leaves deeply lobed with 5–20 narrow lobes per side; capitula many in ± congested
     corymbs; corollas pale blue . . . . . . . . . . . . . . . . . . . . . . . . . . . . . . . . **9.** *bauchiense*
 −   Leaves with ≤ 8 lobes per side; capitula 1–many, or if corollas blue then capitula few . . 10
10.  Phyllaries 5–7 mm long; corollas red, mauve, pink or white . . . . . . . . . . . . . . . . . . 11
 −   Phyllaries 7.5–12.5 mm long; corollas blue, orange-red to brick red, purple or mauve . . .12
11.  Corolla 5.5–8 mm long, only slightly exceeding involucre . . . . . . . . . . . . . . **1.** *radiatum*
 −   Corolla 8–9 mm long, much exserted from involucre . . . . . . . . . . . . . . . . . **10.** *effusum*
12.  Corollas blue, purple or mauve . . . . . . . . . . . . . . . . . . . . . . . . . . . . . . . . . . . . . . . 13
 −   Corollas orange-red to brick red . . . . . . . . . . . . . . . . . . . . . . . . . . . . . . . . **2.** *crepidioides*
13.  Corollas purple or mauve, leaf base cuneate or attenuate into an exauriculate petioloid
     base (upper leaves) . . . . . . . . . . . . . . . . . . . . . . . . . . . . . . . . . . . . . . . . . **11.** *rubens*
 −   Corollas blue, leaf base attenuate into an exauriculate petioloid base (all leaves) . . . . .
     . . . . . . . . . . . . . . . . . . . . . . . . . . . . . . . . . . . . . . . . . . . . . . . . . . . . . **12.** *coeruleum*

1.  **Crassocephalum radiatum** S.Moore in J. Bot. **56**: 227 (1918). —Jeffrey in Kew Bull.
     **41**(4): 907 (1986). —Lisowski, (Asterac. Fl. Afr. Centr. 2) Fragm. Flor. Geobot. **36**
     Suppl. 1: 345, t. 76 (1991). Type: [D.R. Congo:] 'Belgian Congo, Pueto (Mpueto) near
     Lake Moero under trees; [19.v.1908], *Kassner*, 2825' (BM924601 holotype, E00239801,
     HBG504315, P00101550, P00101551).
      *Senecio heteromorphus* Hutch. & B.L.Burtt in Rev. Zool. Bot. Africaines **23**: 42 (1932). Type: [D.R.
     Congo:] 'Belgian Congo: Between Kitendwe and Kasiki, [20-23] June 1931, G. F. *de Witte* 411'
     (BR867813 'Herb. Mus. Cong.' lectotype, BR8678027, BR8678065, BR8678126), here lectotypified.[70]
      *Crassocephalum heteromorphum* (Hutch. & B.L. Burtt) C.Jeffrey in Kew Bull. **41**(4): 906 (1986).

---

[70] The Herbarium of the Congo Museum (Tervuren) was transferred to BR in 1934. There are four
     specimens in BR all marked as isotypes. BR867813 has a determinavit label initialled by 'J.H. & B.L.B.'
     [= J. Hutchinson & B. L. Burtt], a copy of the protologue, the original de Witte collecting ticket with
     location data, together with a later fully displayed dissection of one capitulum; BR8678027, BR8678065
     (a plant with roots), BR8678126 all appear to be 'duplicates' and flowering branches (or roots) with simple
     'Herbarium Musei Congoensis' labels lacking location data, but merely annotated '*de Witte* 411'. It is likely
     that Hutchinson & Burtt were aware of all of the material as it all appears to have come from one plant.
     However, since Hutchinson & Burtt did not specify which of the collections was the 'type' (all of it could
     be considered as the 'holotype') it seems appropriate to consider one sheet as the lectotype – BR867813.

Annual herb 10–240 cm tall, erect. Stems pale green, often speckled or tinged reddish-purple, sparsely crispate-setulose. Leaves sessile, elliptic, narrowly elliptic or lanceolate, 1.2–14 × 0.3–6.5 cm, cuneate or attenuate into exauriculate petioloid base, margins remotely sinuate-serrate to serrate-laciniate, unlobed or shortly to deeply 2-lobed towards base, apex obtuse, apiculate, shortly scattered crispate-setulose. Capitula numerous to rarely solitary in rather lax terminal corymbs, erect, radiate or discoid; pedicels slender, glabrous or almost so; involucre cylindrical, base slightly swollen, 5.5–7.5 × 2–3 mm; calycular bracts 8–12, dark-tipped, 1–3.5 mm long, margins glabrous or ciliate; phyllaries 13 or sometimes 8, green with purple tips, 5–7 × 0.6–1.3 mm, shortly hispid or glabrous. Ray florets 13, 8, 5 or 0, tube 4 mm long, glabrous, rays when present deep pink or pinkish-mauve, 1–3.5 mm long (6–8 mm in Lisowski 1991: 345), 1 mm wide, 3–4-veined. Disc florets c. 106, corollas 5.5–8 mm long, dull brick-red, magenta or white, corolla tube glabrous, gradually expanded above middle, lobes 0.5–0.7 mm long. Achenes 1.5–1.6 mm long, ribbed, short-setuliferous between ribs; pappus setae 4.5–6 mm long.

**Malawi**. N: Chitipa Dist., Misuku Hills, Mughesse, Songwe view, 5600′, 6.vii.1973, *Pawek* 7037 (CAH, K, MAL, MO, UC).

Also in D.R. Congo, Burundi, Tanzania. Grasslands, brush slopes, roadside; c. 1700 m; collected in flower in July.

Conservation Status: One specimen from FZ area, DD.


2. **Crassocephalum crepidioides** (Benth.) S.Moore in J. Bot. **50**(590): 211 (1912). — Andrews, Fl. Pl. Sudan **3**: 21 (1956). —Lind & Tallantire, Common Fl. Pl. Uganda: 182 (1962). —Jeffrey in Kew Bull. **41**(4): 908 (1986). —Lisowski, (Asterac. Fl. Afr. Cent. 2) Fragm. Flor. Geobot. **36** Suppl. 1: 354 (1991). —Jeffrey in Fl. Masc., Composées **109**: 159 (1993). —Agnew & Agnew, Upland Kenya Wild Fl., ed. 2: 219 (1994). Types: 'Sierra Leone, [*G.*] *Don*; and, apparently the same species, Senegal, *Heudelot' Don* s.n. (BM000924608 syntype);[71] *Heudelot* s.n. (?G, ?P00101786, ?P00101787 syntypes).[72]

> *Senecio diversifolius* A.Rich., Tent. Fl. Abyss. **1**: 437 (1848), nom. illeg., non Dumort. (1827) [= *Senecio doria* L.]. Type: [Ethiopia:] 'Crescit in campis humidis circa Adoua [Adowa], mense Novembre florens (*Quartin Dillon*)' *Quartin-Dillon & Petit* s.n. (P00111989, P0011190, P00111991 syntypes).
>
> *Gynura crepidioides* Benth. in Hooker, Niger Fl.: 438 (1849). —Oliver & Hiern, F.T.A. **3**: 403 (1877).
>
> *Gynura polycephala* Benth. in Hooker, Niger Fl.: 437 (1849). Type: [Equatorial Guinea:] 'Fernando Po [Bioko], [xi.1841], *Vogel* [s.n.]' (K000306766 – '[*Vogel*] 139. ... Nov. [18]41', ex Herb. Hookerianum, K000306767 – *Vogel* s.n., ex Herb. Benthamianum syntypes).
>
> *Crassocephalum diversifolium* Hiern, Cat. Afr. Pl. **1**(3): 594 (1898), nom. illeg. superfl., based on *Senecio diversifolius* A.Rich.

Annual herb 25–120(150) cm tall, rarely a short-lived perennial. Stems erect, green, often flecked with dull purplish-red, crispate-pubescent. Leaves sessile, obovate, broadly elliptic, rhombic or ovate, unlobed or pinnato-lyratcly 2–8-lobed, 5–26 × 2–10 cm wide, cuneate or attenuate and slightly decurrent on to exauriculate petioloid base, margins usually coarsely and sometimes rather irregularly sinuate-serrate or sinuate-serrato-bidentate, apex obtuse to acute, scattered crispate-setulose, sometimes with purplish margins or purplish-tinged beneath especially on main veins. Capitula few to numerous in rather dense to fairly lax terminal corymbs, discoid, drooping in bud and at anthesis, becoming erect in flower and fruit; pedicels crispate-pubescent; involucre cylindrical or ± so above a somewhat bulging base, 9–13.5 × 3–5 mm; calycular bracts 6–21, purplish on blackish with darker tips, 2–6 mm long, scattered-pubescent or at least with ciliate margins; phyllaries 13–21, usually 21 or less frequently 13, yellow-green or green with purplish or blackish tips, 8–12.5 × 0.6–1.5 mm, scattered-setulose, or rarely glabrous. Florets numerous, corollas 7–12 mm long, tube often yellow, glabrous, slender, gradually expanded in upper third, lobes 0.5–1 mm long, orange-red or brick-red, rarely yellow and style arms red. Achenes 1.8–2.7 mm long, ribbed, setuliferous between ribs; pappus setae 7–13 mm long.

---

[71] The *G. Don* material was taken as the holotype, BM000924608, Jeffrey (1986: 908) and Jeffrey & Beentje (2005: 610) apparently disregarding Bentham's comment on the *Heudelot* material.

[72] It is unclear where Bentham may have seen the *Heudelot* syntype; nothing in K appears with *Heudelot*'s label or handwriting that I have been able to find. There are two duplicates of *Heudelot* 785 in P, from Senegambia (= Senegal) that may correspond to type material – P00101786, P00101787.

**Zambia**. N: Lake Chila, below Club House, 5000′, 3.v.1955, *Richards* 5505 (K). W: Mwinilunga Dist., in disused garden by R. Lunga, just below R. Mudjanyama, 25.xi.1937, *Milne-Redhead* 3392 (K). C: near Mumbwa, 15°S, 28°E, 1911, *Macaulay* 694 (K). S: Mapanza Mission, 3500′, 10.v.1953, *Robinson* 206 (K). **Zimbabwe**. E: N: Trelawney Tobacco St., 4.v.1943, *Jack* 196 (SRGH, fide M. Hyde). C: Chegutu (Hartley) Dist., Poole Farm, 12.iv.1948, *Hornby* 3134 (SRGH, fide M. Hyde). Chirinda Forest, in forest glade, 3800′, 2.vii.1952, *Wild* 3838 (K, SRGH). S: Zimbabwe Ruins, 1.vii.1930, *Hutchinson & Gillett* 3335 (K). **Malawi**. N: Chitipa Dist., Misuku Hills, along dirt road in Mughesse rain forest, 5500′, 17.ix.1975, *Pawek* 10141 (K, MAL, MO, SRGH, UC). S: Mulanje Dist., Mulanje Mt., Lichenya Path between Likhubula & Sawke Rivers, 15°5'S, 35°31'E, 3.v.1989, *Pope et al.* 2261 (BR, K). **Mozambique**. N: Niassa Dist., arredores de Vila Cabral, iv.1934, *Torre* 73 (LISC). Z: Namuli Moutain, Muretha Plateau, 15°23' 20.2"S, 37°02'16.2"E, 1867 m, 29.v.2007, *Timberlake et al.* 5070 (K). MS: Manica e Sofala: Chimioi, serra de Garuzo, 28.iii.1948, *Barbosa* 1255 (LISC). GI: Sul do Save, Jangamo, no "machongo" Nhacula, 15.ix.1948, *Myre & de Carvalho* 214 (K, LMA). M: Maputo, Matutuíne, Ponta de Ouro, a 3 km for a de Ponta de Ouro, na estrada para Maputo, 3.i.1980, *de Koning* 7865 (K, LM).

Also in Angola, Benin, Burundi, Cameroon, Congo (Brazzaville), D.R. Congo, Ethiopia, Gabon, Ghana, Guinea, Ivory Coast, Kenya, Madagascar, Nigeria, Rwanda, Senegal, South Africa, Sudan, Tanzania, Togo, Uganda; naturalized in large parts of tropical Asia and the Pacific. Forest margins and secondary vegetation, a common weed of disturbed places and cultivation; 0–2100 m; flowering throughout the year.

Conservation Status: LC.

3. **Crassocephalum montuosum** (S.Moore) Milne-Redh. in Kew Bull. **5**(3): 376 (1950) [Feb. 1951]. —Andrews, Fl. Pl. Sudan **3**: 21 (1956). —Jeffrey, Kew Bull. **41**(4): 906 (1986). —Blundell, Wild Fl. E. Afr.: fig. 361 (1987). —Lisowski, (Asterac. Fl. Afr. Cent. 2) Fragm. Flor. Geobot. **36** Suppl. 1: 331, t. 72 (1991). —Agnew & Agnew, Upland Kenya Wild Fl., ed. 2: 219, t. 91 (1994). —Jeffrey & Beentje in F.T.E.A., Compositae **3**: 602, fig. 126 (2005). —Beentje, Fl. Gabon **56**: 43 (2021). Type: [Kenya:] 'Hab. British East Africa, between Machakos and Kikuyu, 5–6000 feet; *G. F. Scott Elliot*, no. 6587' (BM000924604 holotype – mounted with *Scott Elliot* 7729, see var. *minor* below, K000306755).

    *Senecio montuosus* S.Moore in J. Linn. Soc., Bot. **35**: 354 (1902).

    *Senecio montuosus* var. *minor* S.Moore in J. Linn. Soc., Bot. **35**: 355 (1902). Type: [D.R. Congo?:] 'Hab. Ruwenzori Mountain, 9600 feet; *G. F. Scott Elliot*, no. 7729' (BM000924604 holotype – mounted with *Scott Elliot* 6587).

    *Senecio butaguensis* Muschl. in Wiss. Ergebn. Deut. Zentr.-Afr. Exped. (1907-1908), Bot. **2**: 403 (1911). Type: [D.R. Congo:] 'Ruwenzori-West: Butagu-Tal, Übergang vom Bergwald zu den Ericaceen, in einer Höhe von 2700–3000 m ü. m. Ein Halbstrauch von 1 m Höhe mit goldgelben Blüten (blühend Mitte Februar 1908 – [*Mildbraed*] n. 2534)' (B† holotype, BR8678171 – 1 leaf, an inflorescence branch and 2 separate capitula in a capsule, with an 'ex Museo botanico Berolinensis' label for the *Mildbraed* collection).

    *Crassocephalum butaguense* (Muschl.) S.Moore in J. Bot. **50**(590): 211 (July 1912).

    *Gynura lutea* Humbert, (Composées Madagascar) Mém. Soc. Linn. Normandie **25**: 122, 302 (1923). Types: [Madagascar:] 'Exsicc.: *PERRIER DE LA BATHIE* 2928, 3212 (types)' *Perrier de la Bâthie* 2928 (K000306749 – 1 leaf and 2 capitula in a capsule from a sheet in P, P00558759 syntypes); *Perrier de la Bâthie* 3212 (P00558760 syntype).

    *Crassocephalum bumbense* S.Moore in J. Linn. Soc. Bot. **47**(314): 279 (1928). Type: [Angola:] 'Sobato de Bumba, in moist thickets; *Welwitsch*, 3687. ... conspecific are Toro, Mpanga Forest; *Bashawe*, 1010, and Uganda, forest clearing at Kipayo; *Dümmer*, 1043' *Welwitsch* 3687 (BM000924528, K000306750 syntypes); *Bagshawe* 1010 (?BM syntype); Uganda, *Dümmer* 1043 (?BM syntype).[73]

---

[73] Because of Moore's terminology, the *Bagshawe* and *Dümmer* collections are considered syntypes, along with the *Welwitsch* collection.

*Crassocephalum afromontanum* R.E.Fr. in Acta Horti Berg. **9**: 144, t. 6 (1928). Type: [Kenya:] 'Mt. Aberdare: Abhang des Kinangop-Gipfels in dichtem Bambusdickicht, ca. 3000 m ü. d. m. (blühend und früchtend 2.IV.1922 – *Rob. E.* und *Th. C. E. Fries* n. 2727)' (UPS holotype).

*Senecio rufopilosulus* De Wild., Pl. Bequaert. **5**: 116 (1929). Types: [D.R. Congo:] 'Ruwenzori, vallée du Butagu, 13 avril 1914 (*J. Bequaert*, n. 3627. –Ver 2500 m. d'altitude fleurs jaunes); Ruwenzori, vallée du Lanuri, 14 mai 1914 (*J. Bequaert*, n. 4242. –Vers 2500 m. d'altitude; buisson de 1 m 50 à 2 m. de hauteur, à fleurs jaunes); Ruwenzori, 24 mai 1914 (*J. Bequaert*, n. 3861. –Vers 2800 m. d'altitude; atteint 3 m. de hauteur; fleurs jaunes)' *Bequaert* 3627 (BR8678010 lectotype – as labelled by C. Jeffrey, BR8678003), lectotypified by Jeffrey (1986: 906); *Bequaert* 3861 (BR8678034, BR8678041 syntypes); *Bequaert* 4242 (BR8678072, BR8678089 syntypes).

*Gynura montuosa* (S.Moore) Bullock in Bull. Misc. Inform. Kew **1932**(10): 499 (1932).

*Senecio afromontanus* (R.E.Fr.) Humbert & Staner in Bull. Jard. Bot. État Bruxelles **14**: 104 (1936).

*Crassocephalum luteum* (Humbert) Humbert, Fl. Madagasc., Composées **189**: 836 (1963).

Annual or short-lived perennial herb or soft-wooded shrub, erect or sometimes semi-scandent, rather weak-stemmed, 25–240 cm tall; stems green, sometimes tinged red towards base, finely crispate-pubescent, glabrescent. Leaves sessile, ovate, lanceolate, elliptic or obovate, unlobed to deeply and narrowly pinnately or pinnato-lyrately 2–8-lobed, 5.5–41 × 1.5–22 cm, base rounded, cuneate or attenuate and slightly decurrent onto a sometimes purplish tinged petaloid base, margins closely sinuate-serrate to coarsely serrate-bidentate, apex acute to obtuse, often ± attenuate, almost glabrous to scattered crispate-pubescent above and beneath especially on veins, glabrescent. Capitula numerous in congested terminal corymbs, discoid; pedicels densely pubescent to glabrous; involucre cylindrical, 5.5–10.5 × 2–4 mm; calycular bracts 6–13, linear to lanceolate, 1.5–7 mm long, dark-tipped, glabrous or margins ciliate; phyllaries 8–22, usually 13–15, green or yellow-green with dark brown or reddish tips, 5–9.5 × 0.4–1 mm, glabrous or sparsely shortly pubescent. Florets numerous, corollas yellow or sometimes orange, 5.5–9.5 mm long, tube glabrous, gradually expanded above middle, lobes 0.5–1.5 mm long. Achenes 1.7–2.5 mm long, ribbed, rather sparsely short-setuliferous between ribs; pappus setae 5–10 mm long.

**Zambia**. E: Nyika, 26.vi.1966, *Fanshawe 9744* (K, NDO). **Zimbabwe**. E: Mutare (Umtali) Dist., Stapleford Reserve Forest, roadside S of Mt Nusa, 5500′, 28.xi.1955, *Chase 5899* (K, SRGH). **Malawi**. N: Chitipa Dist., Mugesse For. Res., 9°40'S, 33°32'E, 24.v.1989, *Pope et al. 2332* (BR, K, MAL). S: Mt. Mulanje, Muluzi Valley, Chisongeli Forest, 1320 m, 15.ix.1988, *Chapman & Chapman 9302* (K, MO). **Mozambique**. Z: Namuli Mt, Muretha Plateau, 15°23'02.1"S, 37°02'15.6"E, 1898 m, 31.v.2007, *Timberlake et al. 5135* (K). MS: Manica e Sofala, Manica, Macequese, junto da cascata de Ruhalonga, 27.vii.1941, *Torre 3192* (LISC).

Also from Nigeria to Ethiopia and S to Uganda, Kenya, Tanzania, Angola and Madagascar. Evergreen rain forest, especially in forest edges, glades and river banks, also in disturbed places such as pine plantation and farm fallows; may be locally common; 1000–1900 m; flowering mostly between April and September.

Conservation Status: LC.

4. **Crassocephalum torreanum** Lisowski in Bull. Jard. Bot. Natl. Belg. **60**(1–2): 178 (1990). —Lisowski, (Asterac. Fl. Afr. Centr. 2) Fragm. Flor. Geobot. **36** Suppl. 1: 344 (1991). Type: [D.R. Congo:] 'Lualaba Province ('Shaba'), plateau de la Manika, près de Manga Manga', 22.i.1969, *Lisowski et al. 218* (POZG holotype). FIGURE 6.4.**11**.

*Senecio* sp. n° 33 of Mendonça, Contrib. Conhec. Fl. Angola **1** (Compositae): 125 (1943).

Subshrub 40–140 cm tall, branched below. Stems erect or slightly decumbent at base, subcylindrical, striate, bright purple below, green above, upper parts covered by dense to sparse whitish arachnoid pubescence. Leaves subsessile or with petiole 3–8 mm long, slightly winged, glabrescent, lamina oblong, elliptic or ovate, 3.3–10.5 × 0.9–3.1 cm, cuneate, rounded or subcordate, margins serrate-dentate, sometimes slightly revolute, apex obtuse or rarely acute, glabrescent or scattered crispate-setulose, especially on main veins. Capitula 4–9 in dense to lax terminal corymbs, discoid, erect or slightly drooping; pedicels densely lanulose; involucre campanulate, 10–13 × 6–10 mm; calycular bracts 4–10, usually with darker tips, 3–6.5 mm long, densely arachnoid pubescent; phyllaries c. 14, yellow-green, 10–13 × 0.8–1.9 mm, sparse arachnoid at base or glabrescent. Florets numerous, corolla lobes yellow

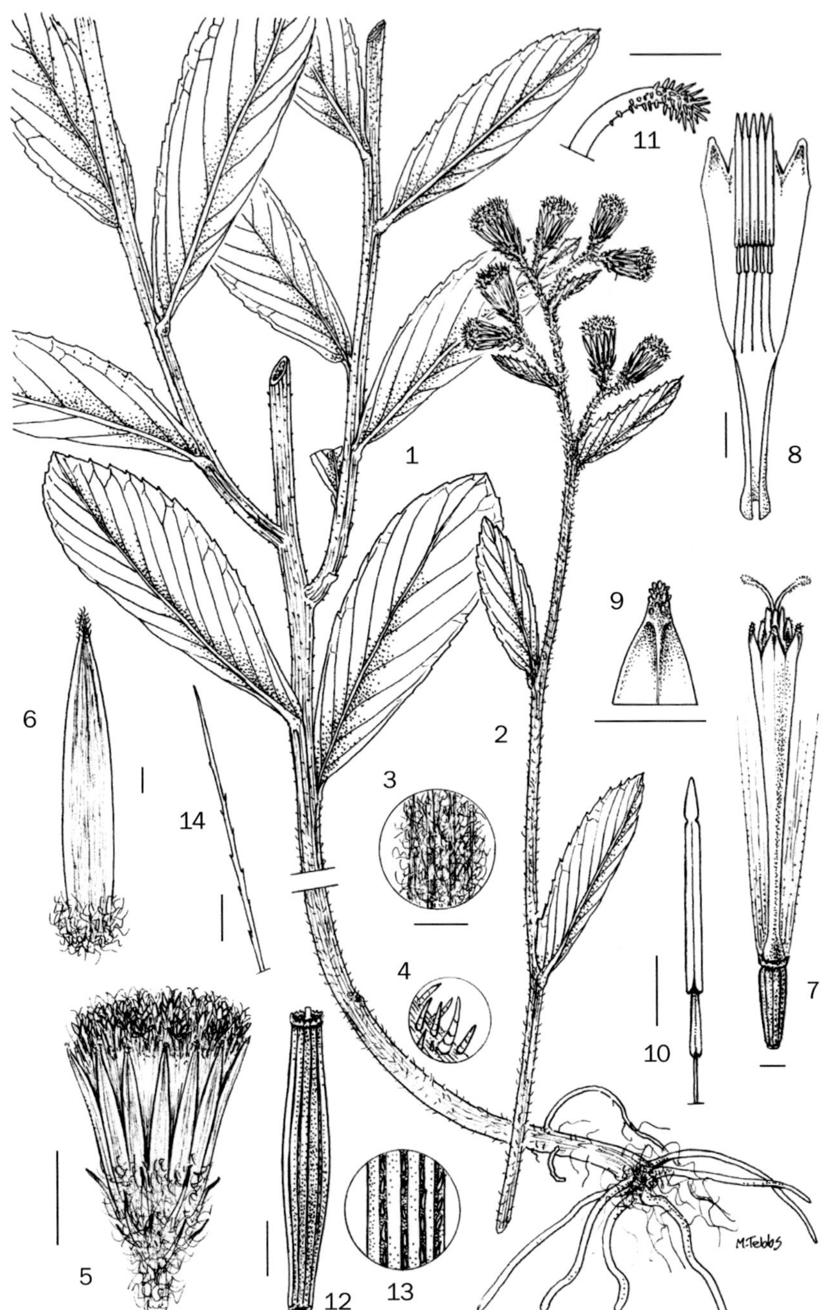

Fig. 6.4.**11**. CRASSOCEPHALUM TORREANUM. 1, leafy base of plant; 2, apex of flowering stem; 3, detail of indumentum of peduncle; 4, detail of indumentum of lower stem; 5, capitulum; 6, phyllary; 7, floret; 8, corolla opened out to show attachment point of filaments; 9, detail of outer surface of corolla lobe; 10, stamen; 11, detail of style arm; 12, achene; 13, detail of setulae in sinuses of ribs of achene; 14, detail of apex of caducous pappus seta. 1–11 from *Milne-Redhead* 3657; 12–14 from *Brummitt* 13990. Scale bars: 1, 2 = 10 mm; 5 = 5 mm; 3, 4, 6–10 = 1 mm; 11, 14 = 0.5 mm. Drawn by Margaret Tebbs.

to orange-yellow, corollas 8.5–12.2 mm long, tube yellow to orange-yellow, glabrous, slender, gradually expanded in upper third, lobes 1–1.2 mm long. Achenes 3.6–4.9 mm long, ribbed, setuliferous between ribs; pappus setae 4–6 mm long.

**Zambia**. N: Mwinilunga Dist., Dobeka Dambo, 45 km W of Mwinilunga, 11°42'S, 24°06'E, 1350 m, 22.i.1975, *Brummitt et al.* 13990 (K).

Also from D.R. Congo and Angola. Wet dambo and river banks; 1200–1350 m; flowering from November to January.

Conservation Status: Four specimens from FZ area from a restricted area, likely NT.

5. **Crassocephalum uvens** (Hiern) S.Moore in J. Bot. **50**(590): 212 (1912). ——Gibbs Russell in Kirkia **10**(2): 502 (1977). —Jeffrey in Kew Bull. **41**(4): 907 (1986). —Lisowski, (Asterac. Fl. Afr. Centr. 2) Fragm. Flor. Geobot. **36** Suppl. 1: 342, t. 75 (1991). Type: [Angola:] 'Huilla. -In swampy places at the river Cacolovar, between Ivantâla and Quilengues; fl. and fr. Feb. 1860; also by the river at Lopollo. [*Welwitsch*] No. 3670.' (BM holotype).

 *Senecio uvens* Hiern, Cat. Afr. Pl. **1**(3): 602 (1898).
 *Senecio telmatophyllus* O.Hoffm. in Warburg, Kunene-Sambesi.-Exped.: 423 (1903), as '*telmatophilus*'. Type: [Angola:] 'Am rechten Kubango-Ufer oberhalb des Kueio, auf moorigem Sumpfboden, 1130 m. ü. m. ([*Baum*] Nr. 354, blühend am 1. Nobember 1899' (B† holotype, BM000924593, HBG505290, M0105358).
 *Senecio kyimbilensis* Mattf. in Bot. Jahrb. Syst. **59**(4, Beibl. 133): 34 (1924). Type: [Tanzania:] 'Nördliches Nyassaland: Station Kyimbila, [Rungwe District], am Wasser im Made-hani-Walde, 2000 m ü. m. (A. Stolz n. 2325, blühend 3. Dezember 1913)' (B† holotype, C10000340, EA, K000306752, P00111863).
 *Senecio vitellinoides* Merxm. in Proc. & Trans. Rhodesia Sci. Assoc. **43**: 70 (1951). Type: [Zimbabwe:] 'Southern Rhodesia: Marondera (Marandellas); wet vlei. leg. *G. Dehn* 5.1.1941. D.n.19.' (M0105357 holotype).
 *Crassocephalum telmatophyllum* (O.Hoffm.) Milne-Redh., comb. inval. in sched.[74]

Perennial herb with short rhizome, lower part of stem creeping but becoming erect and 20–65 cm tall, subscapiform. Stems pale green or slightly reddish-purple tinged, glabrous or crispate-pubescent. Leaves sessile, oblanceolate, 1–7 × 0.2–1.2 cm, base attenuate to exauriculate, margins remotely sinuate-denticulate or sinuate-serrate, apex obtuse, apiculate, sparsely shortly setulose or hispid, bright green above, paler beneath, sometimes margins purplish. Capitula solitary, terminal, erect, long-stalked, discoid; pedicels finely crispate at least in upper part; involucre cylindrical, 8–11 × 3.5–5 mm; calycular bracts about 6, lanceolate, green or purple, 2.5–3 mm long; phyllaries 13–18(21), purple or green tipped purple or brown, 7–10 × 0.8–1.5 mm, glabrous or shortly sparsely crispate-pubescent. Florets numerous, corollas orange-yellow or orange, 7.5–11 mm long, tube glabrous, gradually expanded above middle, lobes 0.8–1 mm long. Achenes 3–4 mm long, glabrous; pappus setae 6.5–10 mm long.

**Zambia**. N: Samfya Dist., c. 5 km W of Samfya, Kasamba Dambo, 29°32'S, 11°22'E, 21.iv.1989, *Pope et al.* 2201 (BR, K). W: Mwinilunga Dist., 100 yards NE of Dobeka Bridge, 11.xii.1937, *Milne-Redhead* 3614 (K). C: Chakwenga Headwaters, 100-129 km E of Lusaka, 27.x.1963, *Robinson* 5775 (K). **Zimbabwe**. N: Sebungwe, Karyangwe Tsetse-Entomologist's camp, c. 2800', 15.xi.1958, *Phipps* 1467 (K, SRGH).W: Hwange (Wankie) Dist., Katsatetsi river, Kazuma range, 1000 m, 10.v.1972, *Gibbs Russell* 1947 (SRGH, fide M. Hyde). C: Harare (Salisbury) Dist., enterprise in vlei, 4500', 1.iii.1950, *Wild* 3231 (K). E: Mare River, Nyanga (Inyanga), 6000', 27.x.1946, *Wild* 1561 (K). S: Masvingo (Victoria) Dist., Oatlands Farm adjacent to [Great] Zimbabwe and near Chipopo river, 12.iv.1973, *Chiparawasha* 711 (SRGH, fide M. Hyde). **Malawi**. N: Dembo, 4000', 5.i.1976, *Phillips* 865 (K, MO). C: Mchinji Dist., 5 km NE of Mchinji, on road to Kasungu,

---

[74] Milne-Redhead provided that determination on Kew's duplicate of *Stolz 2325* but never published it. Subsequently the combination appeared in two checklists: Goodier & Phipps in Kirkia **1**: 44–66 (1960) and Gill in Willdenowia **12**(1): 95–128 (1982).

1215 m, 27.iv.1970, *Brummitt* 10205 (K). S: Zomba Dist., Zomba mountain, Chitengi Dambo, 15°17'S, 35°19'E, 14.v.1989, *Pope et al.* 2285 (BR, K). **Mozambique**. N: Niassa Dist., Vila Cabral, 27.i.1935, *Torre* 751 (LISC). MS: Manica, Chimanimani mountains, S of Namudima mountain, 19°45'47"S, 33°04'56"E, 1603 m, *Osborne et al.* 1182 (BR, K, LMA, SRGH). GI: Inhambane: Zavada, Zandamela, 5.xii.1944, *Mendonça* 3273A (LISC).

Also known from Angola, D.R. Congo, Tanzania. Marshy or riverine grassland, dambo and vlei; 850–2000 m; flowering throughout the year.

Conservation Status: LC.


6. **Crassocephalum vitellinum** (Benth.) S.Moore in J. Bot. **50**(590): 212 (1912). —Andrews, Fl. Pl. Sudan. **3**: 21 (1956). —Lind & Tallantire, Common Fl. Pl. Uganda: 182, fig. 120 (1962). —Maquet in Fl. Rwanda **3**: 652, fig. 201/2 (1985). —Jeffrey in Kew Bull. **41**(4): 907 (1986). —Blundell, Wild Fl. E. Afr.: fig. 423 (1987). —Lisowski, (Asterac. Fl. Afr. Cent. 2) Fragm. Flor. Geobot. **36** Suppl. 1 : 339, t.74 (1991). —Agnew & Agnew, Upland Kenya Wild Fl., ed. 2: 220 (1994). Type: [Equatorial Guinea:] 'Fernando Po [Bioko], [xi.1841], *Vogel* [s.n.]' (K000306764 – ex Herb. Hookerianum holotype, K000306762 – s.n., ex Herb. Benthamianum, K000306763 – '198', ex Herb. Benthamianum syntypes).

 *Gynura aurantiaca* Benth. in Hooker, Niger Fl.: 437 (1849), nom. nud. et nom. illeg., non (Blume) DC. (1838).[75]

 *Gynura vitellina* Benth. in Hooker, Niger Fl.: 438 (1849). —Oliver & Hiern, F.T.A. **3**: 402 (1877).

Annual or perennial herb 30–150(350) cm tall, erect or basally procumbent and rooting at nodes with erect flowering branches, or semi-scandent. Stems green often streaked or tinged with purplish, red or violet, almost glabrous to densely setulose. Leaves sessile, ovate to broadly ovate, elliptic or obovate, 3–12.5 × 1.5–7 cm, base cordate, truncate, cuneate or broadly cuneate and slightly decurrent onto sometimes pinnato-lyrately 2–4-lobed, auriculate or exauriculate, petioloid base, margins sinuate-serrate to sinuate-bidentate or broadly sinuate-lobulate, apex obtuse to acute, finely or usually densely pubescent above and beneath, sometimes with purplish midrib. Capitula solitary or up to 3, terminal, discoid, long-stalked, sometimes nodding; pedicels densely glandular-pubescent, sometimes purplish; involucre broadly cylindrical, 7–12 × 5–10 mm; calycular bracts 10–18, lanceolate, 2.5–7 mm long, glandular-setulose and ciliate; phyllaries usually 21, rarely 13, green or yellow-green with dark tips, 6.5–11 × 0.8–1.1 mm, densely glandular-setulose with hairs sometimes purplish. Florets numerous, corollas orange, less often orange-yellow or yellow, 6.5–10 mm long, tube glabrous, gradually expanded above middle, lobes 0.7–1.5 mm long. Achenes 1.8–2 mm long, ribbed, short-setuliferous between ribs; pappus setae 6–9 mm long.

**Zambia**. B: Mwinilunga Dist., by River Matonchi near garden in damp grassy places at edge of evergreen vegetation, 11.xi.1937, *Milne-Redhead* 3192 (K). W: Kalabo Dist., 2 miles W of Kalabo, edge of swamps bordering river, 13.xi.1959, *Drummond & Cookson* 6436 (K, SRGH). **Malawi**. N: Karonga Dist., Nganda hill, Nyika Plateau, 7600', 6.ix.1962, *Tyrer* 828 (SRGH, fide M. Hyde).

Also in Nigeria, Cameroon, Equatorial Guinea (Bioko), D.R. Congo, Rwanda, Burundi, Uganda, Kenya and Tanzania. Forest margins, damp grassland, edge of swamp bordering rivers; 1000–1500 m; flowering in November.

Conservation Status: Rare in FZ area but wide distribution; LC.


7. **Crassocephalum paludum** C.Jeffrey in Kew Bull. **41**(4): 906 (1986). —Lisowski, (Asterac. Fl. Afr. Cent. 2) Fragm. Flor. Geobot. **36** Suppl. 1: 335 (1991). Type: Uganda, 'Kigezi District, above Lake Bunyongi, Muko' *Morrison* 251b (K holotype).

 *Crassocephalum sp. A* of Maquet in Fl. Rwanda **3**: 652, fig. 201/3 (1985).

---

[75] In the text, Bentham referred to 'the *G. aurantiaca*, described below, is intermediate in habit, ...' without using that name. The description doubtless referred to that of *Gynura vitellina*.

Herb 30–200 cm tall, erect or sometimes straggling. Stems sometimes purplish, pubescent to glabrous. Leaves sessile, occasionally reddish, linear-lanceolate, lanceolate, elliptic or oblanceolate, 4–19 × 0.4–0.7(2.5) cm, cuneate-attenuate into a winged petioloid or non-petioloid exauriculate or usually auriculate base, margins remotely shallowly sinuate-denticulate to coarsely sinuate-dentate or sinuate-serrate, often with ± recurving teeth, apex obtuse to acute, sometimes attenuate, glabrous or densely to thinly pubescent, often ± glabrescent. Capitula 1–10 in congested to lax terminal corymbs, discoid; peduncles of individual capitula pubescent; involucre cylindrical, 7.4–9.7(12) × 0.5–0.9 mm; bracts of calyculus 7–16, lanceolate, 1.5–5 mm long, glabrous or ciliate, dark-tipped; phyllaries 13–21, usually 13, green, 7.4–9.7(12) × 0.4–1.1 mm, glabrous or sparsely pubescent. Florets numerous, corollas yellow, rarely orange, 7–9.5 mm long, tube glabrous, gradually expanded above middle, lobes 0.7–1.5 mm long. Achenes 1.5–1.7 mm long, ribbed, short-setuliferous between ribs; pappus setae 6–10.5 mm long.

**Zambia**. N: Kawambwa Dist., Mantapala, river swamp, 1050 m, 20.iv.1957, *Richards* 9356 (K). C: Kabwe (Broken Hill) Dist., Lukanga Swamp, 19.viii.1958, *Seagrief* CHA 3175 (K, SRGH).

Also in D.R. Congo, Rwanda, Burundi, Sudan, Uganda, Kenya and Tanzania. Floating swamps and river swamps; 1050–2400 m; collected in flower in April and August.

Conservation Status: Rare in FZ area but wide distribution in fairly common habitat; LC.


8. **Crassocephalum picridifolium** (DC.) S.Moore in J. Bot. **50**(590): 212 (1912).[76] — Andrews, Fl. Pl. Sudan **3**: 21 (1956). —Hilliard, Compositae Natal: 369, fig. 13 (1977). —Gibbs Russell in Kirkia **10**(2): 502 (1977). —Jeffrey in Kew Bull. **41**(4): 906 (1986). — Blundell, Wild Fl. E. Afr.: fig. 362 (1987). —Lisowski, (Asterac. Fl. Afr. Centr. 2) Fragm. Flor. Geobot. **36** Suppl. 1: 336, t. 73 (1991). —Agnew & Agnew, Upland Kenya Wild Fl., ed. 2: 220, t. 92 (1994). Type: [South Africa:] '... in Africâ austr.-orient. ad Port-Natal et Omsameulo [, 1.iii.1832,] legit cl. *Drege* [5161]! ... (v. s.)' (G-DC00474273 holotype).

    *Senecio picridifolius* DC., Prodr. **6**: 386 (early Jan.1838). —Oliver & Hiern, F.T.A. **3**: 413 (1877).

    *Senecio picridifolius* sensu Harvey in Harvey & Sonder, Fl. Cap. **3**: 379 (1865), non DC.

    *Senecio acutidentatus* A.Rich., Tent. Fl. Abyss. **1**: 436 (1848). Type: [Ethiopia:] 'Crescit in locis umbrosis circa Kouaietha, in provincia Chiré, altitudine circiter 7,000 pedum supra mare (*Quartin Dillon*).' *Quartin-Dillon & Petit* s.n. (P00101514 – ex Herb. Richard, as 'lectotype', P00101515 – as 'isolectotype' syntypes).[77]

    *Senecio papaverifolius* A.Rich., Tent. Fl. Abyss. **1**: 437 (1848). Type: [Ethiopia:] 'Crescit circa Adoua [Adowa], mense Julio florens (*Quartin Dillon*)' *Quartin-Dillon & Petit* s.n. (P00101516 – ex Herb. Richard, P00101517 – ex Herb. Richard, P00101518 syntypes).[78]

    *Gynura picridifolia* Burtt Davy in Bull. Misc. Inform. Kew **1935**(10): 569 (1935).[79] Type: [South Africa:] 'Transvaal. Pretoria District: Groenkloof, near Pretoria, along water-furrow, [27.iii.1920,] *Burtt Davy* 18809 in Kew Herb.' (K000373090 holotype).[80]

Perennial herb 30–120 cm long or high or scrambling to 180 cm, erect or lower stem prostrate and rooting at nodes, sending up erect branches, sometimes clambering through other plants. Stems

---

[76] This is sometimes given as ×*picridifolium*, a hybrid between *C. paludum* and *C. vitellinum*.

[77] Jeffrey & Beentje (2005) stated that the 'holotype' was in P, although it is unclear why the material indicated is considered the 'lectotype'.

[78] Jeffrey & Beentje (2005) stated that the holotype was in P, probably unaware that there were two specimens from Richard's herbarium, thus no holotype.

[79] Jeffrey (1986: 907) and Jeffrey & Beentje (2005: 606) incorrectly referred to Burtt Davy's plant name as a combination, based upon de Candolle's in *Senecio*. However, Burtt Davy was at pains to point out that he was excluding all material cited by Harvey, except de Candolle's Drège type, and including the material in his new species, *Gynura picridifolia*. Moore (1912), without mentioning material, based his combination upon de Candolle's entity. Unfortunately, the remaining accounts cited by Moore did cite material of another entity – of what Burtt Davy considered his plant.

[80] The date of this publication, as provided in F.T.E.A., as '1925', is incorrect.

green or streaked or tinged with purplish, red or brownish-red, glabrous to densely crispate-setulose with red hairs. Leaves sessile, obovate, oblanceolate, elliptic, lanceolate or ovate, sometimes distinctly pinnately or pinnato-lyrately 2–8-lobed in lower half, 3–15.5 × 0.5–9 cm, cuneate to attenuate into a usually auriculate petioloid base, margins sinuate-dentate or sinuate-serrate to broadly to deeply sinuate-lobulate, especially towards base, apex obtuse to acute, sometimes somewhat attenuate, finely crispate, pubescent, sometimes also setulose on veins beneath, sometimes glabrescent. Capitula 1–5, terminal, lax, discoid; long-stalked, pedicels pubescent; involucre cylindrical, 9–13 × 4–8 mm, base wide and truncate; calycular bracts 8–15, linear to lanceolate, dark-tipped, 2.5–8 mm long, crispate-pubescent or at least sinuate-margined; phyllaries 13–21, usually 21, green with purplish or brownish tips, 8–13 × 0.6–1.8 mm, finely crispate-pubescent or densely crispate-glandular, sometimes hairs tinged purplish, or rarely glabrous. Florets numerous, corollas yellow, orange-yellow or orange, 7.2–11.5 mm long, tube glabrous, gradually expanded above middle, lobes 0.7–1.7 mm long. Achenes 1.5–2.5 mm long, ribbed, short-setuliferous between ribs; pappus setae 7–12 mm long.

**Caprivi Strip**. Mukwe, Popa Falls Rest Camp, island walk, banks of Okavango river, 18°7'17.83"S, 21°35'1.21"E, 990 m, *Kolberg & Tholkes* 2272 (K, WIND). **Botswana**. N: Growing around outer edge of Lediba at Matsaudi a ga mmatelelamma Talelo Island, 18.iii.1973, *Smith* 459 (K, SRGH). **Zambia**. B: Mongu S., 15°15'S, 23°06'E, 1000 m, *Bingham* 10032 (K). N: Mbala (Abercorn) Dist., edge of woodland and Lake Chila, 1500 m, 15.xi.1964, *Richards* 19277 (K). W: Ndola, in long grass by dambo, 27.iii.1954, *Fanshawe* 1020 (K). C: Chinyunyu, 55 miles E of Lusaka, 4000', 15.vi.1956, *King* 382 (K). E: Tigone Dam, Lundazi-Chama, 1.5 miles, 1200 m, 18.x.1958, *Robson & Angus* 151 (K). S: Kabulamwanda dam, 80 miles N of Choma, 3400', 24.iv.1954, *Robinson* 718 (K). **Zimbabwe**. W: Bulawayo Dist., Khami, at stream edge, 4500', 25.iv.1946, *Wild* 1057 (K). C: Rusaki (?Rusape), 03.ii.1921, *Hislop* Z236 (K). E: Chimanimani (Melsetter) Dist., Cashel, Umvumvumvu R., by stream, 3800', 10.xi.1952, *Wild* 3877 (K, SRGH). S: Masvingo (Victoria) Dist., Oatlands Farm adjacent to [Great] Zimbabwe and near chipopo river, 12.iv.1973, *Chiparawasha* 710 (SRGH, fide M. Hyde). **Malawi**. N: Mzimba/Nkhata Bay Dist., N Vipya, Camp Gordon Army Base, 33 miles N of Mzuzu, 6000', 30.vii.1972, *Pawek* 5578 (K). C: Mchinji Dist., 5 km NE of Mchinji on road to Kasungu, 1215 m, *Brummitt* 10199 (K). S: Zomba Dist., Zomba Plateau, by Mlunguzi River near Trout Ponds, 1520 m, 18.iv.1970, *Brummitt* 9951 (K). **Mozambique**. N: Nampula, at the foot of Monte Matharia, 14°58'32"S, 38°08'02"E, 637 m, 21.x.2017, *Osborne et al.* 1281 (K, LMA, LMU). Z: Namuli Mountain, Muretha Plateau, 15°23'29.6"S, 37°02'54.9"E, 1856 m, *Patel et al.* 7310 (K). MS: Manica, Chimanimani mountains, S of Namudima moutain, 19°45'47"S, 33°04'56"E, 1603 m, *Osborne et al.* 1181 (BR, K, LMA, SRGH). GI: Inhambane, Massinga, Rio das Pedras, *de Koning & Hiemstra* 8934 (K, LMU). M: Matola, 10.xii.1897, *Schlechter* s.n. (K).

Also in W Africa from Gambia to Cameroun, D.R. Congo, Sudan, Ethiopia, Uganda, Kenya, South Africa, Tanzania. Swamps or swampy grassland; may be locally common or even mat-forming; 630—2500 m; flowering throughout the year.

Conservation Status: LC.

Most likely a hybrid between *C. paludum* and *C. vitellinum*. As noted by Jeffrey (1986), *C. picridifolium* is more widespread and with a larger spectrum of habitats than its putative parental species. All morphological intermediate forms between *C. paludum* and *C. vitellinum* are found making identification quite challenging.

9. **Crassocephalum bauchiense** (Hutch.) Milne-Redh. in Kew Bull. 5(3): 376 (1950) [6 Feb. 1951]. —Jeffrey in Kew Bull. **41**(4): 908 (1986). —Lisowski, (Asterac. Fl. Afr. Centr. 2) Fragm. Flor. Geobot. **36** Suppl. 1: 354 (1991). Type: Nigeria, Bauchi Plateau, Vom, 3000–4500 ft., 1922, *Young* 158 (K000306765 holotype).

    *Gynura caerulea* Hutch. & Dalziel in F.W.T.A. **2**: 147, 148 (1931), nom. illeg. non *G. coerulea* O.Hoffm. (1893: 86) (= *Crassocephalum coeruleum* (O.Hoffm.) R.E.Fr.).

    *Gynura bauchiensis* Hutch. in F.W.T.A. **2**: 608 (1936), nom. nov. pro *G. caerulea* Hutch. & Dalziel.

Annual herb up to 120 cm tall. Stems erect, shortly crispate-pubescent. Leaves sessile, obovate-lanceolate, deeply pinnate-lyrately-lobed with 5–20 lanceolate sinuate-dentate or sinuate-lobulate lateral lobes, 2–15 × 2–5.5 cm, apex obtuse, minutely acuminate-apiculate, scattered crispate-pubescent especially on veins. Capitula numerous in copious compound terminal corymbs, pedicels crispate-pubescent, rather crowded. Capitula discoid; involucre cylindrical, 7.5–8.5 × 3 mm; calycular bracts 11–18, narrowly lanceolate, 2–3.5 mm long, ciliate-margined; phyllaries 12–18, 7–8 × 0.6–1.1 mm, glabrous. Florets numerous, corollas pale to bright blue, 6–7 mm long, tube glabrous, gradually expanded in upper third, lobes 0.7 mm long. Achenes 2.5–3 mm long, ribbed, very shortly sparsely setuliferous between ribs; pappus setae 6–7 mm long.

**Zambia**. W: Mwinilunga Dist., by R. Kasompa, in garden in open in boggy grassland, 1.ii.1938, *Milne-Redhead* 4435 (K).

Also known from Cameroon, D.R. Congo, Nigeria and Uganda. Boggy grassland; c. 1300–1400 m; collected in flower in February.

Conservation Status: One specimen from FZ area, DD.

10. **Crassocephalum effusum** (Mattf.) C.Jeffrey in Kew Bull. **41**(4): 907 (1986). Type: [Tanzania:] 'Nördliches Nyassaland: Station Kyimbila, [Rungwe Dist.,] Mulinde-Walde, 900 m ü. m., lichter Wald (A. Stolz n. 1456, blühend 22 Juli 1912)' (B† holotype, K000306751, M0105360).

    *Senecio effusus* Mattf. in Bot. Jahrb. Syst. **59**(4, Beibl. 133): 34 (1924).

Annual or short-lived perennial herb 90–150 cm tall, erect. Stems pale green with reddish tinge, sparsely shortly crispate-pubescent. Leaves sessile, elliptic, 3–21 × 0.5–4 cm, attenuate to an exauriculate petioloid base or (especially upper) sessile, margins shallowly and remotely to (especially upper) coarsely sinuate-serrate, unlobed or (especially upper) lyrato-pinnately or pinnately 2–8-lobed, apex ± attenuate and obtuse to acute, shortly crispate-pubescent. Capitula discoid, several to rather numerous in rather lax terminal cymes, erect; stalk of individual capitula glabrous or crispate-pubescent; involucre cylindrical, 6.5–7 × 2.5–3 mm; calycular bracts 9–12, lanceolate, dark-tipped, 1.5–3.5 mm long, ciliate-margined; phyllaries 13, pale green with dark tips, 6–6.5 × 0.6–1.3 mm long, sparsely shortly pubescent. Florets numerous, corollas mauve or pinkish-mauve (label of type says white), 8–9 mm long, tube glabrous, gradually expanded above middle, lobes 1–1.2 mm long. Achenes 1.8–2 mm long, ribbed, short-setuliferous between ribs; pappus setae 6–8.5 mm long.

**Malawi**. N: Nkhata Bay Dist., 3 miles S of junction, N.B. Sec. School, 1800', 31.iii.1974, *Pawek* 8288 (K, MAL, MO, UC).

Also from Tanzania. *Brachystegia* woodland, secondary lowland evergreen forest; 500–1550 m; flowering from February to July.

Conservation Status: Seven specimens from FZ area from a restricted area, likely NT.

11. **Crassocephalum rubens** (B.Juss. ex Jacq.) S.Moore in J. Bot. **50**(590): 212 (1912). — Andrews, Fl. Pl. Sudan **3**: 21 (1956). —Jeffrey in Kew Bull. **41**(4): 907 (1986). —Lisowski, (Asterac. Fl. Afr. Centr. 2) Fragm. Flor. Geobot. **36** Suppl. 1: 350, t.78 (1991). —Jeffrey in Fl. Masc., Composées **109**: 157, t. 52 (1993). Type: 'Floret in caldario, orta ex India Orientali.' Original material not traced.[81]

    *Senecio rubens* B.Juss. ex Jacq., Hort. Vindob. **3**: 50, t. 98 (1777).

    *Senecio cernuus* L.f., Suppl. Pl.: 370 (1782), nom. illeg. superfl. pro *Senecio rubens* B.Juss. ex Jacq.

    *Gynura cernua* Benth. in Hooker, Niger Fl.: 437 (1849), nom. illeg. superfl. pro *Senecio rubens* B.Juss. ex Jacq. —Oliver & Hiern, F.T.A. **3**: 402 (1877).

    *Gynura rubens* (B.Juss. ex Jacq.) Muschl. in Repert. Spec. Nov. Regni Veg. **11**(274–278): 119 (1912).

    *Gynura rubens* (B.Juss. ex Jacq.) Roberty in Bull. Inst. Franç. Afrique Noire, A. Sci. Nat. **16**(1): 73 (1954), nom. illeg., isonym.

---

[81] Original material from Paris Jardin des Plantes sent by Bernard de Jussieu. If no material can be found then the original plate could serve as the lectotype.

Annual herb 20–150 cm tall, erect. Stems green striated with purple, densely to sparsely crispate-pubescent or crispate-setulose. Leaves sessile, obovate, oblanceolate, elliptic or lanceolate, rarely ovate, unlobed or (especially upper) lyrato-pinnately or pinnately 2–8-lobed, 1.2–20 × 0.5–7.5 cm, cuneate or attenuate into an exauriculate petioloid base or (especially upper) sessile, margins remotely sinuate-denticulate to coarsely sinuate-serrate, apex rounded to obtuse or acute, scattered crispate-pubescent at least on veins beneath. Capitula 1–8, long-stalked, usually at first nodding then ± erect, but sometimes erect throughout, discoid; pedicels ± crispate-pubescent and purplish-tinged; involucre cylindrical, 8–13 × 2.5–8 mm; calycular bracts 5–23, purplish or dark green with purple tips, lanceolate or narrowly lanceolate, 3–6(8) mm long, glabrous or margins ciliate; phyllaries 13–25, commonly 21 or 13, pale green to green, often tinged purple especially towards apex and purple-tipped, 7.5–12 × 0.4–2.1 mm, glabrous or sparsely shortly pubescent. Florets numerous, corollas purple or mauve, less often pink or red, 6.2–10.5 mm long, tube glabrous, gradually expanded in upper third, lobes 0.4–1.5 mm long. Achenes 2–2.5 mm long, short-setuliferous between ribs; pappus setae 7–12 mm long.

*C. rubens* has been treated as a single polymorphic sp. including *C. sarcobasis* (Jeffrey 1993; Tadesse in Hedberg *et al.*, Fl. Ethiopia & Eritrea **4**(2): 263, 2004) or considered as two distinct species (Jeffrey 1986; Lisowski 1991), and more recently as two varieties (Jeffrey & Beentje 2005; Beentje 2021). In the FZ area the minor differences between both varieties are not completely consistent and intermediate specimens may indicate hybridization.

Leaves often narrowly and deeply lobed; capitula 1–4, phyllaries commonly 21. .   a) var. *rubens*
Leaves usually rather broadly lobed; capitula 1–12, phyllaries commonly 13. . . b) var. *sarcobasis*

## a) Var. **rubens**

Leaves often narrowly and deeply 2–8-lobed. Capitula 1–4; involucre broadly cylindrical, 9.5–13 × 5–8 mm; calycular bracts 12–23; phyllaries 13–25, commonly 21. Corollas purple, mauve, magenta, pink or red.

**Zambia**. B: near Senanga, 29.vii.1952, *Codd* 7229 (K, PRE). N: Shiwa Ngandu, 5400′, 21.vii.1938, *Greenway* 5458 (K). W: Ndola, 5.iii.1956, *Fanshawe* 2821 (K). C: Great East Rd. between Undaunda and Rufunsa, 135 km E of Lusaka, 15°13'S, 29°25'E, 1150–1200 m, *Kornás* 772 (K). E: Petauke Dam, 850 m, 6.xii.1958, *Robson* 855 (K). **Zimbabwe**. N: Trelawney Tobacco Station, 29.iii.1944, *Jack* 197 (K, SRGH). W: Matobo Dist., Farm Besna Kobiba, 4800′, iii.1957, *Miller* 4202 (SRGH, fide M. Hyde). C: Marondera Dist. (Marandellas), Digglefold, 5000′, 5.vi.1948, *Corby* 120 (K, SRGH). E: woodland above Mutare, on road to Vumba Mountain, 3.viii.1988, *Carte & Coates Palgrave* 2676 (K). S: Masvingo Dist. (Fort Victoria), Zimbabwe, 3500′, 10.x.1949, *Wild* 3010 (K, SRGH). **Malawi**. N: Rumphi Dist., Nyika Plateau, 9 mi. N of Ml., 1700 m, 21.xii.1977, *Pawek* 13321 (K, MAL, MO). C: Dedza Dist., 11 km NW of Dedza on road to Lilongwe, 1370 m, 18.ii.1970, *Brummitt* 8619 (K). S: Mt. Mulanje - Likabula Valley, 860 m, 11.iv.1988, *Chapman & Chapman* 9033 (K, MO). **Mozambique**. N: Nampula, at the foot of Monte Matharia, 14°58'32"S, 38°08'02"E, 637 m, 21.x.2017, *Osborne et al.* 1282 (K, LMA). Z: Entre Mopeia e Morrumbala, a 28.3 km de Mopeia, 5.ix.1949, *Grandvaux Barbosa & Carvalho* 3962 (K, LMA). T: Tete Dist., Moatize, Zóbuè para Metengobalama ao km 21 de Zóbuè (Ef), c. 900 m, 11.i.1966, *Correia* 399 (LISC). MS: Chimanimani Mts, Musapa Gap, 3250′, 6.x.1950, *Sturgeon & Panton* 30738 (K, SRGH).

Also in W Africa from Liberia to Cameroon, Gabon, Sudan, Ethiopia, Uganda, Kenya, Tanzania, South Africa, Madagascar, Comoros, Mauritius and Reunion. Swamps or swampy grassland, grassland, woodland but especially in disturbed areas (cultivations, roadsides) and secondary vegetation; 300–2050 m; flowering throughout the year.

Conservation Status: LC.

b) Var. **sarcobasis** (DC.) C.Jeffrey & Beentje in F.T.E.A., Compositae **3**: 609 (2005). Type: Madagascar, 'in campis cultis prob. Emirnae ins. Madagascar legit cl. *Bojer*. Senecio sarcobasis Bojer! in litt. 1835. ... (v. s. comm. à cl. inv.)' *Bojer* s.n. (G-DCG00147063 holotype, P00558756).

*Gynura sarcobasis* DC., Prodr. **6**: 300 (1838).

*Senecio sarcobasis* Bojer ex DC., Prodr. **6**: 300 (1838), nom. illeg., nom. nud. pro syn. sub *Gynura sarcobasis* DC.

*Crassocephalum sarcobasis* (DC.) S.Moore, J. Bot. **50**(590): 211 (1912). —Jeffrey in Kew Bull. **41**(4): 907 (1986). —Blundell, Wild Fl. E. Afr.: fig. 678 (1987). —Lisowski, (Asterac. Fl. Afr. Cent. 2) Fragm. Flor. Geobot. **36** Suppl. 1: 349 (1991). —Agnew & Agnew, Upland Kenya Wild Fl., ed. 2: 219, t. 92 (1994).

Leaves usually rather broadly 2–8-lobed. Capitula 1–12; involucre cylindrical, 8–12 × 2.5–7 mm; calycular bracts 5–17; phyllaries 13–25, commonly 13; corollas purple, mauve, magenta or pink.

**Zambia**. N: road Old Boma Gardens, Mbala (Abercorn), 5000ʹ, 10.v.1955, *Richards* 5630 (K). W: Ndola, 23.iv.1954, *Fanshawe* 1138 (K, NDO). C: 5 miles E of Lusaka, 4200ʹ, 20.v.1955, *King* 22 (K). S: Kafue Basin, 10 miles N of Choma, 15.iv.1963, *van Rensburg* 1951 (K). **Zimbabwe**. N: vicinity of Umvukwe Mountains, near Darwendale (Darwindale), 20.iv.1948, *Rodin* 4348 (K, UC). W: Matobo Dist., Besna Kobila Farm, 4800ʹ, 28.iii.1963, *Miller 8423* (K, SRGH). C: Mashonaland East (Salisbury) Dist., Domboshawa, 5000ʹ, 8.iii.1950, *Wild* 3245 (K, SRGH). E: Umtali Div., Manica Dist., Odzani River Valley, 1914, *Teague* 171 (K). S: Zimbabwe Ruins, 1.vii.1930, *Hutchinson & Gillett* 3335 (SRGH, fide M. Hyde). **Malawi**. N: Nkhata Bay, Vipya Link rd., ca. 70 mi. SW of Mzuzu, 5800ʹ, 16.v.1971, *Pawek* 4822 (K, MAL). C: Mchinji Dist., c. 7 km W of Namitete, 14°01'S, 33°16'E, 29.iv.1989, *Pope et al.* 2236 (BR, K, LISC, MAL). S: Zomba Plateau, Zomba Dist., 1500 m, 9.vi.1946, *Brass* 16326 (K, NY). **Mozambique**. Z: entre Lioma e Gurué, a 41.7 km do Lioma, 15.ix.1949, *Grandvaux-Barbosa & Carvalho* 4091 (K, LMA). MS: Manica, Sussudenga Dist., Chimanimani foothills, Moribane Forest Reserve, Mbiquisa Forest, 19°44'44"S, 33°19'08"E, 494 m, 28.vi.2015, *Cheek et al.* 17950 (K, LMA).

Also in Sudan, Ethiopia, D.R. Congo, Rwanda, Burundi, Uganda, Kenya, Tanzania, South Africa, Madagascar, Comoros, and Yemen. Damp, marshy places, riverine forest, but especially in disturbed areas (cultivations, roadsides) and secondary vegetation; 200–2165 m; flowering throughout the year.

Conservation Status: LC. Sometimes cultivated.

12. **Crassocephalum coeruleum** (O.Hoffm.) R.E.Fr. in Wiss. Ergebn. Schwed. Rhodesia–Kongo-Exped. 1911–1912, **1**: 342 (1911). —Merxmüller, Prodr. Fl. Südwestafr. **139**: 46 (1967). Type: [Namibia:] 'Standort: Südost-Ondonga, Groot-Fontein [19.iv.1886] (*Schinz* [718]).' (?B†, K000306748, ?Z syntypes).

*Gynura coerulea* O.Hoffm. in Bull. Herb. Boissier **1**(2): 86 (1893).

*Crassocephalum cernuum* (L.f.) Moench var. β *coerulea* (O.Hoffm.) Hiern, Cat. Afr. Pl. **1**(3): 594 (1898), as '*caerulea*'.

Annual herb 45–60 cm tall, erect. Stems corrugated, green, frequently striated with purple, sparsely crispate-setulose or glabrate. Leaves sessile, blade ovate or elliptic, unlobed or (especially upper) lyrato-pinnately, 3.5–13.1 × 1.5–8.3 cm, attenuate into an exauriculate petioloid base, rarely with minute auricles (one specimen with auricles up to 2.5 cm long), margins irregularly sinuate-serrate, apex acute, glabrous to sparsely crispate-pubescent, especially on veins. Capitula 1–4, long-stalked, erect discoid; pedicels sparsely to densely crispate-pubescent, green sometimes purplish-tinged; involucre cylindrical, 9–13 × 4.5–10.1 mm; calycular bracts 8–25, green, frequently with purple tips, linear, 2.7–4.5 mm long, margins ciliate; phyllaries (14–)16, pale green to green, often tinged purple especially towards apex and purple-tipped, 6.5–9.5 × 1–2 mm, glabrous or sparsely shortly pubescent. Florets numerous, corolla blue 7.7–9 mm long, tube glabrous, gradually expanded in upper third, lobes 1.5–2 mm long. Achenes 2–2.6 mm long, densely short-setuliferous between ribs; pappus setae (6–)8–11 mm long.

**Zambia**. B: Mongu-Lealui, Mongu, 10.i.1960, *Gilges* 934 (SRGH, fide M. Hyde). ?N: Fries (1911: 342) found the species at Lake Bangweulu at several localities but I could not find specimens from there. W: Mwinilunga Dist., 60 miles S of Mwinilunga on the Kabompo road, 6.vi.1963, *Loveridge* 743 (SRGH, fide M. Hyde).

Also from Namibia, Angola and one record from Tanzania. On well-drained sandy woodland, on roadside; c. 1000–1300 m; flowering from January to June.

Conservation Status: Four specimens from FZ area. Despite the comment of Fries (1911: 342) that the species was a ruderal at lake Bangweulu, it is rarely collected in FZ area. At present it is best recorded as DD (Data Deficient).

The description is based on specimens from Namibia.

## 104. **EURYOPS** (Cass.) Cass.

**Euryops** (Cass.) Cass. in Dict. Sci. Nat., ed. 2, **16**: 49 (1820). —Nordenstam, Opera Bot. **20**: [1]–409 (1968). —Dyer, Gen. S. Afr. Fl. Pl. **1**: 715 (1975). —Jeffrey in Kew Bull. **41**(4): 931 (1986). —Jeffrey & Beentje in F.T.E.A., Compositae **3**: 566–569 (2005). —Devos *et al.* in Taxon **59**(1): 57–67 (2010).

*Othonna* L. subgen. *Euryops* Cass. in Bull. Sci. Soc. Philom. Paris **1818**: 140 (1818).

*Psilothamnus* DC., Prodr. **6**: 41 (1838).

*Ruckeria* DC., Prodr. **6**: 483 (1838).

*Jacobaeastrum* Kuntze, Revis. Gen. Pl. **1**: 347 (1891), nom. illeg.

*Lasiocoma* Bolus in Trans. S. African Philos. Soc. **16**(4): 390 (1906).

*Thodaya* Compton in Trans. Roy. Soc. S. Afr. **19**(3): 324 (1931).

*Lysichlamys* Compton in J. S. African Bot. **9**: 118 (1943).

*Gamolepis* sensu auct., non Less. (1832).

Shrubs, subshrubs, perennial or annual herbs. Leaves alternate, sessile, lax to very tightly arranged, often imbricate, sometimes rosulate or fasciculate. Inflorescences lateral, terminal or pseudoterminal; peduncles naked, simple. Capitula heterogamous and radiate, rarely homogamous and discoid; involucres hemispherical, campanulate or urceolate, ecalyculate; phyllaries uniseriate, rarely ± biseriate, equal or rarely subequal, variously connate, or rarely free to base; receptacle alveolate, rarely smooth, flat or convex, rarely conical, epaleaceous and glabrous. Ray florets usually present, sometimes absent, female, fertile, ray limb yellow or orange, sometimes discolorous; styles terete or flattened, bifid, style arms linear, glabrous, sometimes apically minutely penicillate; staminodes occasionally present. Disc florets hermaphrodite, fertile or sterile; corolla tubular widening above or campanulate towards limb, yellow, lobes (3)5, with a distinct mid-vein; styles terete or flattened, bifid or sometimes simple; style arms linear-oblong, flattened or ± terete, apically penicillate and truncate to convex or conical; anthers 5, base rounded or obtuse to subacute, apical appendages deltoid or ovate-lanceolate, obtuse to acute, filaments filiform and lacking distinct filament collars. Achenes ribbed, veined, or smooth, sometimes tuberculate or muricate or papillate, body glabrous or setuliferous (often densely so), setulae of twin-hairs, often myxogenic; pappus setae few to many, caducous, or absent.

A genus of c. 100 spp. (see Devos *et al.* 2010, for the most recent estimate, cf. Nordenstam 2006, 2012, both suggesting a larger number), mostly of southern Africa, northwards through tropical Africa to Socotra and the Arabian Peninsula. Two species are native in the Flora area, with a third, *E. chrysanthemoides* (DC.) B.Nord., widely cultivated for hedging and sometimes naturalizing; *Euryops pectinatus* (L.) Cass. and *E. virgineus* (L.f.) DC. are included in the key as they are sometimes cultivated, although no herbarium voucher material has been seen.

*Euryops* is perhaps slightly unusual within the Senecioneae, the 'Senecionoid Group' (Jeffrey 1992: 84), even within the 'Othonnoid complex' (see Jeffrey 1986: 876), or the 'othonnoids' (Jeffrey 1992: 62, 84, 99), now the subtribe Othonninae. The species possess cylindrical filament collars with no discernible balusterform dilation. Otherwise, the genus has the typical shrubby habit, heterogamous and radiate capitula, ecalyculate involucres, largely connate phyllaries, fistulose or alveolate receptacles, 'centrally depressed achene testa cells' (see Jeffrey 1992: 99) and pappus setae (usually caducous, rarely absent), found in many of the related genera. The conspicuous filament collars were considered 'apparently plesiomorphic for the tribe ...', and one of the typical characters found in members of the Senecionoid complex.

1. Achenes epappose . . . . . . . . . . . . . . . . . . . . . . . . . . . . . . . . . . . . . . .**3.** *chrysanthemoides*
 – Achenes pappose . . . . . . . . . . . . . . . . . . . . . . . . . . . . . . . . . . . . . . . . . . . . . . . . . . 2
2. Pedicels lateral, axillary . . . . . . . . . . . . . . . . . . . . . . . . . . . . . . . . . . . . . . . . . . . . . . 3
 – Pedicels pseudoterminal at branch tips and branching points. . . . . . . . . . . . . . . . . . . . 4
3. Leaves cuneate and fan-shaped, apically 3–7-lobed; achenes glabrous; plants
 0.5–3.5 m tall. . . . . . . . . . . . . . . . . . . . . . . . . . . . . . . . . . . . . . . . . . . . . . . . . . *virgineus*
 – Leaves narrow-linear, sometimes tripartite towards apex, rarely 7(10)-lobed; achenes
 densely lanate-setuliferous, setulae mucilaginous; plants 0.2–0.5 m tall . . . . **1.** *subcarnosus*
4. Leaves pinnatipartite or pinnately-lobed, lobes parallel and linear, appressed light- or
 grey-tomentose; achenes glabrous (or sparsely setuliferous); plants 0.5–2 m tall, shrub . .
 . . . . . . . . . . . . . . . . . . . . . . . . . . . . . . . . . . . . . . . . . . . . . . . . . . . . . . . . . . . .*pectinatus*
 – Leaves pinnately-lobed, rarely 3-forked or entire, glabrous; achenes setuliferous; plants
 2–25 cm tall, subshrub with a woody xylopodium and annual flowering stems . . . . . . .
 . . . . . . . . . . . . . . . . . . . . . . . . . . . . . . . . . . . . . . . . . . . . . . . . . . .**2.** *transvaalensis*

The descriptions are largely based upon those in Nordenstam (1968), with appropriate
modification if material allowed.

1. **Euryops subcarnosus** DC., Prodr. **6**: 445 (1838). Type: [South Africa:] '... in Carro legit
 cl. Drège. ... (v. s.)' *Drège* 6099 (G-DC-G00472180 lectotype, P, S-G-9762), lectotypified
 by Nordenstam (1968: 193).

   *Euryops punctatus* DC., Prodr. **6**: 445 (1838). Type: [South Africa:] '... in deserto Carro propè
   Gauritz legit cl. *Ecklon*! ... (v. s.)' (G-DC-G00472137 – 'No. 357/ 99.12 Georg. Mr Eklon 1835'
   holotype).[82]

   *Euryops subcarnosus* [var.] β *indivisus* DC., Prodr. **6**: 445 (1838). Type: [South Africa:] ' ... ad
   Zwartenbergen et Zwart Ruggen (*Drège*!). ... (v. s.)' (G-DC-G00472082 – '6098. Zw: Zwartebergen,
   und Zwart Ruggen. R.III.' holotype).[83]

   *Jacobaeastrum punctatum* (DC.) Kuntze, Revis. Gen. Pl. **1**: 348 (1891).

   *Jacobaeastrum subcarnosum* (DC.) Kuntze, Revis. Gen. Pl. **1**: 348 (1891).

   *Thodaya elongata* Compton in Trans. Roy. Soc. South Africa **19**(3): 324 (1931). Type: [South
   Africa: Laingsburg Division:] 'Collected by Prof. Thoday at Whitehill, Oct. 1921. (Herb. Bolus
   17160).' (BOL139128 holotype).

Nordenstam (1968: 193–210) recognized 4 subspp., although understanding the
intraspecific variation and cytology was necessary to properly understand the subspp. Both
subsp. *subcarnosus* and subsp. *foetidus* have simple leaves, and other than achene length and the
smell of opened capitula (which is impossible in herbarium material!) the two are difficult
to separate. I am following Nordenstam's account in recognizing subsp. *foetidus* as the only
subspecies present in our Flora area.

Subsp. **foetidus** B.Nord., Opera Bot. **20**: 207 (1968). Type: [South Africa:] '*Acocks* 530,
 Barkly West Div., Potfontein, 21.VII.1936' (PRE0197038-0 holotype, BOL139129,
 K000374790).

---

[82] Nordenstam (1968: 193) unnecessarily lectoptyified the name based on this material, and cited
 duplicates in 'L, P, S.' The other material probably corresponds to P0001118 – s.col., s.n., s.loc.,
 P0001127 – 'No. 667. DC./ 99.12', mounted with a further duplicate ex herb. Schultz Bipontinus,
 numbered '357. DC. 99.12', but lacking a separate barcode, S-G-9762, in addition to ?GH00008090.
 A supplementary 'Plantae Capenses' label on the sheet in G-DC, gives 'Georg, Karroogegend
 zwischen Gouritzrivier und Langekloof, (IV, B, c). December.' and indicates that it was an Ecklon &
 Zeyher collection; other material in P may well also be duplicate material of this collection.

[83] Nordenstam (1968: 193) unnecessarily lectotypified the name based on the same material.

Erect shrub, 0.2–0.5 m tall, much-branched, branches ascending or spreading. Leaves usually entire (sometimes 2–3-forked above middle), erect to patent or spreading, filiform to narrowly-linear, 15–40 × c. 1 mm, subterete or flattened, coriaceous to slightly fleshy, apices acute to obtuse, mucronate. Inflorescences of solitary, lateral capitula on juvenile branches or brachyblasts; pedicels ascending, (1)2–7 × c. 0.5 mm, scarcely striate. Capitula heterogamous, radiate, unpleasantly scented (in this subsp.); involucre hemispherical c. 7–12 mm diam.; phyllaries (5)6–11, basally connate, oblong to ovate-lanceolate, 4–6 mm long, usually 3-veined, veins dark, apices acute to obtuse; receptacle ± convex, distinctly alveolate. Ray florets 5–9, limb narrowly-oblong, 4–7 × 1.5–3 mm, 4(6)-veined; style arms linear to tapering, 1–1.5 mm long, apices acute to obtuse. Disc florets c. 15–35, corollas 3.5–4.5 mm long, tube 1.5–2 mm long, limb 2–2.5 mm, lobes 0.8–1 mm long, ovate; style arms 0.5–0.9 mm long, apices truncate. Achenes oblong-obovate, 4–5 × 2–4 mm (incl. setulae), densely lanate-setuliferous, setulae 1–2 mm long, light brown or whitish, ± flexuous, mucilaginous when wetted; pappus setae (1)1.5–2 mm long, white.

**Botswana**. SE: Metsimesau, Serowe [1005 m], 1940, *Miller* B/227 (PRE).

This subsp. is also found in South Africa (Cape Province), although the species is relatively widespread in Namibia and South Africa (Northern Cape Province and Orange Free State); c. 1005 m; flowering from July to September.

Conservation Status: This species was not listed in Raimondo *et al.* (2009). It is probably best regarded as LC (Least Concern).

The occurrence of this species in the Flora area is based upon two collections cited by Nordenstam (1968: 208). The description is based on limited material.

2. **Euryops transvaalensis** Klatt in Bull. Herb. Boiss., sér. 1, **4**(12): 843 (1896). Type: [South Africa:] 'Transvaal: [leg. *Mrs.*] *Saunders*, [Comm.] *Wood*, No 5349 [*sic* = '5449'].' *Wood* 5449 (K000307063 lectotype, NH0007552-0 – Natal Government Herbarium No. 7552, 'Collected by Mrs Saunders'),[84] lectotypified by Nordenstam (1968: 299). FIGURE 6.4.**12**.

    *Euryops multinervis* N.E.Br. ex S.Moore in J. Bot. **41**: 135 (1903). Types: [South Africa:] 'Hab. Open veldt around Johannesburg. [*Rand*] No. 747. (Also Middelburg; *Bolus*, No. 7610 in Herb. Kew.)' *Rand* 747 (BM000924728 lectotype), lectotypified by Nordenstam (1968: 299);[85] *Bolus* 7610 (K000307064 syntype).

Subsp. **setilobus** (N.E.Br.) B.Nord., Opera Bot. **20**: 299 (1968). Type: [South Africa: KwaZulu Natal:] 'Natal. On a grassy hill near Nottingham Road Station, at 4000–5000 ft.[, 26-10-97], *Wood*, 7193.' (K000307061 holotype, BM000924726, MO391401, NH0008564-0).[86] FIGURE 6.4.**12**

---

[84] Since neither of the two duplicates mentioned above were seen by Klatt, it suggests that there may be/have been at least one other in another institution. There is a collection in GH (GH00008086), which has a single flowering stem (with a single-headed scape) mounted on the sheet, and in the packet is a description, in German with the Latin diagnosis that was published in the Bulletin's protologue, in Klatt's hand; Klatt's own label is a copy of the label of the type material in K/NH. The det. slip by Vernon Bates (1985) indicates this is the 'Holotype'. The NH material has had changes made during its history, but whether the single shoot in GH came from this material is unclear; it is clear that the shadows on the sheet represent material no longer there or in their original position on the sheet, and may also account for Klatt's description indicating '1–3 cephalo', when all other shoots have just one capitulum. The 3-headed specimen is not of this taxon, nor of a *Euryops*, based on the branched inflorescence, the presence of several bracteoles in the upper part of the pedicel, the phyllaries and the pinnately-lobed leaves at the base of, and in the lower portion of the scape. It more closely resembles a species of *Senecio*.

[85] Effectively selected by Nordenstam (1968: 299), in citing 'Holotype: BM.'

[86] Nordenstam (1968: 299) unnecessarily selected a lectotype based on the material in K, especially since Brown worked at K.

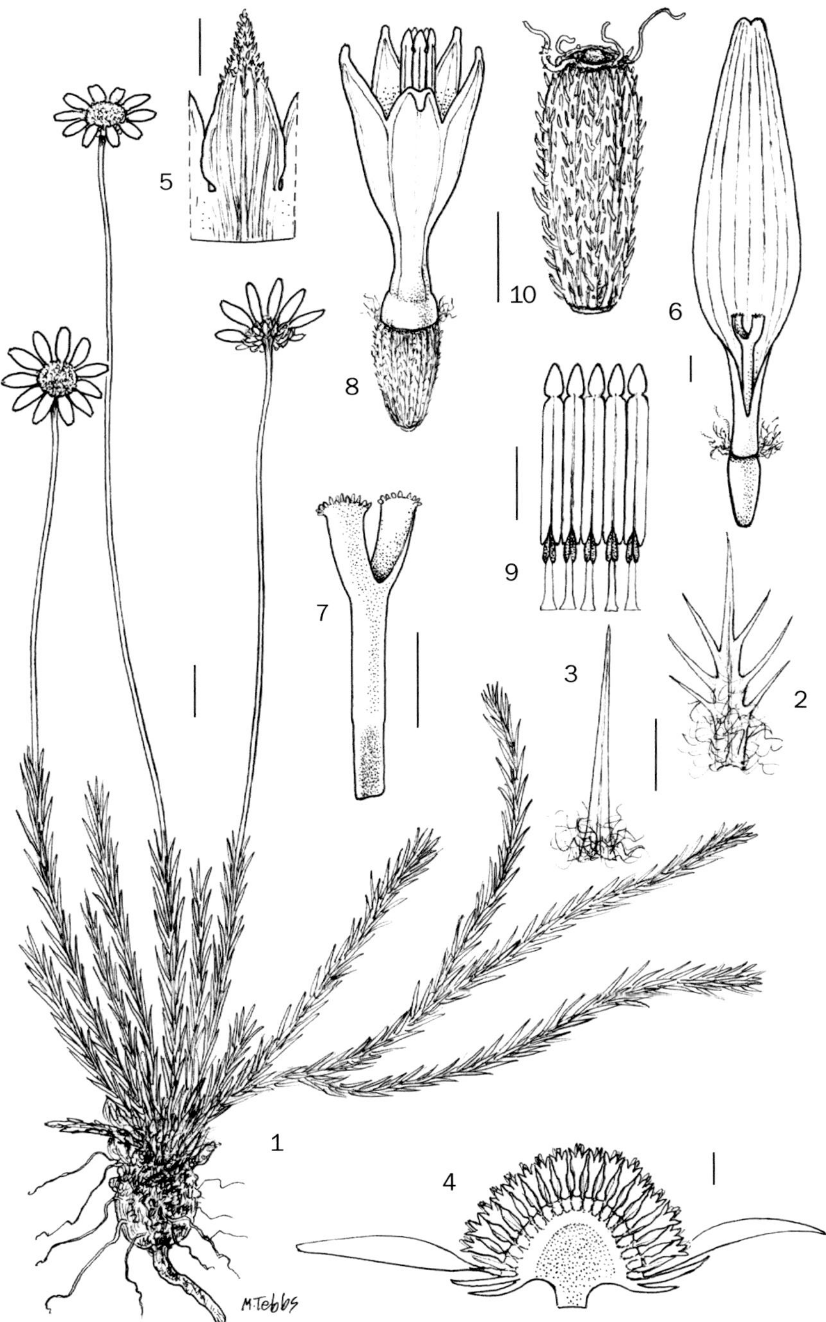

Fig. 6.4.**12**. EURYOPS TRANSVAALENSIS ssp. SETILOBUS. 1, flowering plant; 2, pinnate leaf; 3, simple leaf; 4, l.s. flowering capitulum, showing solid receptacle and convex receptacle; 5, phyllary; 6, ray floret; 7, detail of ray floret style arms; 8, disc floret; 9, anther cylinder, opened out, showing balusterform filament collars; 10, mature disc floret achene. 1 from *Davies* 832; 2, 3 from *Wild* 3554; 4–10 from *Robinson* 1921. Scale bars: 1 = 10 mm; 2, 3 = 3 mm; 4–6, 8, 10 = 1 mm; 7, 9 = 0.5 mm. Drawn by Margaret Tebbs.

*Euryops setilobus* N.E.Br. in Bull. Misc. Info. Kew **1906**(1): 22 (1906).

*Euryops striatus* N.E.Br. in Bull. Misc. Info. Kew **1906**(1): 23 (1906), as '*striata*'. Type: [South Africa:] 'Transvaal. Summit of Saddleback Mountain near Barberton, at 5000 ft., *Galpin*, 988.' (K000307062 holotype, BOL, NH0007213-0, NH0008714-0, PRE0197146-0).

*Gamolepis caudata* Klatt ex Burtt Davy & Pott-Leendertz in Ann. Transvaal Mus. 3(3): 171 (1912), nom. nud.[87]

*Euryops neptunicus* S.Moore in J. Bot. **59**(704): 230 (1921). Type: [South Africa:] 'Transvaal, Kaapsche Hoop; *H. Wager* (Hb. Rogers, 24007).' (BM000924727 holotype).

*Euryops caespitosus* Markötter in Ann. Univ. Stellenbosch **8**(Sect. A, 1): 44 (1930). Type: [South Africa:] 'Witzieshoek. O. V. S. Oct. 1905. Stony grassy hillsides. Alt. ca. 6000' Several inches high. Heads yellow. Leg. *J. Thode*. Herb. Univ. Stell. (No. 5720).' (NBG0200287-0 holotype).

Subshrub with annual stems from a woody xylopodium. Stems simple or few-branched, erect, 2–25 cm tall, densely leafy throughout, glabrous or tomentose in leaf axils. Leaves imbricate, erect to spreading, pinnately-partite, with 2–8 pairs of lobes or 3-forked, rarely entire, 5–20 mm long, rachis 0.5–2(3) mm wide towards base, few (1–5)-veined abaxially, lobes 1–10 mm long, subulate, straight or curved, green. Inflorescences terminal or at branching points, capitula solitary, erect, pedicels 6–25 cm long, glabrous, terete or striate to sulcate. Capitula heterogamous and radiate; involucre hemispherical to broadly campanulate, 6–12 mm diam.; phyllaries 10–16, c. 5 × c. 2 mm, connate ⅓ to ⅔ length, apices acute to acuminate, puberulous-tipped; receptacle convex, shallow-alveolate. Ray florets 8–15, ray limb 4–11 × 1.2–3 mm, elliptic-oblong to oblong, yellow, 4–10-veined; style base swollen, shaft terete, style arms 0.4–0.8 mm long, apices obtuse to truncate; staminodes sometimes present, filiform. Disc florets c. 40–100; corolla 2.5–3.5 mm long, corolla tube base inflated, limb campanulate, lobes narrow-ovate, 0.5–1 mm long; apical anther appendages obtuse to subacute; style shaft terete or flattened, style arms 0.1–0.5 mm long, truncate. Achenes elliptic-oblong, 2–4 × 1–2 mm, 10-ribbed, light- or yellowish-brown, setuliferous; pappus setae few, 0.2–1.5 mm long, white.

**Zimbabwe**. E: Umtali Dist., Himalayas [Mts], Engwa [Farm], 7000'[2134 m], 2.iii.1954, *Wild* 4466 (K, SRGH45775).

A native of South Africa N to Zimbabwe. Occurs in short grassland on quartzitic or schistose rocks, amongst rocks or in rock crevices, or open and frequently burnt grassland; 1000–2200 m; flowering from (July) October–December (March).

Conservation Status: This species was not listed in Raimondo *et al.* (2009). It is probably best regarded as LC (Least Concern).

There are several collections, all from the mountainous eastern areas of Zimbabwe, in Umtali, Melsetter and Nyanga (Inyanga), suggesting that this taxon has a relatively restricted distribution within the Flora area. It is widespread in South Africa, and especially within the Zimbabwian area is the more widespread, and variable, of the two subspp.

3. **Euryops chrysanthemoides** (DC.) B.Nord., Opera Bot. **20**: 365, t. 62/c–g (1968). Types: [South Africa:] '...Cap. Bonae-Spei in coloniae distr. orient. legit. cl. *Burchell* et mecum sub n. 3641 cat. geog. comm. Legerunt etiam cl. *Drège* ad Buffelrivier et *Ecklon* in distr. Uitenhage prope Bosjesman-rivier. ... (v.s.)' *Ecklon* '726' (G-DC-G00450917 – 'N. 726./ 10.9. Uitenhage', mounted with *Burchell* 3641 and another *Ecklon* lectotype,[88] ?G, ?K, ?LE, M0105114 – numbered as '327', and '10.9.', MO391353 – numbered as '327', and '10.9.', ?S, SAM0017050-0 – numbered as '327', and '10.9.', ?W

---

[87] Numbered as taxon number 3192.

[88] Cited by Nordenstam as '*Ecklon & Zeyher* 10.9.', "Uitenhaag, Urwälder von Olifantshoek, zwischen den Mündungen der Flüsse Boschmansrivier und Zondagsrivier, unter 300'."

syntypes);[89] *Ecklon* 7 (G-DC-G00451239 – 'No. 7/ 10.9. Uitenhage' syntype); *Ecklon* '468' (G-DC-G00450886 – 'N. 468/ 10.10. Uitenhage', mounted with *Burchell* 3641 and another *Ecklon* syntype, ?syntype); *Ecklon* '1070'(G-DC-G00450899 – 'N. 1070/ 10.10. Uitenhage' syntype); *Burchell* 3641 (G-DC-G00450904 – mounted with 2 *Ecklon* collections – the other syntypes, K000307144 – there are 2 labelled *Burchell* duplicates on this sheet with one barcode, alongside a barcoded *Bowker* collection, LE syntypes); *Drège* 5103 (G-DC-G00451237 – '5105. Buffelrivier. R. II.', P0000863 syntypes).

*Gamolepis chrysanthemoides* DC., Prodr. **6**: 40 (1838).

Shrub 0.5–2 m tall, nearly completely glabrous, well-branched. Leaves spreading, 3–10 × 1–3 cm, obovate, pseudopetiolate towards base, lamina pinnatilobate, lobes 3–10 pairs, ovate to lanceolate, 2–20 × 5 mm, usually obtuse. Capitula solitary, pedicels 5–20 cm long, somewhat thickened beneath capitula, striate. Involucre broad-campanulate, 10–18 mm diam.; phyllaries ± uniseriate, partially overlapping, 8–15, ovate-lanceolate, 5–8 × 2–4 mm, green, basally connate between ⅕ and ⅓, apices obtuse, puberulous. Ray florets 1–20, corollas yellow, limb (8)12–20 × 2.5–5 mm, 4-veined, widest above middle; style terete, style arms flattened, c. 1 mm long. Disc florets 50–125, corollas 4–5 mm long, yellow; style arms c. 0.8 mm long, apices conical. Achenes oblong-obovate, 3.5–5 × 1–2 mm, 8–11-ribbed, brown to black, glabrous; pappus absent.

**Zimbabwe**. C: Salisbury, cultivated on double stone wall, [1500 m], 8.iv.1949, *Biegel* 3155 (K, SRGH).

A native of South Africa (Cape Province and KwaZulu-Natal). A commonly cultivated plant in Zimbabwe, sometimes naturalizing and established in a few localities; growing in forest margins, clearings and in mixed scrub, but apparently confined to relatively humid areas with a high summer rainfall; ±5–1070 m; flowering throughout the year.

Conservation Status: A widespread, and widely cultivated species. The species was not listed in Raimondo *et al.* (2009). It is probably best regarded as LC (Least Concern).

Uses: An ornamental plant, often used as hedging.

Vernacular name: Honey Daisy.

*Biegel* 3155 was oddly misidentified as *Chrysanthemum frutescens* L. (= *Argyranthemum frutescens* (L.) Sch.Bip.) [Anthemideae]. I have cited this material, as I have not seen any from the Flora area that are obviously naturalized.

## 105. **OTHONNA** L.

**Othonna** L., Sp. Pl. **2**: 924 (1753); Gen. Pl., ed. 5: 396 (1754). —Harvey in Harvey & Sonder, Fl. Cap. **3**: 327–345 (1865). —Dyer, Gen. S. Afr. Fl. Pl. **1**: 716 (1975). — Hilliard, Compositae Natal: 513–514 (1977). —Rowley, Succ. Compositae (1994). — Magoswana *et al.* in Ann. Missouri Bot. Gard. **104**(4): 515–562 (2019). —Magoswana *et al.* in Bothalia **52**(1): a5 [pp. 9] (2022).

*Aristotela* Adans., Fam. Pl. **2**: 125 (1763), p.p.

*Doria* Thunb., Nov. Gen. Pl. **12**: 162 (1800), nom. illeg.

*Calthoides* B.Juss. ex DC., Prodr. **6**: 473 (1838), nom. nud. pro syn. sub *Othonna*.

*Ceradia* Lindl. in Edwards's Bot. Reg. **31**(misc.): 12 (1845).

---

[89] Although Nordenstam noted the location code, and locality, as '10.9', there are several numbered *Ecklon* collections with this location number, and all from 'Uitenhage'. It is therefore unclear as to which of the collections in G-DC that he was referring to, especially as the isolectotype material merely has the taxon number '327' with the locality code of '10.9'; all of the *Ecklon* collections in G-DC are considered syntypes, and none marked as 'lectotype'. HAL0111402 is also '327', but locality coded as '10.10.'; it is also databased as an isolectotype. Second-step lectotypification should solve the issue, since at present the first-step appears incomprehensible, especially against the material in G-DC.

Perennial succulent shrubs, subshrubs or herbs, geophytes, with underground tubers or corms; rootstock crown often densely long-pubescent. Leaves cauline and alternate or rosulate, sessile, short-petiolate or conspicuously long-petiolate, entire or dentate to serrate or lobed, rarely lyrate, terete to flat, usually glabrous, often glaucous, coriaceous or somewhat fleshy. Inflorescence of solitary capitula (and scapose or scapiform) or laxly cymose or corymbose and many-headed, capitula pedicellate, pedicels ebracteolate or few-bracteolate (usually towards capitula). Capitula heterogamous and radiate or rarely homogamous and discoid, or heterogamous and disciform, capitula usually homochromous, sometimes heterochromous; involucre campanulate or cylindrical, ecalyculate; phyllaries uniseriate, connate, at least basally; receptacle convex to conical, epaleaceous, glabrous. Marginal florets, when present, uniseriate, fertile, rays (when present) strap-shaped and 3-toothed, or marginal florets filiform in disciform capitula; style arms linear, stigmatic areas lateral, apices subacute. Disc florets hermaphrodite and functionally male; anthers ecaudate; style arms of fertile hermaphrodite florets (in disciform capitula) truncate and penicillate, of functionally male florets simple (or minutely bifid) and sterile, apices conical or obtuse. Corollas usually yellow, white, pink or purple, rarely magenta. Achenes elliptic-oblong, ± ribbed, ray floret achenes setuliferous, setulae often appressed, myxogenic or non-myxogenic, rarely glabrous, disc floret achenes glabrous; pappus setae of fertile achenes numerous (sometimes described as copious), but often in reduced numbers on achenes of disc florets and then somewhat flattened at base, coarsely capillary, finely barbellate, sometimes united at base, persistent or rarely caducous, sometimes elongating and conspicuous in fruit, white, straw-coloured or pink to reddish or purplish, very rarely pappus absent in disc floret achenes.

About 120 spp. (or as many as 150 according to Hilliard, 1977), most in southern Africa (Cape Province), northwards to Angola and Zimbabwe. Only one sp. has been recorded in the Flora area and is relatively rarely collected.

In *Othonna*'s treatment by Harvey in *Flora Capensis*, I consider the '§' as representing the section, contrary to Magoswana *et al.*'s statement that they were unranked (Magoswana *et al.* 2019). *Othonna natalensis* was place in § 5. *Scapigerae*, as one of six spp.[90] In Magoswana *et al.*'s simple key, *O. natalensis* would be grouped with the 'remaining species of *Othonna*'.

**Othonna natalensis** Sch.Bip. in Flora **27**(40): 771 (1844). Type: [South Africa:] '[*Krauss*] 442. ... Ad radicem mont. Tafelberge, Natal, alt. 1000 [Aug. 1839]' (P0004837 lectotype, BM000924715 – mounted with *Medley-Wood* 576, G00016815, K000306922 – mounted with *Baur* 521, K000306925 – mounted with 2 duplicates of *Cooper* 299, M0105094 – s.n., TUB005791), here lectotypified.[91] FIGURE 6.4.**13**.

---

[90] Harvey (1865) divided the genus into six apparently unranked, as '§', groups of species. Elsewhere in *Flora Capensis* (volume 3) he specified 'Sub-genus' (followed by an upper case Roman numeral) in some genera, but without using any symbol, in other genera infrageneric divisions were simply by using Roman capital numbers (I, II, III, etc.) followed by a Latin name (e.g. in *Cotula* L. – I. *Eucotula*, II. *Pleiogyne*, III. *Discocotula*). In yet other genera in the key he used 'Sect.' (sometimes 'Sec.') followed by an upper case Roman numeral, and a Latin name (e.g. in *Eriocephalus* L. – Sect. I. *Phaenogyne*, Sect. 2. *Cryptogyne*), but in the text the abbreviation was replaced using the section symbol, '§'. Complications only arise in the very large genera, such as *Helichrysum* Mill., where another, unnamed rank, appears between the 'sub-genus' and section. In *Muraltia* DC. (Polygalaceae), in volume 1 of *Flora Capensis*, this corresponds to 'Group'. Equally obvious, is the mix of ranks and name endings used in other families (e.g. the 'Order Caryophylleae' [≡ Family Caryophyllaceae] authored by Sonder, where sub-ordinal ranks are named 'Sileneae', 'Paronychiaceae', etc., and within *Pharnaceum* L., the sections, although numbered, have no Latin name, or in other genera the sections sometimes have English names). In yet other families both Sonder and Harvey have treated families and genera informally in non-taxonomic synoptic treatments (using asterisks and a synoptic phrase to separate groups of species).

[91] Whilst there is material in P (ex herb. Schultz Bipontinus), it is highly probable that there was originally material in B, now destroyed. This can be ascertained from previous works on Krauss collections and the many names published in Walpers' *Repertorium Botanices Systematicae* vol. 2 (Walpers, 1843). My selection of the duplicate in P as lectotype, provides a specimen with *Krauss*'s original label, and determined by Schultz Bipontinus.

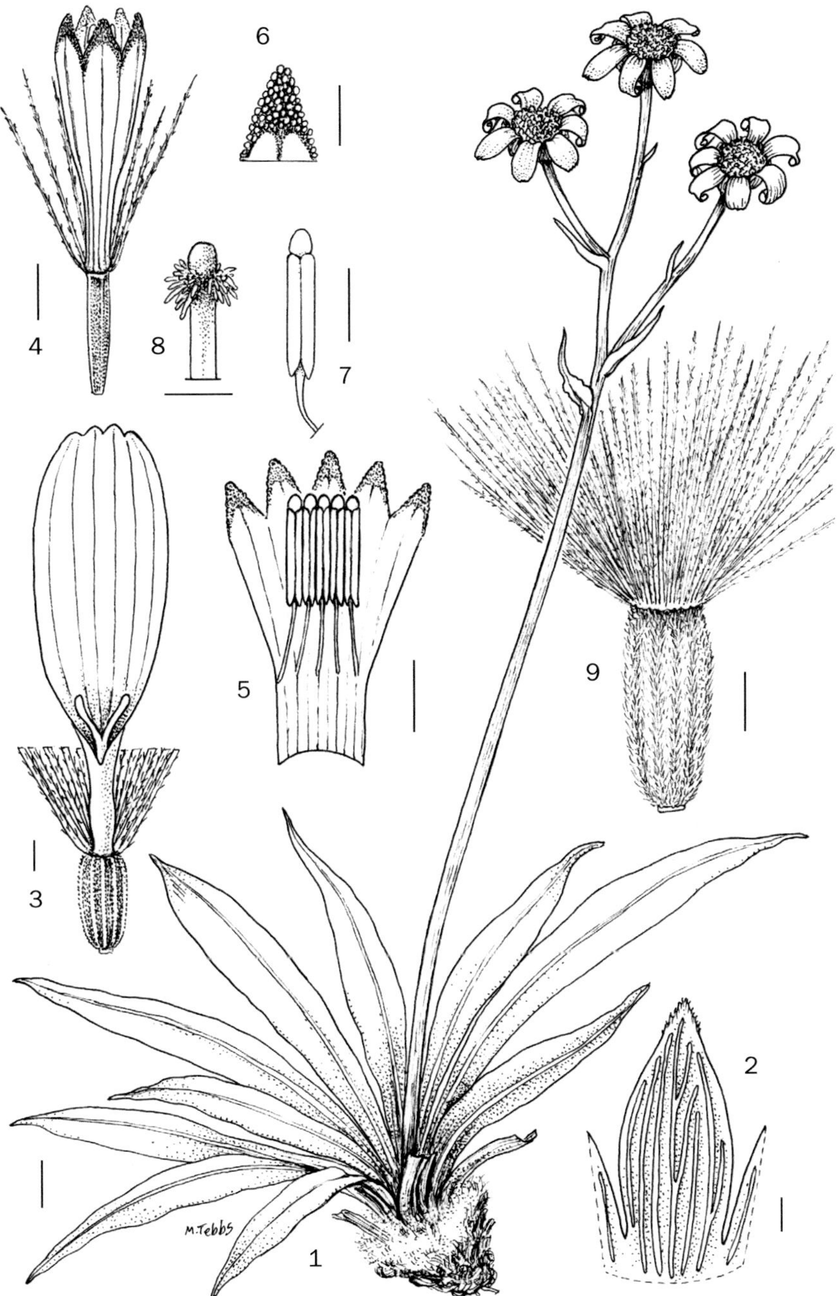

Fig. 6.4.**13**. OTHONNA NATALENSIS. 1, flowering plant; 2, phyllary; 3, ray floret, with part of pappus removed; disc floret; 4, disc floret; 5, disc floret corolla opened out to show attachment point of anther filaments; 6, detail of inner surface of apex of corolla lobe of disc floret; 7, separate stamen, showing absence of filament collar; 8, apical portion of connate style arms of disc floret; 9, mature ray floret achene. 1 from *Wild* 3557; 2–9 from *Venter & Vorster* 97. Scale bars: 1 = 10 mm; 9 = 3 mm; 2–8 = 1 mm. Drawn by Margaret Tebbs.

*Othonna scapigera* Harv., Thes. Cap. **1**: 10 (1859). Type: [South Africa:] 'Near Port Natal, [CBS] *Dr. Sutherland.*' (TCD0001083 holotype, K000306919 – mounted with *Galpin* 1029).

Perennial herb with a woody xylopodiaceous rootstock with few to many rosettes of leaves, plant crown to c. 20+ cm diam., rosette crown conspicuously densely woolly, hairs rufous-tawny, sometimes with remnant burnt petiole/pseudopetiole bases or oldest leaves from previous year's growth. Leaves erect, few to many per rosette, 5–31 × 0.4–5 cm, linear to broad lanceolate-oblong to obovate, larger leaves usually long-pseudopetiolate, ± fleshy, glabrous, glaucous, often reddish-purple towards base (although several herbarium specimens show a remarkable copper-colour all over – leaves, petioles/pseudopetioles, scapes and involucres), midrib prominent and paler than lamina beneath, markedly paler (and whitish green) than lamina above (in fresh material, but in herbarium material midrib usually brown or pale brown), margins entire, flat or rarely revolute (in some fresh material), very rarely dentate (not seen in herbarium material but very rarely in images of live material, and teeth very coarse, and solitary or few per margin), apices obtuse or acute, bases long-attenuate into pseudopetiole in narrower leaves. Inflorescence scapiform, 5–48 cm tall, peduncle/scape striate (at least in herbarium material) simple or few (1–2) -branched, ebracteolate or with few bracteoles, bracteoles 6–14 mm long, scale-like (although largest leaf-like 25–50 × 2.5–6.5 mm) and usually subtending branches, capitula terminal, solitary or few (usually to 3 or 4). Capitula heterogamous and radiate; involucre spherical to hemispherical in bud, becoming campanulate in flower, 15–20 mm diam.; phyllaries 8, 9.5–17 × 2.5–4 mm, lanceolate, united at base, often reddish-tinged from closed bud to post-anthesis, rounded over midrib, phyllaries spreading in fruit. Ray florets female, c. 8, limb c. 15+ × 4–7 mm, oblong and appearing distinctly clawed where limb narrows to corolla tube, bright yellow, conspicuously 5–8-veined when dry, spreading during anthesis and eventually rolled towards end of anthesis of central disc florets, limb apex scarcely 3-toothed. Disc florets ∞, functionally male, corollas c. 4 mm long, lemon- to golden-yellow, corolla lobes c. 0.4 mm long, erect; anther cylinder included within corolla limb, filaments attached to corolla tube in upper half, basal anther appendages scarcely short-sagittate, apical anther appendages obtuse, about as long as wide; style shaft glabrous, style base dilate, but scarcely a distinct node on top of a short nectary, style arms exserted just above corolla lobes, adnate (or minutely bifid) and sterile, apices conical or obtuse. Ray floret achenes 10–10.75 × 4–5 mm, body brown, ribbed, setuliferous, setulae myxogenic and mucilaginous when wetted, disc floret achenes c. 10 mm long, infertile and somewhat shrunken; carpopodium a narrow annulus, inconspicuous on disc floret achenes; pappus setae 5–17(23) mm long (elongating post-anthesis), those of disc florets significantly shorter than on ray floret achenes, conspicuous, persistent, rufescent or tawny coloured, those of ray floret achenes finely barbellate and short-united at base, spreading at maturity, of disc floret achenes slightly coarser barbellate and united at base.

**Zimbabwe**. C: Makoni Dist., Makoni, 1525 m, 18.xii.1946, *Wild* 878 (K, SRGH). E: Melsetter Dist., Pork Pie Mt., 1950 m, 7.x.1950, *Sturgeon & Panton* 30813 (K, SRGH).

Also known from South Africa (Eastern Cape, Free State, Gauteng, KwaZulu Natal, Limpopo, Mpumalana), Lesotho, and Eswatini. Open grassland, apparently frequently burnt, sometimes amongst rocks; 0–c. 2700 m throughout its range; flowering from August to October.

Conservation Status: LC (Least Concern) (Foden & Potter – National Assessment: Red List of South African Plants, vers. 2020.1. 2005) and Raimondo *et al.* (2009), since the species is widespread in South Africa and is present in the Flora area.

Vernacular names: Incama, Incamu, Naka, Natal geelbossie, Phela (Foden & Potter 2005), Natal baboon cabbage, Natal othonna, Natal daisy bush, Elephant's root.

Uses: Along with several other plants found in Lesotho, *Othonna natalensis*, known locally as Phela, is well-known medicinally, and is widely used by locally herbalists to help counter a wide range of conditions and ailments. Its over-collection in some areas has led to it facing local extinction. Veterinary uses, as a vermifuge, have also been reported.

The rust, *Puccinia othonnae* Doidge is hosted by *Othonna natalensis* in Gauteng Province, South Africa (Berndt & Uhlmann in Mycol. Progr. **5**: 154–177, 2006).

## 106. **LOPHOLAENA** DC.

**Lopholoaena** DC., Prodr. **6**: 335 (1838). —Moore in J. Bot. **67**(No. 802): 274–276 (1929). —Phillips & Smith in Trans. Roy. Soc. South Africa **21**(3): 221–238 (1933). —Dyer, Gen. S. Afr. Fl. Pl. **1**: 713 (1975). —Jeffrey in Kew Bull. **41**(4): 932 (1986). —Brusse in Nordic J. Bot. **11**(1): 79–83 (1991). —Jeffrey & Beentje in F.T.E.A., Compositae 3: 569–572 (2005). —Hind in Kew Bull. 80(4): 809–830 (2025).

Shrubs, small trees, or perennial herbs with thickened, sometimes woody, xylopodiaceous rootstock. Stems simple or poorly- to well-branched, young stems green or rarely suffused red, wingless or narrowly winged from short- or long-decurrent leaf bases, glabrous or rarely pubescent, sometimes essentially glabrous but with tufts of hairs in leaf axils. Leaves alternate, distichous or spiralled, sessile or subsessile to distinctly petiolate, base broad-amplexicaul or decurrent, fleshy, narrow-linear to suborbicular, glabrous, or rarely pubescent, obscurely to prominantly veined, venation of a single main vein or pinnate, margins entire or sometimes serrate. Inflorescence of solitary terminal capitula or of dense to lax corymbose panicles of few to many capitula, rachis or pedicel bracteate, capitula occassionally surrounded by upper foliage leaves or large foliaceous bracteoles rarely obscuring involucre. Capitula homogamous, discoid, few-(5–7) to many-(50) flowered; involucre uniseriate, ecalyculate, cylindrical, sometimes very broadly so; phyllaries uniseriate, few (3–10), connate at base, sometimes appearing connate to ⅔ length but margins still free and imbicate, often conspicuously elongating post-anthesis, usually abaxially keeled, very rarely with a dorsal crest or wing above midrib, margins usually scarious and interlocking with marginal groove of thickened portion above midrib of adjacent phyllaries; receptacle flat to slightly convex, areolate, glabrous, epaleaceous. Florets 3–50, hermaphrodite; corollas white, cream, pink or purplish, cylindrical but widening into campanulate limb, exceeding pappus at anthesis but overtopped by pappus in fruit; corolla lobes 5, deltoid, subacute, glabrous, sometimes differently coloured to throat; filament collar balusterform, apical anther appendages ovate to ovate-oblong, anther bases obtuse; style base enlarged, but scarcely nodal, glabrous, style shaft glabrous, filiform, style arms linear to subulate, spreading to recurved, often coiled in age, flattened and papillose marginally above, densely long-papillose abaxially, apices tapering, obtuse, apical appendages filiform papillose, papillae longer at midpoint, noticeably shorter towards apex. Achenes elliptic-oblong, obscurely ribbed, glabrous or densely setuliferous, sometimes body obscured or only sinuses densely setuliferous, setulae of 'twin-hairs', eglandular and non-myxogenic, apices assymetric and scarcely divided, apical cells acute; pappus setae 2–3-seriate, numerous, capillary, finely barbellate, white or straw-coloured, often elongating markedly in fruit, persistent (but sometimes recorded as caducous in some collections and most probably fragile, e.g. as in *L. phippsii*), apical cells acute.

A genus of 18 spp., mostly from southern Africa, but with two from West Africa, and three from Tropical East Africa; six are currently recognized from the Flora area. Jeffrey & Beentje's lower species number for the genus is most probably because post-Phillips & Smith's taxa were not considered. Phillips & Smith (1933) provided a practical infrageneric taxonomy and recognized two sections based on the arrangement of capitula; that infrageneric treatment is not followed here.

1. Inflorescences of lax panicles or of dense corymbs. . . . . . . . . . . . . . . . . . . . . . . . . . . . 2
– Inflorescences usually with solitary capitula or 2 or 3 together and subracemose (with up to 6 capitula) . . . . . . . . . . . . . . . . . . . . . . . . . . . . . . . . . . . . . . . . . . . . . . . . . . . 3
2. Perennial herbs, suffrutices or subshrubs with annual flowering stems and a woody xylopodiaceous rootstock; stems winged; leaves elliptic or slightly obovate, pseudopetiolate; capitula 3- or 4-flowered; involucre c. 2–2.5 mm diam. in flower; phyllaries 3 or 4 . . . . . . . . . . . . . . . . . . . . . . . . . . . . . . . . . . . . . . . . . . . . . . . . . . . . . . . . **1.** *trianthema*
– Shrub or small tree; stems wingless; leaves obovate to elliptic-oblong, conspicuously petiolate; capitula 7-flowered; involucre c. 5 mm diam.; phyllaries 5 . . . . . . .**2.** *brickellioides*
3. Leaves prominently 3–5-veined from base . . . . . . . . . . . . . . . . . . . . . . . . . . . . . . . . . . . 4
– Leaves prominently 1-veined (never more) from base upwards; leaf bases decurrent on stem as distinct wing . . . . . . . . . . . . . . . . . . . . . . . . . . . . . . . . . . . . . . . . . . . . . . . . . . . 5
4. Shrubs or treelet; leaves obovate, apices always rounded or obtuse, (1)3-veined from base; involucre few-bracteolate on subtending pedicel. . . . . . . . . . . . . . . . . . . . . . **3.** *coriifolia*

   – Perennial herbs or suffrutices with annual flowering stems; leaves elliptic or lanceolate, apices acute; pedicels conspicuously bracteolate, bracteoles numerous beneath involucre . . . . . . . . . . . . . . . . . . . . . . . . . . . . . . . . . . . . . . . . . . . . . . . . . . . . . . . . . **4.** *acutifolia*
   5. Capitula 20–25-flowered; shrubs. . . . . . . . . . . . . . . . . . . . . . . . . . . . . . . . . . . . .**5.** *phippsii*
   – Capitula 3–6-flowered; suffrutices or subshrubs with a woody xylopodium . . . .**6.** *whyteana*

1. **Lopholaena trianthema** (O.Hoffm.) B.L.Burtt in Notes Roy. Bot. Gard. Edinburgh **34**(1): 85 (1975). —Cribb & Leedal, Mountain Fl. S. Tanzania: 161, pl. 44B (1982). —Jeffrey in Kew Bull. **41**(4): 932 (1986). —Lisowski, (Asterac. Fl. Afr. Centr. 2) Fragm. Flor. Geobot. **36** Suppl. 1: 436 (1991). Type: [Tanzania:] 'Unyika: Umalila, auf welligem Hochplateau, um 1900 m ([*Goetze*] n. 1352. – Blühend am 21. Oct. 1899).' (E00239833 lectotype, B†, BR6423353, K000307179 – ex E, P00111799), lectotypified by Hind (2025: 815).[92]

    *Senecio trianthemos* O.Hoffm. in Bot. Jahrb. Syst. **30** (3): 437 (1901), as '*S.* (§ *Kleinia*) *trianthemos*'.

Woody perennial herb (often described as a shrub), stems 100–150 cm tall, erect, wingless; rootstock woody (most probably a xylopodium). Leaves somewhat succulent, sessile, elliptic or slightly obovate, 4–10 × 1–4 cm, cuneate-attenuate into a petioloid base, entire, apex obtuse, minutely apiculate, glabrous, somewhat 3-veined. Inflorescences of numerous capitula in dense compound corymbs, ultimate branchlets ± umbelliform, bracteate, bract-axils lanate, pedicellate, pedicels 7–17 mm long. Capitula homogamous and discoid; involucre cylindrical, 8–10(15) × 2–2.5 mm, green, glabrous; phyllaries 3(4), green, c. 8–10 × 2–3 mm at anthesis, growing to 14–15 × 4–5 mm in fruit, eventually spreading in fruit, obtuse and margins laciniate in free apical portions, and apex papillate adaxially. Florets 3–4, corollas c. 12 mm long, limb c. 7 mm long, white, cream to orangish, exserted and exceeding pappus at anthesis, corolla lobes c. 1.2 mm long. Achenes ellipsoid-cylindrical, 6–6.5 × 3 mm, obtusely ribbed, setuliferous in sinuses, more densely so at base of achene and beneath apical callus (described as glabrescent, probably when ribs are seen as glabrous); pappus setae c. 8 mm long in flower, conspicuous and elongating to 12–20 mm long in fruit, very finely barbellate in upper half, but essentially smooth in basal third, persistent, white.

    **Zambia**. N: Abercorn Dist. Track to Vomo Gap., alt. 1500 m, 8.vii.1961, *Richards* 15302 (K). **Malawi**. N: 72 km S of Mzimba, c. 12°20'S, 33°40'E, alt. 1500 m, 31.x.1966, *Gillett* 17521 (EA, K).

    Also in D.R. Congo and Tanzania. Pyrophyte in upland grassland, *Protea* bushland and open sites in light woodland, old cultivation areas, miombo woodland, often on sandy soils; 1400–2150 m; flowering July–September (November).

    Conservation Status: A widespread, moderately well-collected species and locally common or frequent throughout its range. LC (Least Concern). In the Flora area the collections, in Zambia, are all concentrated in the Abercorn [= Mbala] area.

2. **Lopholaena brickellioides** S.Moore in J. Linn. Soc. Bot. **40**(275): 118 (1911). —Phillips & Smith in Trans. Roy. Soc. South Africa **21**(3): 229 (1933). Types: 'Hab. In crags of the Chimanimani Mountains, 7000 ft.; Mt. Pene, 7000 ft.; in fl. Sept., Oct.; [*Swynnerton*] nos. 1802, 6133.' 26.ix.1906, *Swynnerton* 1802 (BM000924557 lectotype, K000307177, SRGH0028757-0), first-step lectotypification by Phillips & Smith (1934: 229),[93] second-step lectotypification by Hind (2025: 816).

---

[92] Lisowski (1991: 436) and Jeffrey & Beentje (2005: 572), stated that the holotype was in B (destroyed), and an isotype in K, apparently unaware of the other three duplicates. I selected the E duplicate as it is a markedly leafy stem apex with a large inflorescence, well-representing the species.

[93] As *Swynnerton* 1802, citing 'Type (BM, K)'.

Large shrub to small tree, to 3 m+, glabrous throughout. Stems laxly branched towards tips, wingless, striate. Leaves alternate to spiralled, obovate to elliptic-oblong, to 2–10 × 1.5–6.7 cm, much reduced upwards, lower gradually tapering at base of lamina, upper cuneate, petiolate, petiole 5–40 mm long, uppermost sessile on rounded or subtruncate base, lamina fleshy, glaucous, midrib prominent on both surfaces, veination actinodromous to brochidodromus, lamina apices rounded, short-apiculate. Inflorescence a relatively lax, many-headed, conspicuously bracteolate panicle to 20 × 15 cm, lowermost bracts leaf-like, to 25 mm long, inflorescence branches conspicuously overtopping subtending bracts, much-reduced and bracteolate towards capitula, capitula sessile to subsessile, pedicels 0–3 mm long. Capitula homogamous and discoid, 7-flowered; involucre c. 10 × c. 5 mm, cylindrical; phyllaries 5, 4–8 × 1.2–1.7(2.2) mm, oblong-linear, flat, ecarinate, tips papillate on margins and abaxial portion of apex, apices acute, penicillate, phyllaries spreading upon achene loss; corollas pale-purple (often described as white on specimen labels, doubtless drying to that colour), tube 9 mm long, as long as pappus setae, corolla limb c. 1.5 mm long, corolla lobes c. 0.8 mm long; anther cylinder conspicuously exserted from throat. Achenes 5-ribbed, 5 × 1.2–1.5 mm, moderately to densely adpressed-setuliferous particularly between ribs, setulae apices scarcely assymetric, acute; pappus setae 5–8 mm long at anthesis (as long as or slightly longer than corolla tube), white.

**Zimbabwe**. E: Mt. Peza, 1580 m, 16.x.1950, *Wild* 30407 (K, SRGH). **Mozambique**. MS: Manica, Sussundenga Dist., Chimanimani mountains, slopes of Mt Peza, 19°44'40.0"S, 33°00'55.9"E, 1567 m, 17.iv.2014, *Timberlake et al.* 5964 (K, SRGH).

Endemic to the Chimanimani (Sussundenga) and Nyanga (Melsetter) Districts of the Chimanimani Mountain range. Forest margins, on quartzitic soils, and amongst rocks; 1565–2133 m; flowering April–October.

Conservation Status: Darbyshire *et al.* (2019) recorded the species as 'NE2+3' – with a 'global range < 10,000 km and 'known from five sites or fewer', although not providing a provisional conservation status statement for this 'near endemic' (NE) species in the Supplementary material 1; a global assessment is still required.

3. **Lopholaena coriifolia** (Sond.) E.Phillips & C.A.Sm. in Trans. Roy. Soc. South Africa **21**(3): 234 (1933). Types: [South Africa:] 'Hab. Magalisberg, *Burke & Zeyer. Zey.*[*her*] 944 [*sic* = '994']. (Herb. Hk, Sd.)' *Zeyher* 944 (S08-6737[94] lectotype, BM000924546), lectotypified by Hind (2025: 822); *Burke* s.n. (K000307233[95]).[96]

 *Othonna? coriifolia* Sond. in Harvey & Sonder, Fl. Cap. **3**: 333 (1865).

 *Othonna bainesii* Oliv. & Hiern in F.T.A. **3**: 423 (1877). Type: 'South Central? South African Gold Fields, *Baines!*' (K000307176 holotype).[97]

 *Lopholaena randii* S.Moore in J. Bot. **41**(484): 133 (1903). Type: [South Africa:] 'Hab. Rocky places in the neighbourhood of Johannesburg. [Rand] No. 746.' (BM000924547 holotype).

 *Lopholaena randii* var. *brachycephala* S.Moore in J. Bot. **67**(802): 275 (1929). Type: [Zimbabwe:] 'Hab. Rhodesia, Makoni, on exposed summit of mountain; *F. Eyles*, 741, in Herb. Mus. Brit.' (BM000924554 holotype, K002655507, SAM).

---

94 '994/ Othonna coriifolia Sond/ Magalisberg/Juli/Zeyher', 'Nur 1 kl. Ex. vorhanden' – in Sonder's hand.

95 The label annotated in ink in two different hands, 'Othonna ?', 'July Macalisberg'[presumably by Burke], and in pencil as 'A. 36' , 'coriifolia Sd' and 'Magalies Berg' in a different hand, the sheet further annotated as 'nr. Pretoria West.'

96 One sheet of type material in K is unnumbered, but is clearly a *Burke* collection (as indicated on the sheet), localized as 'Macalisberg [*sic*]'.

97 The non-specific collection locality (typical of many plant collections made by Baines, and simply based on the title of the expedition) has led to speculation on the sheet as to where it might have been collected. The pencilled suggestion of 'Transvaal, probably' was met with a counter comment, in black ink, of '(NO!)'. The collection was most probably made in Zimbabwe. Phillips & Smith (1933: 235) indicated Mashonaland, a region in northeastern Zimbabwe bordering Zambia and Mozambique.

*Lopholaena dehniae* Merxm. in Proc. & Trans. Rhodesia Sci. Assoc. **43**: 66 (1951). Type: [Zimbabwe:] 'Southern Rhodesia: Marandellas; on a kopje, low bush between rocks. leg. *G. Dehn* 17.8.1941, D.n.345.' (M0105366 holotype).[98]

Well-branched shrub (or treelet) to c. 1(2) m tall, glabrous throughout. Stems to 4 cm diam. at base, branches ascending, often covered with old, withered leaves (missing in many herbarium specimens, but seen in verified habitat photographs). Leaves alternate and scattered, obovate to oblong-obovate, to 20–40(60) × 12–24 mm, coriaceous to subfleshy, base ± conspicuously narrowed into a pseudopetiole, obscurely short-decurrent on stem but not to next node below, essentially (1)3-veined from base, midrib prominent almost throughout and paler than lamina, lamina glaucescent, entire, apex obtuse to rounded. Inflorescences of solitary capitula or capitula in terminal clusters or on short side branches aggregated towards apices of main stems or subterminal side branches. Capitula homogamous and discoid, 14–20-flowered; involucre c. 12 × 5–6 mm, campanulate-cylindrical; phyllaries 5(6), oblong to ovate-lanceolate, 10–13 (but clearly elongating to 14 mm+, post anthesis) × 4–5 mm, yellowish-green, rounded above midrib (and clearly lacking a keel), apices short-acuminate. Corollas white to cream (once described as yellow – *Biegel* 2149, K), c. 16 × c. 1 mm at base and narrowing upwards to throat, throat c. 6 mm long and expanding gradually to distal portion of corolla limb. Achenes 8–9 mm long, fawn-coloured, densely white-setuliferous (Phillips & Smith, 1933: 234 suggested that they were rarely glabrous, but I have not seen this), setulae adpressed, apical cells acute, connate; pappus setae 12–14 mm at anthesis, elongating to 17–20 mm long and markedly exserted from involucre when mature.

**Zimbabwe**. W: Matobo, plentiful along the old Kwanda Rd., 3.x.1971, *Leach & Cannell* 14848 (SRGH). C: Salisbury, Widdycombe Road, 9.ix.1946, *Wild* 1210 (K, SRGH). S: Gutu, Chikwanda Reserve, s. dat., *Robinson* 17 (= SRGH 17329) (SRGH). **Mozambique.** M: Maputo (Lourenço-Marques), Namaacha, 4.vi.1957, *Grandvaux Barbosa & de Lemos* 7589 (LMJ, SRGH).

Widely collected in Zimbabwe, as well as two collections from Mozambique. Also from Angola, Eswatini, South Africa (Cape Province, Transvaal). Wooded grassland, rocky hillsides and often conspicuous amongst rocks and boulders especially when in fruit, and on sandy or poor or over-grazed soils; 760–1950 m; flowering June–November.

Conservation Status: LC (Least Concern) – Foden & Potter (2005), Raimondo *et al.* (2009).

Vernacular names: Chigunguru (Shona), Mugakatombo (Shona), Mulwiradundu (Shone), Nyakatondo (Shona), Leather-leaved fluff-bush, Pluisiesbos (Foden & Potter 2005), Small-leaved fluff-bush, Kleinbaarpluisbos, Pluisbossie.

4. **Lopholaena acutifolia** R.E.Fr., Wiss. Ergebn. Schwed. Rhodesia-Kongo-Exped. 1911-1912, **1**: 341 (1916). —Phillips & Smith in Trans. Roy. Soc. South Africa **21**(3): 232–233 (1933). Type: [Zambia:] 'Northwest-Rhodesia: Bwana Mkubwa, auf mit Gräsern und Kräutern reichlich bewachsener Lichtung in dem Trockenwald (blühend 16. Aug. – [*R. E. Fries*] n. 365).' (UPS046495 holotype, B†). FIGURE 6.4.**14**.

> *Lopholaena alata* P.A.Duvign. in Bull. Soc. Roy. Bot. Belgique **91**(2): 148 (1959). Type: [D.R. Congo:] 'Entre Shinkolobwe et Swambo, forêt à *Brachystegia* sur sol rouge sur calcaire de Kakontwe, *Duvigneaud* 2034 D, VII-1956 (Holotype, BRLU).' (BRLU90020780 – 'Entre Shinko et Mindingi ...', and only dated to '1956' holotype, BR8877468 – 'Entre Shinko et Mindingi ...', and only dated to '1956').[99]

---

[98] Merxmüller (1951: 67) also added the following note concerning Dehn's coloured painting and drawing of dissections on the sheet: 'The drawing shows a plant in fruit, too, but it is doubtful whether it is the same plant.'

[99] Durigneaud's original label determination (from 1957) was 'Derriksia alata n. sp.', the holotype with a handwritten description (in blue ink) for the ined. genus '*Derriksia*' (comparing it with 'certains Notonia africains'), and a second label (in black ink) with Durigneaud's diagnoses as to where the new genus belonged. Both the holotype and isotype were '2034 D1', not simply 2034 D' as in the protologue, the 'D1' written with a much thicker pen. The label annotation suggests that this may have been a *Duvigneaud & Timperman* collection.

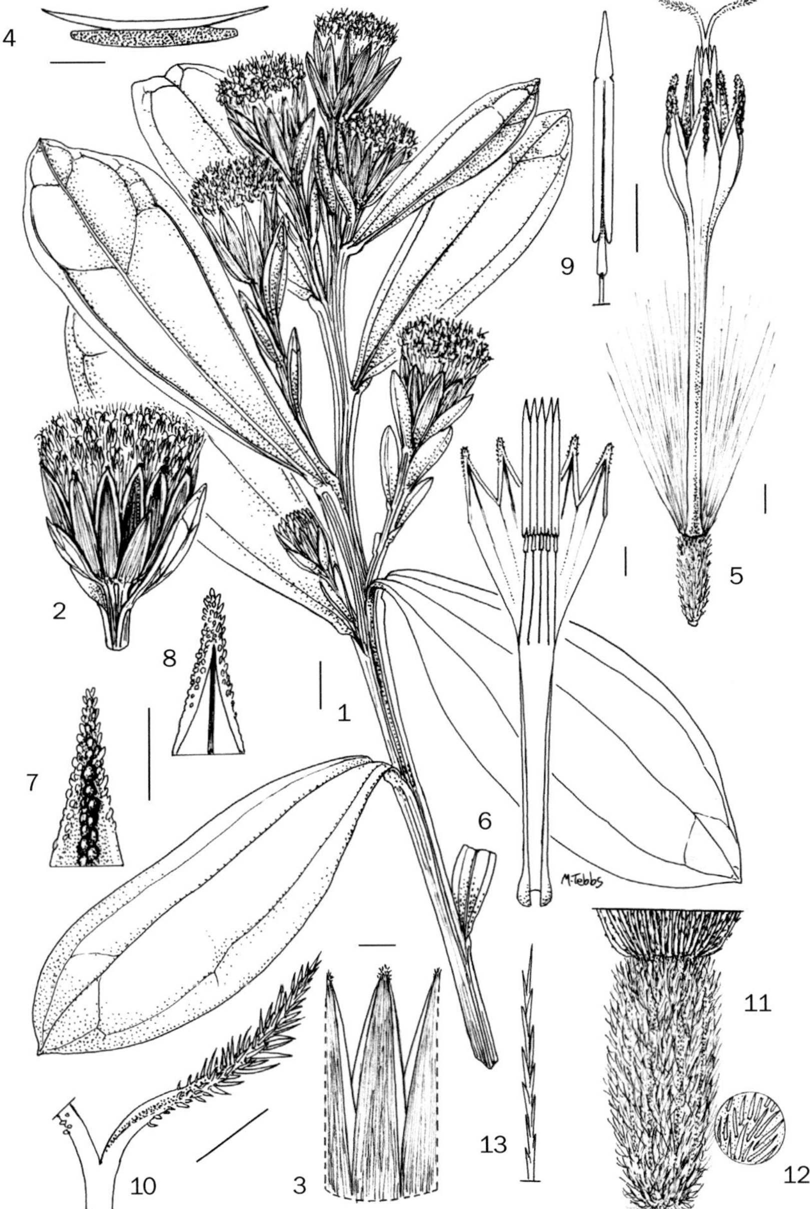



Fig. 6.4.**14**. LOPHOLAENA ACUTIFOLIA. 1, apex of flowering stem showing long-decurrent leaf bases and winged stem, as well as multi-bracteate pedicels; 2, capitulum; 3, detail of interlocking phyllaries; 4, detail of cross section of phyllary, adaxial surface uppermost, showing characteristic slots beneath thickened middle of phyllary where adjoining phyllary margins interlock; 5, floret; 6, corolla opened out to show attachment point of filaments; 7, detail of outer surface of corolla lobe; 8, detail of inner surface of corolla lobe; 9, stamen, showing only upper portion of filament; 10, detail of style arm; 11, achene and base of pappus setae; 12, detail of surface of achene showing moderately dense 'twin-hairs'; 13, detail of apex of pappus seta. 2, 3, 5–8, 11–13 from *Fanshaw* 1383; 1, 4, 9–10 from *Fanshaw* F3316. Scale bars: 1 = 10 mm; 2 = 5 mm; 3, 11 = 2 mm; 4 – 10 = 1 mm. Drawn by Margaret Tebbs.

Perennial herb or subshrub, 30–40 cm tall, glabrous. Stems rigid, erect, branched, branches erect, striate, leafy. Leaves sessile, lanceolate or narrow-lanceolate, 6–10 × 1–2.5 cm, base cuneate and decurrent into c. 2 mm wide wing, lamina glabrous, light green, margins entire, apex acute. Capitula terminal to main branch and solitary in upper leaf axils, pedicels 1–3 cm long, conspicuously bracteolate, uppermost bracteoles shorter than, or ± overtopping phyllaries, remainder numerous and clothing/obscuring pedicel; phyllaries glabrous, flat, c. 10, 14–17 × 2–5 mm, lanceolate to linear-lanceolate, ± glaucous, margins hyaline, apices acute. Florets c. 30+ per capitulum, corollas c. 15 mm long, limb expanding significantly to throat, 3–4 mm long (to base of corolla lobes), corolla lobes c. 2 mm long; anthers c. 4 mm long, apical anther appendage c. 1.2 mm long acute; style c. 16 mm long, style arms c. 3 mm. Achenes c. 9.5 mm long, c. 3 mm broad, ribbed, ribs rounded and glabrescent, densely setuliferous in sinuses between ribs (Fries described the ovary as glabrous); pappus setae 8–10 mm long (at anthesis), 13–17 mm at fruiting stage, white.

**Zambia**. W: Luanshya, 19.vii.[19]54, *Fanshawe* F1383 (K).

Rarely collected in the Copperbelt area of Zambia through to the D.R. Congo where it is known from few collections; open woodland, grassy steppe savannahs on sand and on copper-bearing soils; c. 1200–1400 m; flowering July–August.

Conservation Status: Although now considered a relatively widespread species, more field data needs to be collected to ascertain population size, especially as the few collections available give no indication of population size. Clear threats exist in mining areas where populations exist on copper and cobalt rich deposits. At present it is best recorded as DD (Data Deficient).

Duvigneaud's protologue of *L. alata* is supplemented, in his Fig. 5, by a small sketch of a single capitulum (with numerous subinvolucral bracts) subtended by a solitary leaf. The leaf venation is not accurate, as both the type material of *L. alata* and the *Fanshawe* collections from the FZ area show that it is very distinctly 3-veined to the base (with a prominent midrib) and the leaf apex apiculate, together with very distinctly winged stems and pedicels, which are far from clear in the figure. *Lopholaena alata* and the *Fanshawe* collections are considered conspecific with *L. acutifolia*, hence the synonymy (see Hind 2025).

5. **Lopholaena phippsii** D.J.N.Hind in Kew Bull. 80(4): 811 (2025). Type: [Zimbabwe:] 'Melsetter. Chimanimani. Point 71. Crevices of rock pavement, at summit. Shrub 2′to 4′high. Leaves glaucous, involucrae green, corolla off-white, A sepia, Quite common. 8000′. 16.3.1957. *J.B.Phipps* 646.' (K000374763 holotype, CAS-SRGH 28776). FIGURE 6.4.**15**.

Shrub, 0.6–1.20 m tall. Stems few-branched, wingless, glabrous. Leaves alternate, sessile to pseudopetiolate, lamina obovate, 18–60 × 13–37 mm, lamina glabrous, glaucous, single-veined and lacking any visible secondary venation, midrib prominent, base pseudopetiolate/sessile and short-decurrent on stem, but stems not conspicuously winged, apex rounded or obtuse, margins entire. Inflorescences usually of few, terminal, loose clusters of 3–6 capitula, each pedicel with 2 or more highly-reduced leaf-like bracteoles just beneath involucre, c. 12 × 4–7 mm wide, 1-veined; pedicels apparently increasing in length post-anthesis (see *Whellan* 2101, SRGH) and continuing to branch with additional capitula produced, eventualy appearing cymose. Capitula homogamous and discoid, 20–25-flowered; involucre broadly cylindrical to campanulate; phyllaries uniseriate, 7 or 8, valvate, 10–12 × 2.5–4.7 mm, margins scarious, phyllary apices slightly incurved at anthesis; receptacle solid, scarcely convex, epaleaceous, glabrous. Florets all hermaphrodite and fertile, corollas actinomorphic, 'off-white', corolla tube c. 5 × 0.5 mm at slightly dilated base, 0.3 mm diam. beneath limb, corolla limb c. 2 mm long to base of corolla lobes, corolla lobes 5, c. 1.5 × c. 0.5 mm wide at base, apices thickened; anther cylinder scarcely exserted from corolla throat, apical anther appendages narrowly triangular, apices acute, basal anther appendages rounded; filament collar balusterform, filaments attached to corolla tube in lower part of upper half of corolla tube just below base of expanded portion of limb, glabrous; style shaft glabrous throughout, style base expanded into basal node, glabrous. Achenes cylindrical and narrowing in lowermost portion towards base, c. 4 × c. 1.2 mm, glabrous, c. 12-ribbed, ribs rounded; carpopodium inconspicuous (in submature achenes); pappus setae c. 8 mm long (mature length unknown, but on sub-mature achenes setae appear fragile, breaking off at apical callus of achene), barbellate, cells finely acute.

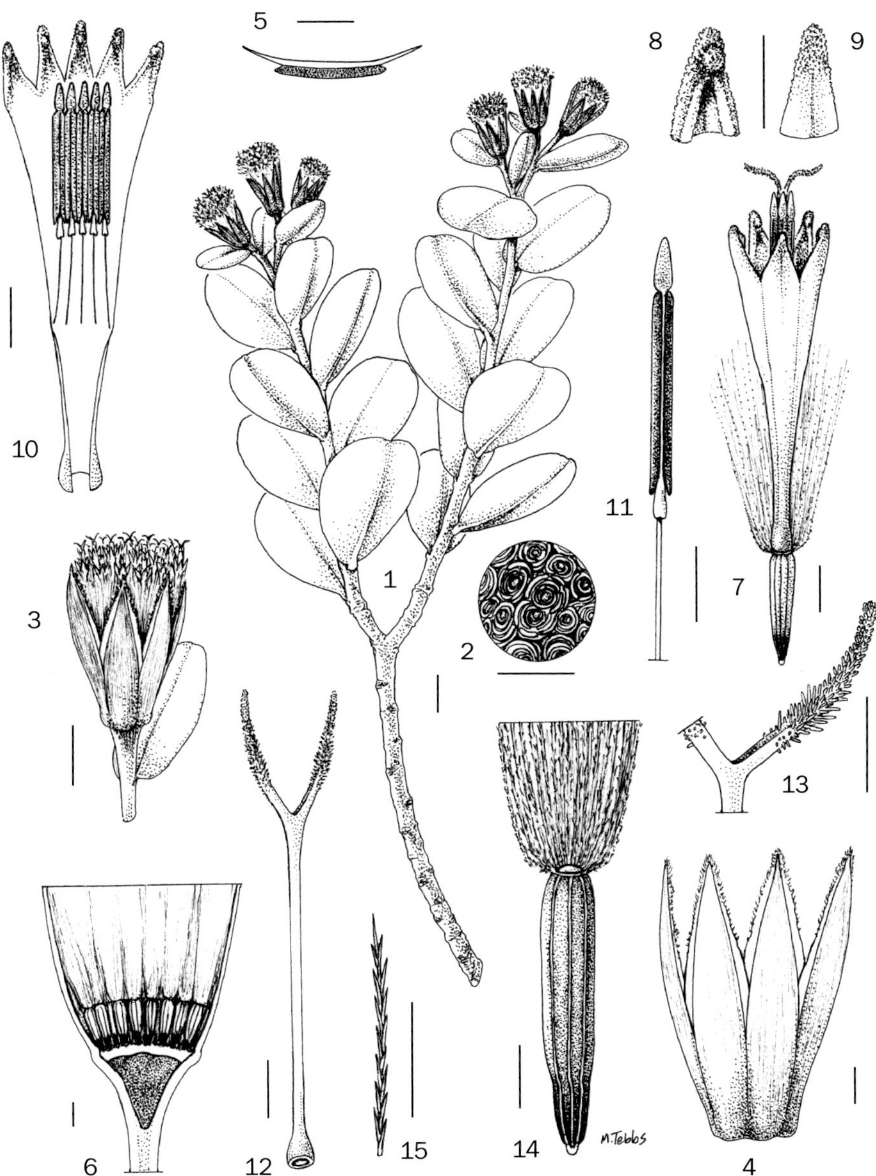

Fig. 6.4.**15**. LOPHOLAENA PHIPPSII. 1, habit of flowering portion of side shoot/main stem; 2, detail of lower leaf surface; 3, capitulum; 4, phyllaries, showing interlocking bases (when dry); 5, t.a. phyllary; 6, l.s. capitulum; 7, floret; 8, corolla lobe – adaxial surface; 9, corolla lobe – abaxial surface; 10, floret opened out showing attachment point of anther filaments; 11, stamen; 12, style; 13, detail of style arm; 14, achene and base of pappus setae; 15, detail of apex of pappus seta. All from *Phipps* 646. Scale bars: 1 = mm; 3 = 5 mm; 2, 4, 5, 6, 9, 10, 11 = 1 mm; 7, 8, 12, 13 = 0.5 mm. Drawn by Margaret Tebbs.

**Zimbabwe**. E: Melsetter, Chimanimani, point 71, 8000′, 16.iii.1957, *Phipps* 646 (K, SRGH).[100]

The species is only known from three collections made on the summit of Mount Binga, and is considered co-endemic to both Zimbabwe and Mozambique; in crevices of rock pavement on the mountain summit; c. 2438 m; flowering March–May.

Conservation Status: This species is currently only known from three collections in Zimbabwe. Whilst it is highly unlikely that there is any threat to the habitat at the summit of Mount Binga, especially because of access issues, the population size may very well be restricted, although recorded as 'Quite common' by Phipps, and no mention was made of colony size by Weiste or Whellan. It is felt best to record it as DD (Data Deficient) until further fieldwork can be undertaken in the type locality, on both sides of the border between Zimbabwe and Mozambique, to determine the population size, and any obvious potential threats to the species.

6. **Lopholaena whyteana** (Britten) E.Phillips & C.A.Sm. in Trans. Roy. Soc. South Africa **21**(3): 236 (1933). Type: [Malawi:] 'Hab. Milanji, 6000 ft. [*Whyte*] No. 62.' (BM000924553 – 'Oct. 1891' holotype, K000307234 – s.n., mounted with *Clonnie* 21).

　　*Othonna whyteana* Britten in Trans. Linn. Soc. London, Bot. 4(1): 21 & pl. IV, Figs 1 & 2 (1894).

　　*Senecio dolichopappus* O.Hoffm. in Bot. Jahrb. Syst. 30(3–4): 438, Taf. XVIII (1901). Types: [Malawi:] 'Usafua: Beya-Berg, auf trockenen Abhängen mit kurzem Gras, um 2700 m ([*Goetze*] n. 1075. – Blühend am 28. Juni 1899). [Tanzania:] Umalila: Uwurungu-Berg, auf rasigen Abhängen, um 2200 m ([*Goetze*] n. 1462. – Mit jungen früchten am 18. Nov. 1899).'[101] *Goetze* 1462 (BM000924551 lectotype, B†,[102] BR8876454, E00239834, L3654794, P00122002), first-step lectotypification by Phillips & Smith (1933: 237),[103] second-step, re-selected[104] by Hind (2025: 823); *Goetze* 1075 (B†, BM000924552, BR8877130, E00239835, P00122001 syntypes).

　　'*Lopholaena dolichopappa* (O.Hoffm.) S.Moore in J. Bot. **41**(484): 134 (1903)'.[105]

　　*Lopholaena dolichopappa* (O.Hoffm.) S.Moore in Bull. Herb. Boissier, sér. 2, **4**(10): 1021 (1904).

　　*Lopholaena dolichopappa* (O.Hoffm.) Prain, Index Kew. Suppl. **3**: 106 (1908), nom. illeg., isonym.

　　*Lopholaena pauciflora* Thell. in Vierteljahrsschr. Naturf. Ges. Zürich **66**: 242 (1921). Type: [Tanzania:] 'TROPISCHES AFRIKA: Nyassa-Hochland, Station Kyimbila, 1913, *A. Stolz* 2267.' (Z000003630

---

[100] The label was annotated as 'Lopholaena sp. = G.H. 9136', and Torre provided a det. slip indicating '*Lopholaena* aff. *disticha* (N.E.Br.) S.Moore' – which the material clearly isn't. It is now clear that this is a reference to a collection of *Miss Weiste* 9136 in SRGH, made in 1942.

[101] Phillips & Smith (1933: 237) discussed the differences between the two *Goetze* collections, eventually (Phillips & Smith 1933: 236) choosing *Goetze* 1075 as the type collection of *L. brevipes*, q.v. Hind (2025: 822).

[102] According to Phillips & Smith (1933: 237).

[103] Phillips & Smith (1933: 237): *Goetze* 1462, 'Be [= B], BM'.

[104] Second-step lectotypification by Jeffrey (1986: 932): 'lectotype B†; isolectotype BM – chosen here'. Jeffrey & Beentje (2005: 570) repeated Jeffrey's earlier statement. Choosing a destroyed specimen as lectotype is considered contrary to the Code, since it must be an extant specimen, effectively permitting a repeated second-step lectotypification.

[105] Although this record appeared in *Index Kewensis* (see Prain, Index Kew. Suppl. **3**: 106. 1908), Moore made no such combination in the *Journal of Botany*, not even mentioning the basionym, *Senecio dolichopappus*. The combination was effectively made by Prain, albeit an isonym; the name also appeared in Phillips & Smith (1933: 236).

lectotype,[106, 107] HBG504804, L3654793, M0105365, WAG0000615, Z000003631[108]), first-step lectotypification by Phillips & Smith (1933: 238), second-step lectotypification by Hind (2025: 823).

*Lopholaena brevipes* E.Phillips & C.A.Sm. in Trans. Roy. Soc. South Africa 21(3): 236 (1933). Type: [Tanzania:] 'South Tropical Africa.–Tanganyika: Usafua, on dry slopes covered with short grass on Beya Mountain, 2700 m., June, *Goetze* 1075! Type (Be[= B], BM).' (BM000924552 lectotype, B†,[109] BR8877130, E00239835, K000307180, P00122001),[110] lectotypified by Hind (2025: 823).

Subshrub (rarely a well-branched small shrub) to 30–75 cm tall from a perennial woody rootstock. Stems several, trailing or erect, becoming somewhat woody with age, often yellowish green, glabrous, densely leafy, often with leafy short-shoots subtended by a conspicuously large main stem leaf, main stem internodes to 30 mm, short-shoot internodes very short. Leaves alternate, sessile, obovate-lanceolate, narrow- or very narrowly-oblanceolate to spatulate, to almost linear, 15–60 × 3–15 mm, somewhat fleshy in some material (sometimes described as succulent), attenuate into petioloid base or pseudopetiolate, single-veined, glabrous except for a few hairs in leaf axils, margins entire, apex obtuse or minutely apiculate. Inflorescences of solitary terminal capitula (on most axillary side shoots at stem apices); pedicels 1.2–8 cm long, erect, ebracteolate (or rarely with 1 or 2 scale-like bracteoles), gradually and slightly expanded below capitula, spreading or reflex to nodding at anthesis, and erect, often thickening post-anthesis. Capitula homogamous and discoid; involucres cylindrical, 9.5–15 × 2.5–3 mm; phyllaries 4 or 5, green or brown, 10–12 × 2–3 mm (significantly enlarging in fruit to 20–25 × 5 mm), oblong or lanceolate, glabrous, often purplish, margins white, finely shortly-ciliate, acute to shortly acuminate, mature phyllaries eventually spreading after achene loss; receptacle flat to ± convex, glabrous, alveolate. Florets 3–6, corollas c. 16 mm long, white, sometimes tinged lilac, exserted and exceeding pappus at anthesis, corolla limb 5–7 mm long; anthers purple. Achenes obovate-cylindrical, 7.5– 8.5 × 2.8–3.5 mm, broadly shallowly ribbed, glabrous (described by Jeffrey & Beentje, 2005: 579, as 'glabrescent with age'); pappus setae c. 10 mm in flowering capitula and scarcely exceeding phyllaries, markedly elongating in fruit to 25–40 mm long, exceptionally dense and paint-brush-like, dirty-white, pale brown or ± straw-coloured, sometimes shiny.

**Zambia**. N: Muchinga Province, Mutinondo Wilderness Area, Mayense Main camp. 12°27'03"S, 31°17'23"E, 1456 m, 27.vii.2012, *Merrett* LM862 (K). **Malawi**. N: Chitipa Dist., Nyika Plateau, Nganda Peak, 8500 ft, 9.i.1974, *Pawek* 7919 (K, CAL). S: Luchenya Plateau, Mlanje Mt, Mlanje Dist., 2100 m, 3.vii.1946, *Brass* 16645 (K, NY, SRGH).

Also in Tanzania. Upland grassland, often growing in rock crevices, where it may be common after burning, and bushland with *Protea* and *Tarchonanthus*, miombo woodland on sandy loamy soil; 1450–2750 m; flowering sporadically throughout the year.

Conservation Status: A widespread species, and moderately common throughout its range; it is best considered as LC (Least Concern) – see also Jeffrey & Beentje (2005: 572).

The modified synonymy of this species is the result of examining type and much other material. The significant overlap of characters and variable leaf shape suggest a widespread variable species that should no longer be split.

A species and name strangely absent in African Pl. Database version 4.0.0 (retrieved 2022), although present under its currently proposed synonymy (albeit with limited synonymy) under *L. dolichopappa*.

---

[106] Annotated by Thellung on a det. label as 'Orig.', det. by C. Jeffrey as the holotype.

[107] It can be argued that Thellung described the species based on the material in Z, where there are two sheets, both of which may well have been used in writing the protologue. It is clear Thellung has annotated both sheets, one on the main label, and the other with a separate det. slip. Phillips & Smith (1933: 238), whilst specifying the type was in 'Herb. Mus. Bot. Univ. Zurich', stated they had not been able to 'examine the type' and were clearly unaware that there were two sheets. I took this as first-step lectotypification bearing in mind the presence of other syntype material. Second-step lectotypification was carried out by Hind (2025: 823), and is based on the material labelled as the holotype by Jeffrey (in late August 1984).

[108] Det. as an isotype by C. Jeffrey, but databased as the holotype.

[109] According to Phillips & Smith (1933: 236).

[110] Although Jeffrey (1986: 932) stated '(syntypes B† BM!; isosyntypes K).', this is not considered as any form of lectotypification, and would still be incorrect, especially as there is only one duplicate of *Goetze* 1075 in K.

Fig. 6.4.**16**. STENOPS HELODES. 1, habit (× ⅔); 2, part of older inflorescence (× ⅔); 3, lower surface of leaf (× 1); 4, capitulum (× 6); 5, ray floret corolla (× 12); 6, ray floret showing corolla tube, style arms and base of ray limb (× 20); 7, disc floret (× 16); 8, achene (× 24). 1, 3 from *Richards* 16354; 2, 4–7 from *Sanane* 241; 8 from *Nutt* s.n. Drawn by Juliet Williamson. From Flora of Tropical East Africa.

## 107. **STENOPS** B.Nord.

**Stenops** B.Nord., Opera Bot. **44**: 73[–75, Fig. 36] (1978). —Nordenstam in Compositae Newslett. **23**: 1–2 (1993). —Jeffrey & Beentje in F.T.E.A., Compositae **3**: 598 (2005). —Nordenstam in Kubitzki, Fam. Gen. Vasc. Pl. **8**: 237–238 [2006](2007).

*Pseudocadiscus* Lisowski in Bull. Jard. Bot. Nat. Belg. **57**(3–4): 468 (1987). —Lisowksi, (Asterac. Fl. Afr. Cent. 1) Fragm. Flor. Geobot. **36** Suppl. 1: 237–238 (1991).

Glabrous annual, subcarnose semi-aquatic or aquatic herbs. Leaves alternate, cauline, sessile, simple, linear to narrowly elliptic-oblong, margins entire or slightly denticulate. Inflorescence few-headed, laxly corymbose-paniculate or of solitary terminal capitula. Capitula radiate, heterogamous, small; involucre ecalyculate; phyllaries uniseriate, connate to about middle; receptacle conical, epaleaceous. Ray florets female, corollas white or pink to mauve. Disc florets hermaphrodite, corollas yellow; apical anther appendages acute, anther bases obtuse, ecaudate; filament collar conspicuously balusterform; style arms apically truncate with short sweeping hairs, stigmatic areas separate. Achenes 5-angled, glabrous; pappus absent.

Two spp. from tropical Africa (D.R. Congo, Zambia and Tanzania); only one sp. present in the Flora area.

**Stenops helodes** B.Nord., Opera Bot. **44**: 73, t. 36 (1978). —Nordenstam in Compositae Newslett. **23**: 1 & Fig. 21 (1993). —Jeffrey & Beentje in F.T.E.A., Compositae **3**: 598 (2005). Type: '*Sanane* 241, Tanzania, Ufipa Distr., road to Katandalo Mission, 6500 ft, 31 vii 1968' (K000306836 holotype). FIGURE 6.4.**16**.

Annual herb, slightly fleshy, glabrous. Stem erect or decumbent, 20–40 cm long, somewhat fleshy, often rooting at lower nodes. Leaves spreading to ascending, linear to narrowly elliptic or narrowly lanceolate, 1–8 × 0.1–0.5 cm, somewhat fleshy, base slightly tapering and slightly decurrent, midrib usually ± evident in herbarium material and inconspicuously submarginally veined from base (at least on broader leaves), margins entire or rarely with 1 or 2 pairs of minute teeth, apex acute. Inflorescence lax, sometimes of simple solitary headed branches, or a few-branched corymb, pedicels 2–10 cm long, bracteolate, bracteoles scale-like, c. 1 mm long. Involucre campanulate, 1.5–2 × c. 5 mm; phyllaries 11–13, connate to about ⅔, apex acute. Ray florets (6)8(10), corollas white, pink or pale mauve, tube 1–2 mm long, ray limb oblong, apically 3-toothed, 4–5.5 × 1.5–2.5 mm, papillate above. Disc florets c. 50–80, corollas yellow, ± 2 mm long, pubescent. Achenes oblong, ± 1 × 0.3–0.4 mm.

**Zambia**. N: Abercorn Dist., Lumi Marsh, 1680 m, 30.v.1957, *Richards* 9959 (K).

Apparently restricted to the Lumi Marsh on the Zambia–Tanzania border – also in Tanzania. In moist grassland or in wet mud; 1500–1950 m – the lowest elevation was recorded on *Richards* 21529 and is specified as '1500 ft.' (= 457 m) but this I believe to be in error for 1500 m; flowering in May and October, although it is likely that the plants can be found flowering at any time throughout the year if the right conditions exist.

Conservation Status: This species is only known from the Lumi Marsh in the Northern Province, Zambia and neighbouring Tanzania, from very few collections. It is most likely, because of the habitat, that the species is under-collected, although it may well be rare; it is best recorded as DD (Data Deficient).

Specimen labels indicate that the capitula smell very sweet, and of honey.

## 108. **EMILIA** Cass.

**Emilia** Cass. in Bull. Sci. Soc. Philom. Paris **1817**: 68 (1817)

*Senecio* L. subgen. *Emilia* (Cass.) O.Hoffm. in Engler & Prantl, Nat. Pflanzenfam. **4**, 5(74): 297 (1892)

Annual or perennial herbs; stem usually erect, little-branched, mostly glabrous. Leaves alternate, usually cauline, simple, sometimes rosulate. Capitula often corymbose, sometimes solitary, discoid or radiate. Involucre ecalyculate, phyllaries free or connate at base, within a single capitulum some wider and some narrower, multi-veined, with scarious margins; receptacle epaleate. Florets of various colours; ray florets present or absent; disc florets tubular or narrowly infundibuliform. Anthers obtuse or slightly sagittate at base, appendaged at apex; style branches truncate to obtuse, sometimes with appendages of fused papillae. Achenes ellipsoid to cylindrical, 5-ribbed; pappus setae uniseriate, rarely pappus absent.

Around 116 spp., paleotropical, mostly in Africa S of the Sahara. 18 spp. for our area, plus two possible ones; this is a very difficult genus with not that many characters to distinguish the taxa.

1. Capitula radiate, ray floret corollas yellow or orange . . . . . . . . . . . . . . . . . . . . . . . . . . 2
– Capitula discoid, ray florets absent . . . . . . . . . . . . . . . . . . . . . . . . . . . . . . . . . . . . . . . . 4
2. Capitula solitary . . . . . . . . . . . . . . . . . . . . . . . . . . . . . . . . . . . . . . . . **6**. *discifolia*
– Capitula usually several in corymbs, rarely solitary . . . . . . . . . . . . . . . . . . . . . . . . . . 3
3. Upper leaves amplexicaul at base . . . . . . . . . . . . . . . . . . . . . . . . . . . . . **1**. *abyssinica*
– Upper leaves obtuse to cuneate at base . . . . . . . . . . . . . . . . . . . . . . . . . . . . . . . . . . 22
4. Corollas yellow or orange[111] . . . . . . . . . . . . . . . . . . . . . . . . . . . . . . . . . . . . . . . . . . 5
– Corollas purple, mauve, red, pink or white . . . . . . . . . . . . . . . . . . . . . . . . . . . . . . . . 9
5. Upper leaves with auriculate/semi-amplexicaul base; capitula solitary or in groups . . 6
– Upper leaves with attenuate, cuneate or obtuse leaf base; capitula always solitary[112] . 7
6. Median and upper leaves at least 4 × as long as wide; basal leaves falling soon . . . . . .
. . . . . . . . . . . . . . . . . . . . . . . . . . . . . . . . . . . . . . . . . . . . . . . . . . . . . . . **4**. *coccinea*
– Median and upper leaves 1–3 × as long as wide; basal leaves persistent . . . **18**. *vanmeelii*
7. Capitula in groups; pappus absent . . . . . . . . . . . . . . . . . . . . . . . . . . . . . **9**. *hindii*
– Capitula solitary; pappus present . . . . . . . . . . . . . . . . . . . . . . . . . . . . . . . . . . . . . . . 8
8. Achenes glabrous; phyllaries 4–7 mm long . . . . . . . . . . . . . . . . . . . . . . **2**. *basifolia*
– Achenes with hairy ribs; phyllaries 9–13 mm long . . . . . . . . . . . . . . . . . . . **10**. *hockii*
9. Capitula solitary or occasionally in pairs on long peduncles . . . . . . . . . . . . . . . . . . 10
– Capitula usually several together[113] . . . . . . . . . . . . . . . . . . . . . . . . . . . . . . . . . . . . . 12
10. Leaves ovate or elliptic, clustered near base of plant, narrowing to base . . . . . . . . . 11
– Leaves broadly reniform-spatulate to obovate, upper leaves with auriculate amplexicaul base; corollas red, orange or yellow . . . . . . . . . . . . . . . . . . . . . . . . . . . **18**. *vanmeelii*
11. Leaves 2.5–6.3 × 1–2.4 cm; corollas bright orange-red or crimson . . . . **7**. *fonszambesiaca*
– Leaves 1.5–5 × 0.2–0.6(1) cm; corollas lilac or mauve . . . . . . . . . . . . . . **17**. *transvaalensis*
– Leaves 2.5–5 × 0.4–0.7 cm; corollas white . . . . . . . . . . . . . . . . . . . . . . . . **17b**. *sp. B*
12. Median/upper leaves with cuneate or obtuse leaf base . . . . . . . . . . . . . . . . . . . . . . 13
– Median/upper leaves with auriculate/semi-amplexicaul base . . . . . . . . . . . . . . . . . . 15
13. Corollas white; achenes slightly hairy . . . . . . . . . . . . . . . . . . . . . . . . . . . . **8**. *fugax*
– Corollas purple, pink or white with purple/mauve tips; achenes glabrous . . . . . . . . 14
14. Capitula broader (5–7 mm in diameter at tips of phyllaries); florets 26–40, corollas purple, mauve or pink; pappus setae 2.5–5 mm long . . . . . . . . . . . . . . . . **11**. *integrifolia*
– Capitula very narrow (2–4.5 mm in diameter at tips of phyllaries); florets 9–14, corollas pale pink or whitish with mauve/purple tips; pappus setae 5–9 mm long . . . . . . . . . .
. . . . . . . . . . . . . . . . . . . . . . . . . . . . . . . . . . . . . . . . . . . . . . . . . . . . . . **16**. *tenellula*
15. Phyllaries 11–21[114] . . . . . . . . . . . . . . . . . . . . . . . . . . . . . . . . . . . . . . . . . . . . . . . 16
– Phyllaries 4–10 . . . . . . . . . . . . . . . . . . . . . . . . . . . . . . . . . . . . . . . . . . . . . . . . . . . 17
16. Median and upper leaves at least 4 × as long as wide; basal leaves falling soon . . . . . .
. . . . . . . . . . . . . . . . . . . . . . . . . . . . . . . . . . . . . . . . . . . . . . . . . . . . . . . **4**. *coccinea*
– Median and upper leaves 1–3 × as long as wide; basal leaves persistent . . **18**. *vanmeelii*

---

[111] *E. coccinea* and *E. vanmeelii* are keyed out both ways.

[112] Except in MR4302, which lacks a pappus.

[113] Very small/young plants sometimes with solitary capitula in *E. coccinea* and *E. parnassifolia*.

[114] *E. coccinea* is included in both leads, it can have 8–21 phyllaries.

17.  Basal leaves persistent and sessile, circular to broadly ellipsoid . . . . . . . **12**. *parnassifolia*
 –   Basal leaves not very persistent, narrowly to broadly ovate or spatulate to obovate  . 18
18.  Stem prostrate or decumbent, or leaning on other plants; corollas mauve or magenta; plants of floodplains or bogs in spray zone . . . . . . . . . . . . . . . . . . . . . . . . . **13**. *protracta*
 –   Plants erect or nearly so; corollas red, orange, yellow, white, purple or pink; plants of drier habitats. . . . . . . . . . . . . . . . . . . . . . . . . . . . . . . . . . . . . . . . . . . . . . . 19
19.  Corollas bright orange, less often red or orange-yellow  . . . . . . . . . . . . . . . **4**. *coccinea*
 –   Corollas white, pink, purple or (in *sonchifolia*) red-purple to pink . . . . . . . . . . . . . . . . 20
20.  Stem and leaves glabrous; phyllaries 4–7; Botswana, Zambia and Zimbabwe . . . . . . . . . . . . . . . . . . . . . . . . . . . . . . . . . . . . . . . . . . . . . . . . . . . . . **14**. *schinzii*
 –   Stem and leaves pubescent on lower surface, at least near base; phyllaries 8–10; Malawi, Mozambique. . . . . . . . . . . . . . . . . . . . . . . . . . . . . . . . . . . . . . 21
21.  Phyllaries 2.3–3.5 mm long; pappus setae 1.5–4 mm long; disc florets 3–5.5 mm long . . . . . . . . . . . . . . . . . . . . . . . . . . . . . . . . . . . . . . . . . . . . . . . . . . **5**. *decipiens*
 –   Phyllaries 7–9.5 mm long; pappus setae 5.4–8 mm long; disc florets 6–7.6 mm long . . . . . . . . . . . . . . . . . . . . . . . . . . . . . . . . . . . . . . . . . . . . . . **15**. *sonchifolia*
22.  Leaf margins with a few spaced teeth; ray florets 6–10; pappus setae 2.5–2.7 mm long . . . . . . . . . . . . . . . . . . . . . . . . . . . . . . . . . . . . . . . . . . . . **3**. *brachycephala*
 –   Leaf margins entire; ray florets 15–16; pappus setae 3.3 mm long . . . . . . . . . **3b**. *sp. A*


1.  **Emilia abyssinica** (Sch.Bip. ex A.Rich.) C.Jeffrey in Kew Bull. **41**(4): 911 (1986). — Lisowski, (Asterac. Fl. Afr. Cent. 2) Fragm. Flor. Geobot. **36** Suppl. 1: 361 (1991). — Mapaura & Timberlake, Checkl. Zimb. Vasc. Pl.: 24 (2004). —Jeffrey & Beentje in F.T.E.A., Compositae 3: 575 (2005). Type: Ethiopia, Adowa, *Schimper* 67 (P holotype, K, M, P).

    *Senecio abyssinicus* Sch.Bip. ex A.Rich., Tent. Fl. Abyss. **1**: 438 (1848).


### Var. **abyssinica**

Annual herb 10–120 cm tall; stems erect, unbranched (except for inflorescence) to well-branched, usually sparsely pubescent, especially towards base, sometimes glaucous; root white, simple. Leaves sessile, pale green, sometimes called fleshy and/or glaucous, broadly spatulate (especially lower leaves) to obovate, 0.8–7.5 × 0.6–4 cm, base attenuate into a briefly semi-amplexicaul petioloid base in upper leaves, becoming more sessile and more cuneate or broadly obtuse, margins sinuate-dentate to coarsely sinuate-serrate in upper part of leaf, apex rounded to obtuse, sparsely hairy to almost glabrous. Capitula radiate, (1)3–24 in lax to dense terminal corymbose panicles (very small plants have fewer capitula than larger ones); peduncles of individual capitula 4–35 mm long, glabrous; involucre 4–6.5 × 1–4 mm, glabrous; phyllaries light green, (5)8–11 mm long, with narrow scarious margins, glabrous but with papillate apex, reflexing in fruit. Ray florets (4)6–10, corollas yellow, tube 2–3.5 mm long, glabrous or with a few hairs at apex, ray limbs 1.2–2 × 0.5–0.8 mm, 4-veined. Disc florets not or hardly exserted, 12–20, corollas yellow or orange, narrowly infundibuliform, 3–4 mm long, tube glabrous, lobes 0.2–0.5 mm long. Achenes blackish, 1–2.5 mm long, shortly hairy on angles; pappus setae white, 3–4 mm long.

**Zambia**. B: Mankoya Agricultural Station, 18.vii.1961, *Angus* 2979 (FHO, K). N: Chilongwelo, no date, *Richards* 2319 (K). W: Kitwe, 15.xii.1967, *Mutimushi* 2403 (K, NDO). C: Protea Hill Farm, 13 km SE of Lusaka, 21.iii.1993, *Bingham* 8929 (K). E: Lundazi, 31.v.1954, *Robinson* 789 (K). S: 40.9 km from Choma along road to Namwala, 8.iii.1997, *Zimba et al.* 1035 (K, MO). **Zimbabwe**. N: Trelawney Dist., 4000 ft, 17.iii.1943, *Jack* 9850 (K). C: Harare, Cranborne, 12.i.1946, *Wild* 699 (K, SRGH). E: Umtali, Zimunya Reserve, *Chase* 5523 (K, SRGH). **Malawi**. N: Mzuzu, Marymount, 6.viii.1970, *Pawek* 3678 (K). C: Dedza, Chongoni Forest, 4.vi.1973, *Salubeni* 1892 (K, MAL, SRGH). S: 6 km from Cholo on Limbe road, 11.iii.1970, *Brummitt* 9022 (K, MAL). **Mozambique**. Z: 9.4 km from Ile to Nampevo, 18.ix.1949, *Grandvaux Barbosa & Carvalho* 4140 (COI, K, LMA).

Also in Nigeria, Cameroon, Central African Republic, D.R. Congo, Rwanda, Burundi, Sudan, Ethiopia, Djibouti, Uganda, Kenya and Tanzania. A weed of cultivation including gardens, in ruderal sites in grassland or open woodland; may be locally common on sand or damp sites; (500)1000–1650 m.

Conservation Status: I have seen 60 specimens from our area; widespread in a common habitat, no threats known, which makes this Least Concern (LC).

A second variety is restricted to northern Tanzania.

2. **Emilia basifolia** Baker in Bull. Misc. Inform. Kew **1898**(139): 154 (1898). —Jeffrey in Kew Bull. **41**: 913 (1986). —Lisowski, (Asterac. Fl. Afr. Cent. 2) Fragm. Flor. Geobot. **36** Suppl. 1: 368 (1991). —Da Silva *et al.*, Prelim. Checkl. Vasc. Pl. Mozamb.: 31 (2004). —Jeffrey & Beentje in F.T.E.A., Compositae **3**: 580 (2005). Type: Malawi, Zomba, *Whyte* s.n. (K holotype).

Annual herb 12–60 cm tall including capitulum and its long stalk; leafy stems 2–36 cm long, prostrate or erect, pubescent in lower part; stem giving rise to a single or several inflorescence peduncles Leaves often dense near base of plant, greyish or cream-green, sessile, obovate or elliptic, 1.3–4.5 × 0.5–2.4 cm, base obtuse (in lower leaves) to cuneate (in upper leaves), margins ± sinuate, obscurely denticulate to sometimes dentate, apex obtuse, soft-pubescent on both surfaces with long white hairs. Capitula solitary and terminal on erect (sometimes reddish) stalk 6–30 cm long, discoid, glabrous; involucre cylindrical, 5.5–7 × 3.5–10 mm; phyllaries 9–13, 4.3–7 mm long, sometimes with reddish apex, glabrous, papillate at apex. Florets 25–60, corollas orange or yellow, narrowly infundibuliform with slightly wide base, 3.5–6 mm long, of which lobes 1.5–1.7 mm long. Achenes 1.5–2.7 mm long, glabrous; pappus setae 3.3–4.5 mm long.

**Zambia**. N: Mbala [Abercorn], 10.vii.1964, *Mutimushi* 847 (K, NDO). W: Kitwe, 8.xii.1953, *Fanshawe* 543 (K, NDO). C: Mkushi Dist., David Moffat's farm, 19.ix.1993, *Bingham* 9648 (K). **Malawi**. N: Mzuzu, Katoto, 10.vi.1972, *Pawek* 5447 (K). C: Dedza mountain, 24.x.1956, *Banda* 310 (K, MAL, SRGH). S: Zomba mountain, Chitengi dambo, 14.v.1989, *Pope, Smith & Goyder* 2284 (K, MAL).

Also in D.R. Congo and Tanzania. In swampy grassland and dambos, by stream-banks, may be mat-forming; 1150–1850 m.

Conservation Status: I have seen 24 specimens from our area (as well as 3 from Tanzania); the habitat is fairly common, but prone to conversion to agriculture. As there are more than fourteen locations, I provisionally assess this as Least Concern (LC).

A specimen (*Bingham* 9685) labelled as an 'aquatic on canal bank' is a particularly vigorous plant, with densely leafy stems to 30 cm long and the largest root system I have seen for this species.

3. **Emilia brachycephala** (R.E.Fr.) C.Jeffrey in Kew Bull. **41**(4): 918 (1986). —Lisowski, (Asterac. Fl. Afr. Cent. 2) Fragm. Flor. Geobot. **36** Suppl. 1: 362 (1991). —Mapaura & Timberlake, Checkl. Zimb. Vasc. Pl.: 25 (2004). Type: Zambia, Broken Hill, *Fries* 222 (UPS holotype).

*Senecio brachycephalus* R.E.Fr., Wiss. Ergebn. Schwed. Rhodesia-Kongo-Exped. 1911-1912, **1**: 344 (1916).

Annual herb 15–45 cm high, erect, unbranched to more usually branched from near base, glabrous (or slightly hairy in young leaves). Leaves with lower ones reddish brown or maroon underneath, shortly petiolate or pseudopetiolate, petiole 0.5–1 cm long, winged; lamina of lowermost leaves spatulate to broadly obovate, 1.5–2 × 0.8–1 cm, base cuneate, margins with a few spaced teeth, apex rounded and apiculate, upper leaves sessile, linear-oblong to narrowly elliptic, 1.5–5(9) × 0.1–1 cm, entire or with remote teeth, apex acute, gradually morphing into filiform bracts, all leaves glabrous. Capitula heterogamous, radiate, in a lax corymb or panicle with side branches usually overtopping central ones, these in branched plants forming a super-corymb; peduncles of individual capitula 1.2–5 cm long; involucre campanulate, 3.5–4.7 × 3–4 mm; phyllaries (9)12–14, linear-oblong, 3.5–4.4 mm long, margins narrowly hyaline, apex acute, glabrous or apex minutely papillose, mostly 3-veined; receptacle flat. Ray floret (6)8–10, corollas yellow, ray limbs 2–3.5 × 0.9–1.5 mm, irregularly 2-toothed at apex; disc floret 23–42, corollas yellow, infundibuliform, 2.7–3.2 mm long of which lobes 0.5–0.7 mm and acute, papillate at apex; anthers linear, 1 mm long; style branches 0.7 mm, truncate and papillate at apex, without apical appendage. Achenes oblong, 1.2–1.7 mm long, angled, glabrous when young or sparsely hairy when mature (in ray florets only?); pappus of white bristles (1.7)2.5–2.7 mm long.

**Zambia**. N: Shiwa Nganda, 4.vi.1956, *Robinson* 1578 (K). W?: near Kafue River, 2.vii.1907, *Allen* 500 (K). C: Serenje, 14.iv.1973, *Phiri* 748 (K, UZL). S: Bowood siding, v.1909, *Rogers* 8056 (K). **Zimbabwe.** All the material under this name for Zimbabwe I have seen is *E. discifolia*, distinct in the solitary heads (rather than in the compound inflorescences of *E. brachycephala*, in which the central is usually overtopped by the lateral).

Also in D.R. Congo (upper Katanga). On sand in woodland especially on roadsides where it may be common; 1250–1650 m.

Conservation Status: I have seen 18 specimens from our area; despite its relative rarity this occurs in a common habitat, and I deem it Least Concern (LC).

### 3b. **Emilia sp. A** (= *Pope et al.* 2124)

The few leaves are broadly ovate to almost circular with an entire (not toothed) margin; phyllaries 8–9, 4.5–5.8 mm long, pilose; ray florets 15–16 (not 6–10), with a lamina of about 4 × 1.7 mm; disc florets about 20, 5.3 mm long; achenes 1.8 mm long, slightly hairy, with white pappus setae 3.3 mm long.

**Zambia**. N: 5 km E of Lumangwe falls, Kalungwishi R., 14.iv.1989, *Pope, Smith & Goyder* 2124 (K)

The habitat is high rainfall miombo on loose sandy soil.

This specimen keys near *E. brachycephala* but looks quite different. This specimen is unlike any of the other rayed species for the area of the D.R. Congo, of F.T.E.A. or of F.Z.; it is possibly new, and more material is required to make sure.

4. **Emilia coccinea** (Sims) G.Don in Sweet, Hort. Brit. ed. 3: 382 (1839). —Jeffrey in Kew Bull. **41**(4): 913 (1986); in Kew Bull. **52**(1): 209 (1997). —Da Silva *et al.*, Prelim. Checkl. Vasc. Pl. Mozamb.: 31 (2004). —Mapaura & Timberlake, Checkl. Zimb. Vasc. Pl.: 25 (2004). —Jeffrey & Beentje in F.T.E.A., Compositae **3**: 583, fig. 122 (2005). —Mapaya & Cron in Plant Syst.Evol. **302**: 703–720 (2016). Type: from a plant cultivated at Vauxhall, London (†); Sims in Bot. Mag. **16**: t. 564 (1803), iconotype, in practice lectotypified by Jeffrey (1997: 209).

    *Cacalia coccinea* Sims in Bot. Mag. **16**: t. 564 (1803).

    *Emilia flammea* auct., non Cass. —Da Silva *et al.*, Prelim. Checkl. Vasc. Pl. Mozamb.: 31 (2004).

    *Emilia caespitosa* Oliv. in Trans. Linn. Soc. **29**: 100 (1873). —Oliver & Hiern in F.T.A. **3**: 405 (1877). —Jeffrey in Kew Bull. **41**(4): 913 (1986). —Lisowski, (Asterac. Fl. Afr. Cent.2) Fragm. Flor. Geobot. **36** Suppl. 1: 383 (1991). —Jeffrey in Kew Bull. **52**(1): 210 (1997). —Jeffrey & Beentje in F.T.E.A., Compositae **3**: 581 (2005). Type: Tanzania, Bukoba Dist., Karagwe, *Grant* 464 (K holotype).

    *Emilia sagittata* auct., non DC. —Oliver & Hiern in F.T.A. **3**: 405 (1877). —Da Silva *et al.*, Prelim. Checkl. Vasc. Pl. Mozamb.: 31 (2004)

    *Emilia macaulayae* Garab. in Bull. Misc. Inform. Kew **1924**: 140 (1924). Type: Zambia, near Mumbwa, *Macaulay* 659 (K holotype).

    *Emilia humbertii* Robyns in Bull. Jard. Bot. État Bruxelles **17**: 102 (1943). Type: D.R. Congo, mountains to SW of Lake Edward, *Humbert* 8276 (BR holotype).

    *Emilia humbertii* var. *angustifolia* Robyns in Bull. Jard. Bot. État Bruxelles **17**: 102 (1943). Type: D.R. Congo, near Nzulu, Gahogo, *de Witte* 1399 (BR holotype).

    *Emilia lisowskiana* C.Jeffrey in Kew Bull. **52**(1): 208 (1997). Type: Nigeria, Ibadan, *Bernardi* 8819 (LE holotype).

Annual herb 4–120 cm high; stem erect, sparsely pubescent at least towards base, or glabrous. Leaves subsucculent, light green, often glaucous, sometimes purplish beneath or near margins, spatulate, obovate or elliptic, upper also ovate or oblong, 0.8–19.5 × 0.4–6 cm, cuneate or attenuate into a winged exauriculate petioloid base in lower leaves, upper leaves cordate, hastate or sagittate and semi-amplexicaul, margins remotely sinuate-dentate or subentire, apex rounded to obtuse, rarely acute, apiculate, varyingly pubescent beneath at least on midrib, glabrescent. Capitula discoid, 1–6 in terminal subumbelliform to lax corymbs; capitulum stalk glabrous or with a few hairs; involucre broadly cylindrical, 4–10 × 2.5–5 mm; phyllarics 8–21, 4–9 mm long, glabrous to densely setiferous or hirsute at least towards apex. Florets 24–40, corollas bright orange, occasionally red or orange-yellow, 5.3–12.5 mm long, shortly hairy; lobes 1.5–2.8 mm long, glabrous or shortly hairy in upper part; style arms with awl-shaped appendage of fused papillae. Achenes 2–4 mm long, glabrous or short-hairy; pappus setae 3–7 mm long.

**Zambia**. W: Ndola, 3.v.1954, *Fanshawe* 1151 (K, NDO). C: Broken Hill, 17.viii.1953, *Fanshawe* 7942 (K, NDO). S: Kafue Gorge below dam, 1.vii.1972, *Kornaś* 1920 (K). **Zimbabwe**. N: Darwin, Noro Dam, 9.iv.1970, *Loveridge* 1801 (K, SRGH). C: S Marondera (Marandellas), vii.1931, *Myres* 451 (K). E: Chimanimani Mts, next Haroni-Makurupini forest, 29.v.1969, *Muller* 1188 (K, SRGH). S: Zaka (Ndanga) Dist., 19.ix.1965, *Loveridge* 1424a (SRGH, fide M. Hyde). **Malawi**. N: Chitipa [Fort Hill], Tanganyika plateau, no date (pre-1897), *Whyte* s.n. (K). C: Dedza, Chongoni Forestry School, 18.i.1967, *Salubeni* 503 (K, MAL, SRGH). S: Michiru Hills, 12.vii.1988, *Banda & Tawakali* 3307 (K, MAL). **Mozambique**. N: Metangula, i.1947, *Seddon* 10 (K). Z: km 29 on Mocuba–Milange road, 20.v.1949, *Grandvaux barbosa & Carvalho* 2753 (K, LISC). MS: Chimanimani Mts, close to Nhamadzi River, alt. 648 m, 14.iv.2014, *Mapaura et al.* 633 (BR, LMA, K, SRGH).

Also in Sierra Leone, Liberia, Ivory Coast, Ghana, Togo, Nigeria, Cameroon, Equatorial Guinea, Central African Republic, D.R. Congo, Rwanda, Burundi, South Sudan, Ethiopia, Uganda, Kenya, Tanzania and Angola. In miombo, on road- and pathsides, in grassland, in abandoned fields and pine plantations; 100–1450 m.

Conservation Status: Widespread and in common habitats, so Least Concern (LC); for our area I have seen 74 specimens.

I have provisionally included *E. caespitosa* in the synonymy of *E. coccinea*. These two taxa have long been confused, and Jeffrey (1986, 1997) made a strong case for separating them; this separation was upheld in Jeffrey & Beentje (2005). However, Mapaya & Cron (2016) did a thorough morphological phenetic analysis of the complex and argue convincingly, in my opinion, that the two spp. are virtually indistinguishable, and are possibly a single heterogeneous species. Mapaya & Cron do not (yet) unite the two, awaiting phylogenetic work using molecular markers or population work involving AFLP or microsatellites. For pragmatic reasons I am uniting the two taxa here, as they cannot really be keyed out in any consistent manner. This is not the final solution, possibly, but for the time being it feels like the most practical one. In the same study (Mapaya & Cron, 2016) it turns out that *E. lisowskiana* is not supported as distinct from *E. coccinea*, either, this time without any doubt.

5. **Emilia decipiens** C.Jeffrey in Kew Bull. **41**(4): 917 (1986). —Jeffrey & Beentje in F.T.E.A., Compositae **3**: 594 (2005). Type: Tanzania, Songea Dist., Matagoro Hills just S of Songea, *Milne-Redhead & Taylor* 9868 (K holotype).

Annual herb 10–60 cm tall; stems ascending, unbranched to fairly densely branched from near base, reddish-purple and shortly pubescent in lower part. Leaves sessile, bright green, glaucous and ± purple-tinged beneath, those on lower part of stem ovate or broadly ovate, 2–7 × 1.2–3.3 cm, cordate and decurrent onto a narrow exauriculate petioloid base (up to 5 cm long), margins sinuate-dentate, apex obtuse, pubescent beneath; those in upper part of stem elliptic to oblong, lower of these often somewhat pinnate-lyrately lobed, 0.8–12 × 0.2–5.5 cm, base expanded and auriculate, margins subentire to sinuate-dentate, apex obtuse. Capitula few (3–8) in lax corymbs (but due to branched nature of plant forming large super-corymbs), discoid; peduncles of individual capitula slender, glabrous; involucre cylindrical, 3.5–4 × 1–2 mm; phyllaries 8–10, corollas bright green tinged purple, 2.3–3.5 mm long, glabrous. Florets 10–16, corollas white, pink or purple (in bud?), 3–5.5 mm long, of which lobes 0.8–1.7 mm long. Achenes narrowly ellipsoid, 1.3–1.5 mm long, hairy; pappus setae 1.5–4 mm long.

**Malawi**. C: Dedza, Giwan Hill, 15.iii.1961, *banda* 414 (K, MAL, SRGH). S: Chiradzulu mountain, above Lisao Forest, 13.iii.1977, *Brummitt, Seyani & Patel* 14861 (K, MAL). **Mozambique**. N: 28 miles E of Ribaue, 23.v.1961, *Leach & Rutherford-Smith* 10984 (K, SRGH). Z: Namuli mountain, small peak 5 km W of highest peak, 27.v.2007, *Harris et al.* 210 (K, LMA).

Also in southern Tanzania. Among rock outcrops on poor stony soil, on shallow soil over rock or in disturbed sites (roadsides, tree plantations); 600–2200 m.

Conservation Status: I have seen 20 specimens from our area, as well as three specimens from Tanzania; habitats like shallow soil over rocks do not seem particularly threatened (and neither do the disturbed sites), so I would assess this as Least Concern (LC).

There is large variation in floret length – plants from around Dedza seem to have smaller florets than those from elsewhere. The ratio pappus length/corolla length seems, however, to be consistent.

6. **Emilia discifolia** (Oliv.) C.Jeffrey in Kew Bull. **41**(4): 912 (1986). —Lisowski, (Asterac. Fl. Afr. Cent. 2) Fragm. Flor. Geobot. **36** Suppl. 1: 363, t. 79 (1991). —Mapaura & Timberlake, Checkl. Zimb. Vasc. Pl.: 25 (2004). —Jeffrey & Beentje in F.T.E.A., Compositae **3**: 576, fig. 121 (2005). Type: Uganda, *Grant* s.n. (K holotype).

    *Senecio discifolius* Oliv. in Trans. Linn. Soc. **29**(2): 100 (1873). —Oliver & Hiern in F.T.A. **3**: 410 (1877).

Annual or perhaps rarely short-lived perennial herb 2–90 cm tall, much branched or not branched at all; stems erect or decumbent, pale green, sometimes bluish, sparsely to densely white pubescent especially when young and in axils in lower part, glabrous or almost so in upper part. Leaves pale to bluish green, sometimes glaucous, subsucculent, sessile, spatulate or rotundate-spatulate, uppermost sometimes oblanceolate, 0.8–10.5 × 0.15–5.6 cm, base rounded or cuneate into a winged petioloid base, margins sinuate-dentate in upper half, apex subtruncate or rounded and at least 3–5-toothed, sparsely to densely pubescent, sometimes glabrescent. Capitula terminal, solitary, long-stalked, radiate; capitulum stalk glabrous; involucre cylindrical, 3.5–8 × 4.5–8 mm; phyllaries 6–13, bright green, sometimes tinged reddish, 3–7.5 mm long, glabrous. Ray florets 8–26, most often 13 or 18, corollas yellow or orange, tube 2.7–5.5 mm long, hairy at apex, ray limbs 3.5–7 × 1.5–3 mm, 4-veined. Disc florets 16–40, corollas bright yellow to orange, often slightly darker in colour than ray florets, 3.5–6.5 mm long, tube glabrous, lobes 0.5–0.7 mm long. Achenes 1.5–3 mm long, shortly hairy; pappus setae white, 3.5–7 mm long.

**Zambia**. C: Livingstone–Lusaka road, alt. 930 m, 29.iii.1961, *Richards* 14910 (K). S: Siamambo, 19.xii.1963, *Mutimushi* 543 (K, NDO). **Zimbabwe**. N: Mtoko, Mudzi Dam, 16.ii.1962, *Wild* 5673 (K, SRGH). W: near the outspan for the World's view, Matopo Hills, 10.iii.1910, *Rogers* 5259 (K). C: Silverbow, 15.vi.1957, *Chase* 6525 (SRGH, fide M. Hyde). E: Chmanimani Mts, Martin F.R., 16.xi.1967, *Mavi* 650 (K, SRGH). S: Victoria, Makoholi experimental station, 13.xii.1977, *Senderayi* 114 (K, SRGH).

Also in Sudan, Ethiopia, Uganda, Kenya, Tanzania, D.R. Congo, Rwanda and Burundi. Miombo, where it may be common in rocky or disturbed sites, often on poor soils or shallow soil over rock (and then may be quite small); 900–1700 m; flowering before the rains (fide Burtt).

Conservation Status: I have seen 52 specimens from our area; widespread in a common habitat, Least Concern (LC).

Our material has, in general, smaller leaves than specimens from further North.

7. **Emilia fonszambesiaca** Beentje, sp. nov. This species is slightly reminiscent of *Emilia transvaalensis* (Bolus) C.Jeffrey but is distinct in its larger leaves, corolla colour (bright orange to crimson, rather than lilac or mauve) and larger length of corolla lobes. It is geographically far separate from the distribution of *E. transvaalensis* as well. Type: Mwinilunga Dist., 15 km N of Kalene Hill, 14.xii.1963, *Robinson* 6039 (K holotype). FIGURE 6.4.**17**.

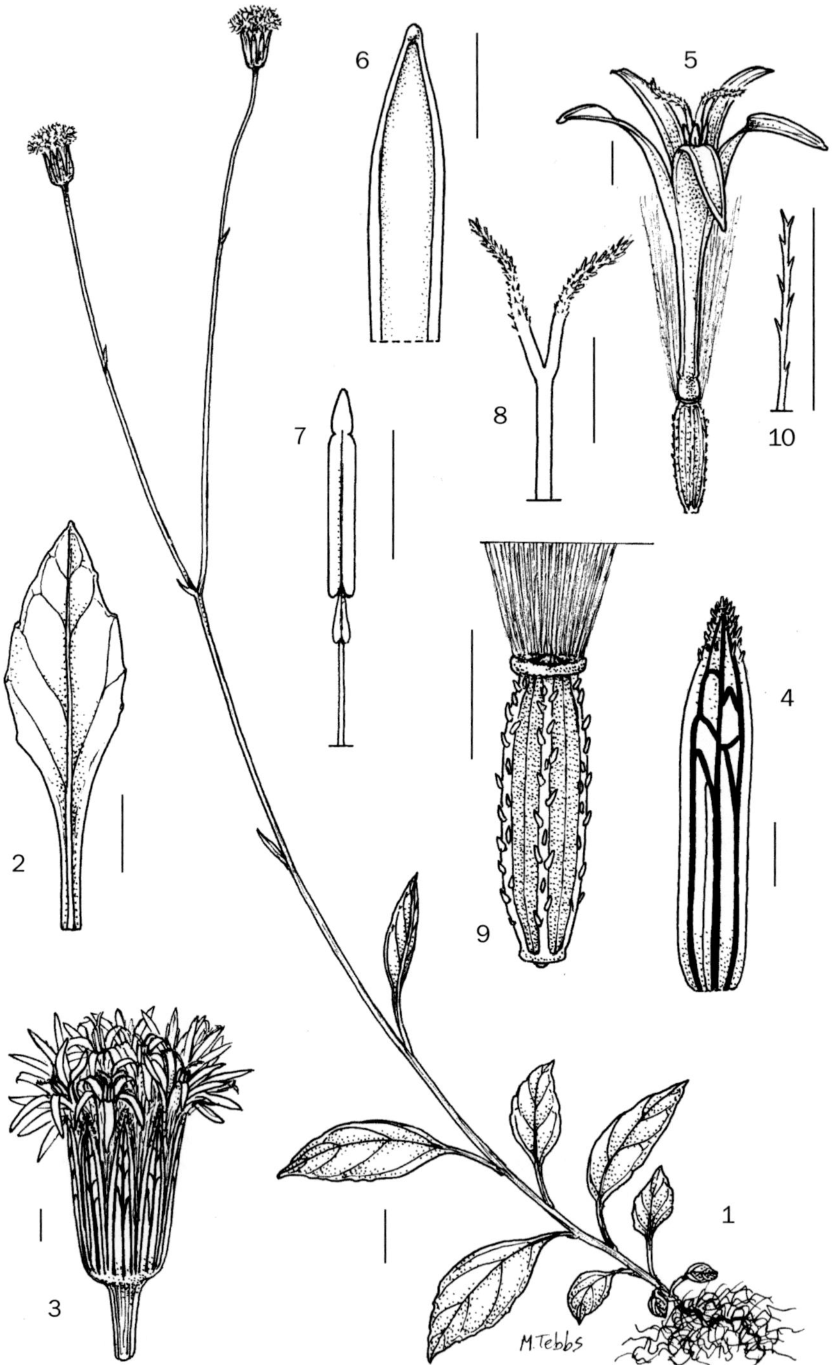

Fig. 6.4.**17**. EMILIA FONSZAMBESIACA. 1, habit; 2, lower leaf surface; 3, capitulum; 4, phyllary; 5, floret; 6, corolla lobe, adaxial surface; 7, stamen; 8, style arms; 9, achene; 10, detail of apex of pappus seta. All from *Hooper & Townsend* 269. Scale bars: 3–10 = 1 mm; 2 = 5 mm; 1 = 10 mm. Drawn by Margaret Tebbs.

Annual herb 15–55 cm tall; stems erect, unbranched except for inflorescence (which is sparsely branched), sparsely pilose to glabrous. Leaves slightly ovate to elliptic, 2.5–6.3 × 1–2.4 cm, base attenuate into a petioloid base, margins entire, apex acute, slighty pilose when young to glabrous. Capitula discoid solitary or 1–2 together; peduncles of individual capitula 5–11 cm long, glabrous; involucre 6–6.3 × 5–6 mm, glabrous; phyllaries probably pale green, 8, 6–6.3 mm long, with narrow scarious margins, glabrous but with papillate apex. Ray florets none; disc florets 35–40, corollas bright orange-red or crimson, narrowly infundibuliform,7–7.4 mm long of which tube 4–4.3 mm, lobes 3–3.6 × 0.6–1 mm. Achenes 2.5–2.8 mm long, with short stubby hairs along angles; pappus setae 3.4–4 mm long.

**Zambia**. W: Mwinilunga Dist., 15 km N of Kalene Hill, 14.xii.1963, *Robinson* 6039 (K); 14 km N of Kalene Hill, 21.ii.1975, *Hooper & Townsend* 269 (K).

Not known elsewhere. In open sandy woodland or on hill slopes with bare granite outcrops and massive boulders, miombo woodland near, at about 1250 m.

Conservation Status: on recent satellite imagery this area still seems to have quite some original vegetation. However, with only two specimens known, this is probably best considered as Endangered (EN D1).

Kalene Hill is in NW Zambia, on the border with D.R. Congo and Angola; it is the site of the source (Latin: *fons*) of the Zambesi River.


8. **Emilia fugax** C.Jeffrey in Kew Bull. **41**(4): 916 (1986). —Jeffrey & Beentje in F.T.E.A. Compositae **3**: 591 (2005). Type: Tanzania, Kigoma Dist., Lake Tanganyika, Kibwesa Point, *Newbould & Jefford* 1667 (K holololotype).

Annual herb 18–50 cm high, delicate; stems smooth, unbranched or branched, glabrous or sparsely hairy. Leaves cauline (sometimes a few basal, minute stalked almost round leaves present, but less than 5 mm long) sessile or lowermost slightly stalked, narrowly elliptic, 1–10 × 0.1–1.1 cm, base obtuse, margins entire or occasionally slightly dentate, apex attenuate and obtuse, glabrous or sparsely hairy. Capitula few in lax cymes, discoid; peduncles of individual capitula very slender, 13–43 mm long; involucre 4–6.5 × 2.5–3.5 mm ; phyllaries 7–8, 3–5.4 mm long, with narrow scarious margin, glabrous but for penicillate apex or sparsely hairy, green with 3 yellowish veins. Florets 14–21, corollas white in F.Z. area, apparently pale yellow in Tanzanian type, 4–5.2 mm long, lobes 0.8–1.4 mm long. Achenes 1.6–3.2 mm long, very shortly hairy; pappus setae 2–3.5 mm long.

**Zambia**. N: Kambole, 20.ii.1957, *Richards* 8288 (K). W: Mwinilunga, 18 km E of Kalene Hill, 16.xii.1963, *Robinson* 6104 (K).

Also in Tanzania. In bogs and damp ground near streams, in miombo where it sometimes occurs in roadsides; 1150–1800 m.

Conservation Status: I have seen eight specimens from our area, and one from Tanzania; the habitat requirement information is rather mixed; Data Deficient (DD).

'Rapid ephemeral'; the Tanzanian specimen, in full flower, was collected ten days after rain. The Tanzanian *Emilia leucantha* C.Jeffrey is very similar.

*Bullock* 3944 from **Zambia**. N: Sumbawanga–Mbala at 6000 feet, 4.vi.1951, is similar to this species but has larger capitula: the phyllaries are 5.7–6 mm long and the pappus is 4.3–4.8 mm long. In other respects it conforms to the above description.

Two specimens from Zambia (B): Mongu area key here (though neither gives a flower colour) but look more solid overall, not as delicate as *fugax* usually does: the capitulum peduncles are much longer, there are more and longer disc florets (6.3–6.6 mm long, 27–35 in number). **Zambia**. B: Mongu, 20.ii.1966, *Robertson* 6850 (K); Mulundu local forest, Kaoma–Mongu km 120, 27.iv.1993, Bingham 9239 (K). I am treating these two specimens as "near *fugax*".


9. **Emilia hindii** Beentje, sp. nov. In the yellow- and orange-flowered discoid species of the genus, quite distinct in the sub-corymbose inflorescence branching; in the entire genus, also distinct in the lack of pappus. Type: Mwinilunga Dist., S of Matonchi Farm, 23.i.1938, *Milne-Redhead* 4302 (K holotype). FIGURE 6.4.**18**.

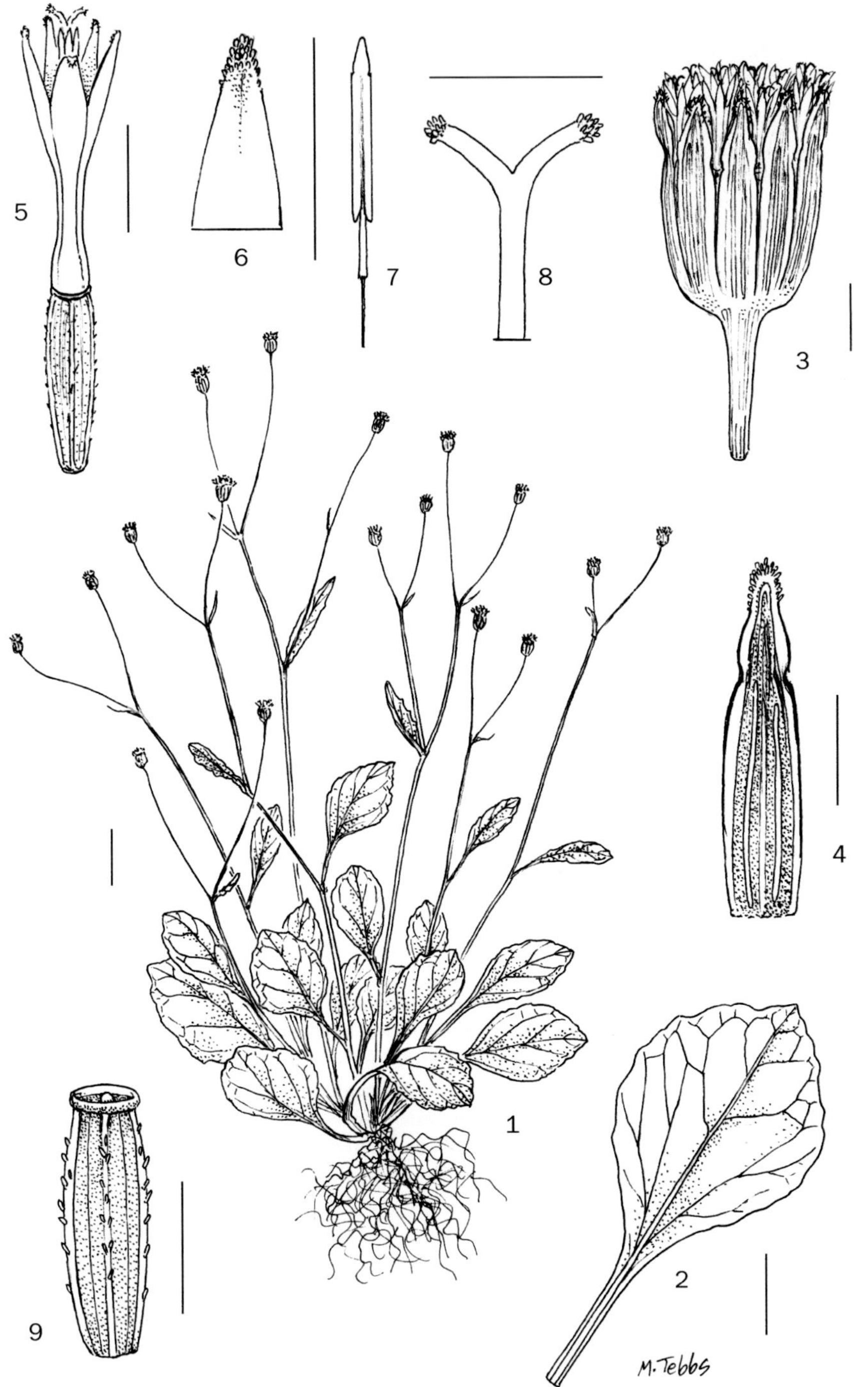

Fig. 6.4.**18**. EMILIA HINDII. 1, habit; 2, lower leaf surface; 3, capitulum; 4, phyllary; 5, floret; 6, corolla lobe, abaxial surface; 7, stamen; 8, style arms; 9, achene. All from *Townsend* 363. Scale bars: 8 = 0.4 mm; 5, 7–9 = 0.5 mm; 3, 4 = 1 mm; 2 = 5 mm; 1 = 10 mm. Drawn by Margaret Tebbs.

Annual herb 5–25 cm tall; stems erect, unbranched (in very young plants, sometimes flowering with a single head) to well-branched, sparsely pilose on younger parts. Leaves dull green above, bright purple below, broadly ovate to subcircular, 1–5 × 1.4–3.3 cm including a petolioid base, base attenuate into a briefly semi-amplexicaul petioloid base (uppermost leaves almost auriculate), margins serrate, apex obtuse, sparsely hairy to almost glabrous. Capitula discoid, few together in lax sub-corymbose panicles (very small plants have fewer capitula than larger ones); peduncles of individual capitula 15–38 mm long, glabrous; involucre 2–3 × 2–4 mm, glabrous; phyllaries pale yellow-green, 9–10, 2.1–3 mm long, with narrow scarious margins, glabrous but with papillate apex. Florets not or hardly exserted, c. 18, corollas bright yellow, narrowly infundibuliform, 1.5–1.7 mm long, tube glabrous, lobes 0.3–0.4 mm long. Achenes brown, 1–1.1 mm long, glabrous to shortly hairy on angles; pappus completely absent, even in bud.

**Zambia**. W: Mwinilunga Dist., S of Matonchi Farm, 23.i.1938, *Milne-Redhead* 4302 (K); 7 km N of Kalene Hill, 17.iv.1965, *Robinson* 6626 (K); 8 km N of Mwinilunga on road to Kalene Hill, 24.ii.1975, *Hooper & Townsend* 363 (K).

Not known elsewhere. On rather bare sandy soil by roadside or in old compound, or at edge of miombo woodland; around 1300 m.

Conservation Status: restricted to the Mwinilunga area, in a fairly common habitat. However, with only three specimens known, this is probably best considered as Vulnerable (VU D1).

I have looked for similar material from Angola and D.R. Congo, but there is nothing like these three specimens; of course the inflorescence branching and the lack of pappus are most distinctive.[115] This must be a narrow endemic peculiar to the Mwinilunga-Kalene Hill area, and a new species. The specific epithet honours Dr D.J. Nicholas Hind, who first put the three specimens together with a note saying "this needs serious checking!"

10. **Emilia hockii** (De Wild. & Muschl.) C.Jeffrey in Kew Bull. **41**(4): 912 (1986). —Lisowski, (Asterac. Fl. Afr. Cent. 2) Fragm. Flor. Geobot. **36** Suppl. 1: 365 (1991). —Jeffrey & Beentje in F.T.E.A., Compositae **3**: 579 (2005). Type: D.R. Congo, Lubumbashi, *Hock* s.n. (BR holotype).

    *Senecio hockii* De Wild. & Muschl. in Bull. Soc. Royal Bot. Belgique **49**: 231 (March 1913).
    *Senecio rogersii* S.Moore in J. Bot. **51**: 185 (Apr. 1913). Type: D.R. Congo, Lubumbashi [Elisabethville], *Rogers* 10176 (BM holotype).

Scapigerous perennial herb with creeping woody rhizome and annual (?) shoots 3–15 cm without scapes; stems unbranched or slightly branched, pubescent. Leaves cauline but sometimes looking sub-radical, slightly fleshy and sometimes called glaucous, grey-green or bluish above, sessile, narrowly obovate or elliptic, 1.18 × 0.2–2.8 cm, base attenuate, margins entire or with a few remote teeth in upper half, apex obtuse to rounded, scattered-pubescent or finely puberulous, glabrescent. Scapes 1–4(8) per plant, glabrous; capitula terminal, solitary, discoid; peduncles of capitula at anthesis (5)7–24 cm long; involucre cylindrical, 11.5–14 × 5.5–12 mm; phyllaries 10–14, dark green, 9.5–13 mm long, glabrous or sparsely hairy, 3–5-veined, often veins branched in upper half of phyllary. Florets 26–60, corollas yellow or orange (called pinkish-mauve in *Fanshawe* 440), 8–11 mm long, tube glabrous, lobes 0.5–1.5 mm long. Achenes 3.4–6 mm long, hairy on angles to almost glabrous; pappus setae 7.5–10 mm long.

**Zambia**. N: Kambole area, 10 m from Kambole–Mbale road, 29.viii.1956, *Richards* 5986 (K). W: Solwezi, 16.x.1953, *Fanshawe* 440 (K, NDO). **Malawi**. N: Nyika Plateau, Zovo, Chipolo, 19.xii.2010, *Ayami et al.* 1040 (FRIM, K). C: Ntchisi F.R., 25.iii.1970, *Brummitt & Evans* 9394 (K, MAL).

Also in D.R. Congo and Tanzania. Grassland, regularly burned grassland, roadsides, dambo edges; once called locally common (in a dry dambo); (1200)1500–2300 m

---

[115] There are two *Emilia* taxa from D.R. Congo without pappus: *E. bampsiana* Lisowski and *E. petitiana* Lisowski; but these both differ from our material.

Conservation Status: I have seen 16 specimens from our area; widespread in a common habitat; Least Concern (LC).

*Greenway & Brenan* 8144 from Zambia, Solwezi to Chingolo, has quite small heads (phyllaries 8 mm long, disc florets 6 mm long, pappus setae 5–6 mm long) with white corollas and (in three of the seven plants collected) inflorescences with two capitula on a single stalk. In other respects this is very similar to normal *E. hockii*; I believe it is an aberrant form.

11. **Emilia integrifolia** Baker in Bull. Misc. Inform. Kew **1895**: 69 (1895). —Jeffrey in Kew Bull. **41**(4): 917 (1986). —Lisowski, (Asterac. Fl. Afr. Cent. 2) Fragm. Flor. Geobot. **36** Suppl. 1: 370 (1991). —Da Silva *et al.*, Prelim. Checkl. Vasc. Pl. Mozamb.: 31 (2004). —Jeffrey & Beentje in F.T.E.A., Compositae **3**: 592 (2005). Type: Zambia, Fwambo, *Carson* 102 (K lectotype), lectotypified by Jeffrey (1986: 917).

    *Emilia limosa* (O.Hoffm.) C.Jeffrey in Kew Bull. **41**(4): 917 (1986). —Lisowski, (Asterac. Fl. Afr. Cent. 2) Fragm. Flor. Geobot. **36** Suppl. 1: 369 (1991). —Mapaura & Timberlake, Checkl. Zimb. Vasc. Pl.: 25 (2004). —Jeffrey & Beentje in F.T.E.A., Compositae **3**: 593 (2005). —Cron in Phytotaxa **159**(3): 197 (2014). Type: Angola, on the Cubango [Kubango] near Cobi [Kobi], *Baum* 907 (B† holotype, HBG, K), syn. nov.

    *Senecio limosus* O.Hoffm. in Warburg, Kunene–Sambesi-Exped.: 422 (1903).

Annual or short-lived perennial herb 10–125 cm tall (*Robson & Fanshawe* 622 states 'with stolons', *Robinson* 1239 states 'rootstock slightly tuberous'); stems slender, erect or basally decumbent and then erect, hardly branched, silvery-pubescent to glabrous, sometimes purplish distally. Leaves cauline, sessile, subsucculent and sometimes glaucous, basal leaves spatulate to oblanceolate or narrowly elliptic, 1.1–9 × 0.2–1.1 cm, attenuate into petioloid base, obscurely sinuate-denticulate, apex obtuse to acute; median and upper sessile, narrowly oblanceolate, narrowly elliptic, elliptic, lanceolate or linear, 3–16 × 0.2–0.7 cm, base attenuate, subentire or sinuate-serrate to sinuate-denticulate, apex obtuse to acute; all leaves scattered-pubescent to glabrous. Capitula 1–14 in lax to congested or subumbelliform terminal corymbs, discoid; stalk slender, glabrous or shortly sparsely pubescent; involucre initially cylindrical but becoming conical, 4–8 × 2–7 mm; phyllaries 6–9(13), usually 8, green tinged purplish or reddish in distal part, 4.5–8 mm long, glabrous or shortly pubescent at least towards apex. Florets 26–40, corollas bright to pale purple, mauve or pink, 4–8 mm long, lobes 1–2.7 mm long, those of outer florets larger. Achenes 1.2–3.1 mm long, glabrous; pappus setae white, 2.5–5 mm long.

**Botswana**. SE: Swaneng near Serowe, 5.iii.1975, *Mitchison* s.n. (K). **Zambia**. N: Mambwe [Mwambe], 1500 m, 9.vii.1970, *Sanane* 1249 (K). W: Mwinilunga, Kalenda dambo, 8.x.1937, *Milne-Redhead* 2660 (K). C: Mkushi Dist., David Moffat's farm, Munchiwemba dambo, 19.ix.1993, *Bingham* 9686 (K). E: Lundazi, 17.x.1967, *Mutimushi* 2231 (K, NDO). **Zimbabwe**. C: Harare Dist., Cleveland Dam, 4500', 3.iii.1950, *Wild* 3237 (K, SRGH). E: Nyanga (Inyanga), Gairesi Ranch on Mozambique border, 5800', 17.xi.1956, *Robinson* 1926 (K, NRGH, SRGH). **Malawi**. N: Mzimba Dist., Champira, Malembo, 22.iv.1974, *Pawek* 8458 (K, MAL, MO). C: Mchinji Dist., 7 km W of Namitete, 29.iv.1989, *Pope, Smith & Goyder* 2231 (BR, K, MAL, LISC).

Also in D.R. Congo, Uganda, Kenya, Tanzania, Angola, South Africa and Madagascar. Dambos, swamps, peat bogs and marshes, wet sandy ground, moist grassland, sometimes in shallow standing water; may be locally common; 900–2250 m.

Conservation Status: Widespread in a range of common habitat types, and with a considerable altitude range: I have seen 56 specimens from our area; Least concern (LC).

The types of *Emilia integrifolia* and *E. limosa* are distinct in that the first has many smallish heads per inflorescence, while the second has rather large solitary heads. The size differences are not that extreme: the type of *E. integrifolia* has phyllaries about 5 mm long, while in the type of *E. limosa* phyllaries are 7 mm long. Other characters are fairly similar, from habit to leaves to achenes, corolla colour (assumed from dry florets) and number. However, with the larger amount of specimens now available (57) assigning specimens to one or the other taxon has become more difficult.

Jeffrey (1986) and Jeffrey & Beentje (2005), dealing with East African specimens, distinguished the two taxa on inflorescence alone: "capitula 3−14 in each inflorescence, not very laxly to densely corymbose/capitula 1−4 in each inflorescence, lax". Lisowski (1991), writing on Congolese specimens, keys these two taxa out as follows: "Leaves (at least the median and upper) auriculate and sagittate at base … *E. limosa*; Leaves not auriculate or sagittate at base … *E. integrifolia*." This does not work on any of the specimens I have seen, and that includes the types. Cron (2014: 197) writing on southern African specimens felt that these two taxa are very similar or the same, and cites several intermediate specimens from the FZ area, including multi-headed inflorescences with large-size heads and few-headed inflorescences with smaller heads.

The size differences between *Emilia integrifolia* and *E. limosa* cannot be upheld; there is a continuous range from small to large, and this is not necessarily linked to number of heads per inflorescence – though the tendency to do so is certainly there. The 'number of heads per inflorescence' character is stronger, but also shows a continuous variation. I am unable to link habitat with one or more of the character variables. With more than 20% of the specimens I have seen 'intermediate' in one way or another between the two taxa, I feel that this is a single variable species, and *E. limosa* is hereby sunk into the synonymy of *E. integrifolia*.

12. **Emilia parnassiifolia** (De Wild. & Muschl.) S.Moore in J. Linn. Soc., Bot. **47**: 284 (1925). —Lisowski, (Asterac. Fl. Afr. Cent. 2) Fragm. Flor. Geobot. **36** Suppl. 1: 385 (1991). Type: D.R. Congo, Kundelungu, *Kassner* 2606 (BR lectotype, HBG, K, P),[116] lectotypified by Lisowski (1991: 385).

*Senecio parnassiifolius* De Wild. & Muschl. in Bull. Soc. Royal Bot. Belgique **49**: 234, t. 20 (1912) as '*parnassiaefolius*'.

Erect annual herb 20–45(65) cm high; stem simple or slightly branched, sub-quadrangular, sparsely leafy, sparsely hairy or glabrous. Leaves membranous, often purple on lower surface, sessile, basal and near-basal circular to broadly ellipsoid, 1–6.8 × 1–4.7 cm, base cordate-amplexicaul, margins entire, apex rounded; median and upper becoming more elongated to ovate and with acute apex, but sometimes (as in *Pope et al.* 2166) still almost circular though with a more acute apex, 1.7–4.2 × 0.8–3.4 cm; lamina glabrous or with a few hairs on lower midrib. Capitula homogamous, in lax few-headed terminal cymes, rarely solitary, discoid; peduncles of individual capitula 7–120 mm long; involucre campanulate, 7–8 × 4–5 mm; phyllaries 7–9, often reddish, linear, 5–6.8 mm long, with narrow or broad hyaline margins, acuminate and papillate at apex, glabrous or puberulous. Ray florets absent; disc florets 15–25, corollas red or purple-red (once described as orange-red), narrowly infundibuliform, exserted, 7–8 mm long, lobes 1–1.5 mm long, obtuse, papillate at apex; anthers yellow, linear, 1.8–2.3 mm long; style branches 1 mm long, with a terminal triangular-filiform appendage 0.5 mm long. Achenes oblong, 2.8–3.1 mm long, ribbed, outer curved, hairy on angles; pappus setae white, 2.5–4.3 mm long.

**Zambia**. B: Kabompo near boma, 23.iii.1961, *Drummond & Rutherford*-Smith 7255 (K, SRGH). N: Kalungwishi River, Kundabwika Falls, 17.iv.1989, *Pope, Smith & Goyder* 2166 (BR, K, LISC, NDO).

Also in D.R. Congo. In open woodland, bushland, often in rocky sites, sandy soil near rivers or lakes; 750–1200 m.

Conservation Status: I have seen 11 specimens for our area; Lisowski cites 12 from D.R. Congo, where the species reaches 1600 m altitude. This is not a common species, but the habitat types seem fairly general. Provisionally I would say this is Least Concern (LC), but the distribution might be fairly restricted.

---

[116] Note that the protologue states the Kassner number is 2686, but all duplicates show the number 2606.

13. **Emilia protracta** S.Moore in J. Bot. **43**: 48 (1905). —Mapaura & Timberlake, Checkl.
    Zimb. Vasc. Pl.: 25 (2004). —Cron in Phytotaxa **159**(3): 201 (2014). Type: Zimbabwe,
    Victoria Falls, 914 m, May 1904, *Eyles* 119 (BM holotype, SRGH).

Prostrate or decumbent herb, short-lived, up to 60 cm tall where supported by surrounding vegetation; stems reddish or purple, hollow and somewhat succulent or even 'inflated' (fide *Greenway*), weakly rooting at nodes where touching soil, glabrous (rarely uppermost leaves puberulous). Leaves slightly succulent (drying membranous), green with maroon edges, purple below, sessile, ovate to lanceolate, 2–8 × 0.8–2 cm, base cordate to auriculate and semi-amplexicaul, margins undulate and/or minutely toothed, apex acute. Capitula discoid, axillary, 2–3 together on short erect peduncles, peduncles of individual capitula 3–34 mm long, densely hairy near apex when young, glabrescent; involucre 5.5–7(8) × 3–4.5(7) mm; phyllaries 7–8, pale green with purple papillate tips, 5–7 × 1–3 mm, with narrow scarious margins. Florets 10–25, corollas mauve or magenta, long-exserted, narrowly infundibuliform, 4.5–7.5 mm long, lobes 0.6–1.4 mm long; all florets sticking together with upper parts after anthesis; anthers 1.2–1.8 mm long including acute anther appendage, filament collars balusterform 0.3–0.8 mm long; style branches 0.5–0.8 mm long, apex shortly penicillate with central tuft of papillae. Achenes narrowly cylindrical, 2.7–5 mm long, minutely hairy on ribs; pappus setae white, 2.3–3 mm long.

**Zambia**. B: Sesheke Dist., before ii.1911, *Gairdner* 502 (K). W: Mongu Dist., Lealui W., 21.ii.1999, *Bingham & Luwiika* 11927 (K, UZL). S: Namwala, 28.iv.1964, *van Rensburg* 2904 (K). **Zimbabwe**. W: Victoria Falls, rainforest, 7.vi.1979, *Ncube* 56 (K, SRGH).

Also in NE Namibia. In bogs in the spray zone or on moist sand at Victoria Falls where it is described as 'fairly common' or 'locally common', elsewhere in floodplains (often with *Oryza*) and flowering where the water has receded; 600–1000 m.

Conservation Status: Rare and ephemeral, with few populations known: Victoria Falls (13 collections seen by me; Cron cites another seven), Namwala on the floodplain, Mongu on the floodplain and Sesheke (1910) 'in shallow water'. Cron saw another three specimens from Namibia, all outside the Caprivi Strip; and three more from Zambia, from Molwe Dambo, Lochinvar N.P. and the Zambezi floodplain at Senenga Dist. (Cron 2014). A habitat specialist, growing on the banks of large rivers subject to flooding; changes in flood regime would pose a threat. A large population is protected in the Victoria Falls National Park, Zimbabwe, with the most recent Zambian collection from 1962 (fide Cron). Cron suggested that this species is data deficient and needs further investigation. I suggest there are about ten populations, of which two are in protected areas, and I suggest an assessment of Least Concern (LC) until a clear threat to any of the populations becomes known; at which point this would move to Vulnerable (VU).

14. **Emilia schinzii** (O.Hoffm.) Cron in Phytotaxa **159**(3): 202 (2014). Type: Namibia,
    Olukonda, 2.i.1886, *Schinz* 719 (K lectotype, GRA, Z), lectotypified by Cron (2014: 202).
    *Senecio schinzii* O.Hoffm. in Bull. Herb. Boissier **1**(2): 87 (1893).
    *Othonna rosea* Klatt in Bull. Herb. Boissier **3**(8): 424 (1895), nom. illeg., non Harv. (1865). Type:
    Namibia, Olukonda, Ondonga, 18.ii.1893, *Rautanen* 44 (GH hololotype, M).
    *Othonna polycephala* Klatt in Bull. Herb. Boissier **4**(6): 471 (1896), nom. nov. for *O. rosea*
    *Othonna ambifaria* S.Moore in J. Bot. **37**: 403 (1899). Type: Zimbabwe, near Sashi River, January
    1898, *Rand* 110 (BM holotype).
    *Senecio dinteri* Muschl. ex Dinter in Repert. Spec. Nov. Regni Veg. **23**: 233 (1926), nom. nud. in syn.
    *Emilia ambifaria* (S.Moore) C.Jeffrey in Kew Bull. **41**(4): 919 (1986). —Mapaura & Timberlake,
    Checkl. Zimb. Vasc. Pl.: 25 (2004).

Annual or short-lived perennial herb, 12–75(130) cm tall, erect, sparsely leafy, branching slightly near base of stem or unbranched; stems striate, glabrous. Leaves grey-green and glaucous, narrowly lanceolate to narrowly elliptic, 2–10 × 0.2–2.5 cm, base amplexicaul, margins entire or less often remotely dentate, apex acute or narrowly obtuse; glabrous. Capitula 4–12 together in lax panicles, sometimes corymbose, discoid; peduncles of individual capitula 5–46 mm long; involucre conical to campanulate, 5–8.5 × 4–7 mm; phyllaries (4)5–7, green with purple or reddish tips, (3.2)4.3–7.8 mm long, with narrow scarious margins, glabrous. Florets 8–30, corollas lilac, pale purple or white, exserted, corolla infundibuliform with a very slender tube and an abruptly widened upper part, 5–8.7 mm long, lobes 1–1.3 mm long, with glandular apex; anthers 2.5–3 mm long, bases rounded to slightly caudate, filament collars balusterform; style arms 0.7–1 mm long, apex rounded to conical with central tuft of sweeping hairs and fringing papillae. Achenes cylindrical, (red-)brown, 3.2–5.5 mm long, shortly and densely white-hairy on ribs; pappus setae white, 4.5–7 mm long.

**Caprivi Strip**. Eastern Caprivi, c. 8 km from Katima Mulilo on road to Linyanti (1724AD), 26.xii.1958, c. 914 m, *Killick & Leistner* 3092 (PRE, SRGH, WIND, cited by Cron 2014). **Botswana**. N: Maun–Narraghe Valley road mile 9, 6.v.1967, *Lambrecht* 179 (K, SRGH). SW: 27 km on Jwaneng–Sekoma road, 1180 m, 11.iii.1989, *Terry et al.* 72 (GAB, K). SE: Kwaneng Dist., Matlolakgang ranch, 16.ii.1977, *Hansen* 3035 (C, GAB, K, SRGH). **Zambia**. Cron (2014) cites a specimen from C: Blue Lagoon N.P., 27.ii.1973, *Osborne* 370 (SRGH) which I have not seen. S: Kafue Hoek pontoon, 21.xi.1959, *drummond & Cookson* 6748 (K, SRGH). **Zimbabwe**. W: Wankie Dist., Kazuma depression, 25.i.1974, *Gonde* 19/74 (K, SRGH).

Also in Angola, Namibia and South Africa. Wooded grassland, edge of pans or sites with impeded drainage, alluvial flats, roadsides; on sand or hard loam, or in cracks of limestone pavement; 900–1200 m.

Conservation Status: Least concern (LC) according to Foden & Potter (2005) and Raimondo *et al.* (2009).

The basal tube of the disc florets characteristically widens very abruptly into the distal lobed part of the corolla, unlike *Emilia limosa* (O.Hoffm.) C.Jeffrey which widens gradually (Cron 2014). Head size on a single plant can vary considerably.

The corolla colour can be pale purple or white; in Zimbabwean populations it seems to be mostly white (over 90% white in one site), while in South Africa pale purple is more usual.

A specimen from **Zambia**, W: Mwinilunga, Kalenda Plain, 16.iv.1960, *Robinson* 3597 (K) is close to *E. schinzii* but looks smaller in most parts, including achenes 1–2.7 mm long and pappus setae 3 mm long. More material from this area would be interesting.

15. **Emilia sonchifolia** (L.) DC. in Wight, Contr. Bot. India: 24 (1834). —Oliver & Hiern, F.T.A. **3**: 405 (1877). —Jeffrey in Kew Bull. **41**(4): 917 (1986). —Lisowski, (Asterac. Fl. Afr. Cent. 2) Fragm. Flor. Geobot. **36** Suppl. 1: 387 (1991). —Da Silva *et al.*, Prelim. Checkl. Vasc. Pl. Mozamb.: 31 (2004). —Jeffrey & Beentje in F.T.E.A., Compositae **3**: 596, fig. 123 (2005). Type: Sri Lanka, *Hermann* s.n. (BM lectotype), lectotypified by Grierson in Dassanayake, Revis. Handb. Fl. Ceylon **1**: 252 (1980).

*Cacalia sonchifolia* L., Sp Pl. **2**: 835 (1753).

Annual herb with basal leaf rosette, 10–60 cm tall; stems erect or semi-erect, pubescent in lower part. Leaves sessile, spreading, subsucculent, often glaucous and/or purplish beneath and on margins, 1.8–9.5 × 0.5–6.5 cm, lower spatulate, broadly ovate-reniform to ovate, base cordate, subtruncate or cuneate and decurrent onto an entire or toothed lyrate-pinnately lobed exauriculate petioloid base, margins sinuate-serrate, apex rounded to obtuse, varyingly pubescent especially towards base; median and upper ovate, lanceolate, oblanceolate or obovate or lyrato-pinnately lobed towards auriculate, semi-amplexicaul base, margins sinuate-dentate, apex obtuse, apiculate. Capitula discoid, 2–10 in lax terminal corymbs; peduncles slender, glabrous; involucre cylindrical, sometimes slightly swollen at base, 7.5–11 × 1.5–2 mm; phyllaries 8–9, glabrous except sometimes towards apex or scattered-pubescent throughout, 7–10.5 mm long. Florets 20–44, corollas reddish-purple, mauve, lavender, pink or pale pink, sometimes described as white but this may refer to pappus, 6–7.6 mm long, lobes 0.7–1 mm long. Achenes 2.7–3.3 mm long, shortly hairy; pappus setae 5.5–8 mm long.

**Malawi**. S: Malosa Mountain Forest, 24.vi.1989, *Balaka & Tawakali* 2112 (K, MAL). **Mozambique**. Z: Macuze, 28.viii.1949, *Grandvaux Barbosa & Carvalho* 3859 (K, LISC). MS: Beira & Villa Machado, 18.ix.1911, *Rogers* 4828 (K).

Pantropical. A weed of cultivation, in lawns, maize fields and on track sides, may be locally common; 50–700 m.

Conservation Status: I have seen 12 specimens from our area; widespread and in a common habitat: Least concern (LC).

16. **Emilia tenellula** (S.Moore) C.Jeffrey in Kew Bull. **41**(4): 919 (1986). —Mapaura
& Timberlake, Checkl. Zimb. Vasc. Pl.: 25 (2004). —Cron in Phytotaxa **159**(3): 203
(2014). Type: Zimbabwe, Matopo Hills, bog near 'View', 1219 m, x.1905, *Gibbs* 203
(BM holotype).

   *Senecio tenellulus* S.Moore in J. Linn. Soc. Bot. **37**: 449 (1906).

Annual erect herb 20–45(60) cm tall, glabrous; stems erect, slender, unbranched (except for
inflorescence) or branched slightly near base, stoloniferous in very wet conditions, sparsely leafy mainly
towards base. Leaves sessile, slightly fleshy, linear to narrowly linear-oblanceolate, 30–65 × 1–3.5 mm,
base narrow, margins entire, revolute when dry, apex acute; one-veined, glabrous but cobwebby in
axils. Capitula in lax corymbs of 3–7 on long peduncles occasionally bracteate, homogamous, discoid;
peduncles of individual capitula 1–5.8 cm long; involucre 7–12 mm long, 2.5–4.5 mm diameter at
apex, phyllaries 5–8, linear-oblong, 7–11.5 × 1–1.7 mm, obtuse, glabrous but with papillose apex,
margins scarious. Florets 9–14, shortly exserted, corollas pale pink or whitish tinged mauve or purple
at tips, narrowly infundibuliform, 6–9.5 mm long, lobes 1–1.5 mm long; anthers 1–1.5 mm long; style
arms 0.5–1 mm long. Achenes brown with paler ribs, narrowly ellipsoid, 2.9–4 mm long, glabrous;
pappus setae white, 5–9 mm long.

**Botswana**. N: island in Ngogha River just downstream of Xaenga, 2.x.1975, *P.A. Smith*
1463 (K, SRGH). **Zambia**. W: just W of Dobeka Bridge, 27.x.1937, *Milne-Redhead* 2979 (K).
**Zimbabwe**. W: Besna Kobila farm, v.1957, *Miller* 4385 (K, SRGH).

Also in Angola. In shallow water (standing or slow flow); 900–1450 m.

Conservation Status: I have seen five collections from our area (and two from Angola); the
habitat is specific, but I do not have data on threats. This will have to remain Data Deficient
(DD) until more information comes to light.

The long, narrow capitula are distinctive and help to distinguish this species it from *Emilia
integrifolia*, which grows in a similar habitat.

Two specimens are similar to this species but differ in smaller involucres (5.2–5.5 mm
long, with 5 phyllaries), shorter florets (5.2–5.8 mm long), shorter achenes (1.8 mm long)
and shorter pappus setae (3–3.8 mm long). The leaves in these two specimens are 10–44 ×
1–5 mm. The specimens are both from **Zambia** N: Chishimba Falls, growing on rocks in the
permanent spray zone at an altitude of 1320 m: 15.x.1960, *Robinson* 3983 (K) and 17.x.1967,
*Simon & Williamson* 1065 (K, SRGH). I have these two down as "*Emilia* cf. *tenellula*".

17. **Emilia transvaalensis** (Bolus) C.Jeffrey in Kew Bull. **41**(4): 919 (1986). —Da Silva *et
al.*, Prelim. Checkl. Vasc. Pl. Mozamb.: 31 (2004). —Mapaura & Timberlake, Checkl.
Zimb. Vasc. Pl.: 25 (2004). —Cron in Phytotaxa **159**(3): 204 (2014). Type: South Africa,
Transvaal, prope Thermas "Warm Bath" dictas, i.1906, *Bolus* 12034 (BOL lectotype, K,
PRE), lectotypified by Cron (2014: 204).

   *Senecio thermarum* Bolus in Trans. S. African Philos. Soc. **16**: 389 (1906), nom. illeg., non
   Phil. (1864).
   *Senecio transvaalensis* Bolus in Trans. Roy. Soc. South Africa **1**: 163 (1909).

Herb or subshrub, annual or occasionally perennial, 15–45 cm tall; stems branching, cobwebby,
especially around axils of younger leaves, glabrescent. Leaves closely clustered, glaucous grey-green
or pale green, slightly fleshy, linear-elliptic or narrowly oblanceolate, 15–50 × 2–5.5(10) mm, base
clasping and sessile, margins entire or sometimes dentate distally, revolute when dry, apex acute to
obtuse; cobwebby when young, especially on lower surface, glabrescent. Capitula solitary on long
peduncles discoid, homogamous; peduncles of individual capitula (5)8–24 cm long, glabrous; involucre
campanulate, 4.5–10 mm in diameter with 15–35 florets; phyllaries 8, green with pale scarious margins,
4.5–7 × 1.5–2.2(3) mm, glabrous. Florets 15–35, corollas pale pink, lilac or mauve, infundibuliform,
5–7.5 mm long including glandular lobes 1–1.7 mm long; anthers c. 2 mm long; style branches c. 0.8 mm
long. Achenes dark brown to black, cylindric, 3.5–4.5 mm long, 5-angled with short white duplex hairs
along angles, mucilaginous when wet; pappus setae white, 3–5 mm long, caducous.

**Botswana**. C: 50 km SE of Sephophe on road to Zanzibar, 841 m, 11.ii.2006, *Farrington et al.* MSB 298 (K, NPGRC). **Mozambique**. M: Namaacha, Changalane, 17.x.1983, *Zunguze, Boane & Dungo* 625 (K, LMU).

Also in South Africa. In rocky grassland or scattered tree grassland, often on stony ground or among rocks; (?0)100–850 m. "Liked by butterflies" (fide *Farrington et al.* MSB 298).

Conservation Status: Least concern (LC) according to Foden & Potter (2005) and Raimondo *et al.* (2009). A widespread species in the northern and north-eastern parts of South Africa; I have seen four specimens from our area.

This species is characterised by its grey-green, semi-succulent leaves and discoid capitula with pale pink or lilac corollas.

17b. **Emilia sp. B** (= *Brummitt et al.* 13947)

Erect herb to 40 cm tall from underground rootstock; stems glabrous. Leaves narrowly elliptic, 2.5–5 × 0.4–0.7 cm, base long-attenuate to stem, margins entire, apex obtuse to acute; glabrous or very sparsely pilose; basal leaves purplish beneath. Capitula discoid, solitary on long peduncles (> 10 cm) or in pairs; involucre 7.5–9.5 × 7–8 mm; phyllaries 8, 7.5–9.4 mm long, with narrow scarious margin and papillate apex. Florets c. 36, corollas white, 6.8 mm long, lobes 1.5–1.6 mm long. Achenes 2 mm long, glabrous; pappus setae white, 4.5 mm long.

**Zambia**. W: Chitunta dambo at km 25 Mwinilunga–Kalene Hill road, 21.i.1975, *Brummitt, Chisumpa & Polhill* 13947 (K).

Grassy dambo; 1390 m.

Conservation Status: on recent satellite imagery this area still seems to have quite some original vegetation. However, with only a single specimen known, and dambos prone to agricultural use, this is probably best considered as Endangered (EN D1).

This taxon is reminiscent of *E. transvaalensis*, but has white corollas. This will have to remain an unidentified taxon; it does not look like anything from the adjacent Flore d'Afrique Centrale area and might be new. The area where it occurs is that of *E. fonszambesiaca* but it differs considerably from that taxon.

18. **Emilia vanmeelii** Lawalrée in Bull. Jard. Bot. État Bruxelles **19**: 225 (1949). —Lisowski, (Asterac. Fl. Afr. Cent. 2) Fragm. Flor. Geobot. **36** Suppl. 1: 381 (1991). —Jeffrey in Kew Bull. **52**(1): 210 (1997). —Jeffrey & Beentje in F.T.E.A., Compositae **3**: 582 (2005). Type: Tanzania, Lac Tanganyika, Utinta, ii.1947, *van Meel* 918 (BR holotype).

Annual herb 20–100 cm high; stems erect, unbranched or branched from base, sometimes purple near base, pubescent in lower part, rarely glabrous. Leaves slightly fleshy, lower often purplish beneath and on margins, broadly reniform-spatulate to obovate, 1.5–13 × 1.4–7 cm (in longer leaves lamina as long as petioloid base), decurrent onto a narrow exauriculate petioloid base to sessile and cordate, margins subentire to undulate-dentate, apex rounded or obtuse; median and upper obovate, oblong, ovate or lanceolate, 3–14 × 1.5–7.5 cm, base auriculate and semi- to fully amplexicaul, margins shallowly sinuate-dentate, apex rounded in median to acute in upper leaves, sparsely pubescent, especially beneath. Capitula solitary or 2–4 in terminal corymbs (but close branches may give impression of more), dense to lax, discoid; peduncles glabrous or shortly pubescent; involucre broadly cylindrical, 5–10 × 3–10 mm; phyllaries 11–21, 4.5–9.3 mm long, green with brown tips, glabrous or pubescent. Florets 26–48, corollas red, orange or less often yellow, 6–10.5 mm long, lobes 1–1.8 mm long. Achenes 2–3.5 mm long, sparsely hairy or glabrous; pappus setae 3–6 mm long.

**Zambia**. N: Muchinga escarpment, Ntumbachushi Falls, 19.iv.1989, *Pope, Smith & Goyder* 2180 (BR, K, LISC, MO, NDO, SRGH). C: Luangwa Valley, S of Lion Plain, 21.iii.1966, *Astle* 4691 (K). E: Chadiza, 11.ii.1957, *Angus* 1514 (K, SRGH). **Malawi**. N: Mizimelo area, Chisengo, Chitipa, 11.iii.2007, *Chapama, Likunda & Salubeni* 599 (K, MAL). C: Lilongwe Nature Sanctuary forest, 3.v.1984, *Patel, Seyani & Banda* 1475 (K, MAL). S: Dedza plateau,

W of Dembeke Mission, 25.iv.1971, *Pawek* 4680 (K, MAL). **Mozambique**. N: Marrupa, 20 km on Lichinga road, 16.ii.1981, *Nuvunga* 514a (K, LMU). T: 27.5 km on Fingoe–Vasco da Gama road, 27.vi.1949, *Grandvaux barbosa & Carvalho* 3322 (K, LMJ).

Also in D.R. Congo and Tanzania. Woodland, more common in disturbed situations; 650–1900(2500?) m.

Conservation Status: I have seen 61 specimens from our area; widespread in a common habitat, so Least Concern (LC).

*E. vanmeelii* taxon was distinguished from *E. coccinea* by Jeffrey (1997) on the basis of glabrous achenes, 'or at most with a few hairs near the apex'; Jeffrey added that specimens of this taxon with hairy achenes may represent the result of introgression from *E. coccinea*. In his key to the species, Jeffrey keyed out *E. coccinea* as having the achenes hairy throughout – but in the description of *E. coccinea* the achenes are said to be 'shortly hairy at least in the upper part'. In the same article the two taxa are differentiated by the median and upper stem leaves, about 3 times as long as wide in *E. coccinea*, but about 1.5 times as long as wide in *vanmeelii*; in the text Jeffrey adds the taxa "may be distinguished (although sometimes with difficulty) by the shape of the cauline leaves". A third taxon *E. caespitosa*, has glabrous achenes, according to this article, and leaves 3–4 times as long as wide. As I have already (provisionally) added *E. caespitosa* to the synonymy of *E. coccinea* (see above) the character of the hairyness of the achenes, already weak as seen by the descriptions in Jeffrey (1997), becomes null. This leaves only the cauline leaves character. Much of the Mozambique material of *E. vanmeelii* has cauline leaves 2–2.5 times as long as wide. The same is true for some specimens from Zambia, and material from Malawi shows a range of 1.7–3 times as long as wide for upper leaves, though some of the cauline leaves are almost circular. Still, material from the combined taxa *E. coccinea* and *E. caespitosa* has consistently narrower leaves, and I keep (reluctantly) *E. vanmeelii* as a species in its own right. While I do not like to distinguish a species on the base of leaf length/width ratio alone, the fact that the habitat seems consistently linked to miombo woodland (more consistently so than *E. coccinea*) tipped the scales. It is possible that the basal leaves of *E. vanmeelii* are more persistent than those of *E. coccinea*.


*Uncertain species*

**Emilia violacea** Cronquist in Bull. Jard. Bot. État Bruxelles **22**: 309 (1952). —Jeffrey in
    Kew Bull. **41**(4): 915 (1986). —Lisowski, (Asterac. Fl. Afr. Cent. 2) Fragm. Flor. Geobot.
    **36** Suppl. 1: 371 (1993). —Jeffrey & Beentje in F.T.E.A., Compositae **3**: 589 (2005).
    Type: Burundi, Muhovuozi Plains, *Michel & Reed* 982 (BR holotype).

Erect or decumbent annual herb 14–80 cm tall; stems light green, smooth, glabrous. Leaves sessile, lanceolate, 2.2–13 × 0.4–2 cm, base sagittate, auriculate and semi-amplexicaul, margins minutely obscurely to clearly sinuate-denticulate, especially on auricles, apex obtuse, apiculate, glabrous. Capitula 2–4 in lax, divaricate terminal cymes, discoid; peduncles of individual capitula slender, glabrous; involucre cylindrical, 5–7 ×2 mm; phyllaries 8 or 13, 4.5–7 mm long, glabrous. Floret corollas pale mauve or violet, 5.5–6.5 mm long, tube glabrous, lobes 1.2–2 mm long, those of outer florets larger. Achenes 2.8–4 mm long, glabrous; pappus setae 2–4 mm long.

It is just possible this specimen occurs in Zambia; a specimen from there: N: Mbala district [Abercorn], 1949–1951, *Bullock 1222* (K) looks rather like this taxon and does not resemble anything else. There are no label data apart from "District: Abercorn" and the date range. The pubescent achenes are 6.2–6.3 mm long, which does not agree with *E. violacea*, and the leaves have small (1.5 mm) auricles at their base; otherwise it conforms to the description of *E. violacea*.

This species is known to occur in D.R. Congo, Burundi and Tanzania. In moist grassland or swamp grassland; 950–1150 m.

## 109. **EMILIELLA** S.Moore

**Emiliella** S.Moore in J. Bot. **56**(668): 225 (1918). —Torre in Garcia de Orta, Sér. Bot. **2**(2): 85–88 & Tabs. I–III (1975). —Lisowski, (Asterac. Fl. Afr. Cent. 2) Fragm. Flor. Geobot. **36** Suppl. 1: 409 (1991). —Nordenstam in Kubitzki, Fam. Gen. Vasc. Pl. **8**: 238 [2006](2007). —Hind & Frisby in Kew Bull. **69**(4)-9550: 1–7 (2014). —Mapaya & Cronn in Taxon **70**-12417: 1–12 (e-published Dec. 2020); in Taxon **70**(1): 127–138 (2021).

Annual herbs. Stems simple or branched at base, glabrous or pubescent, hairs simple, eglandular. Leaves alternate, simple, sessile to pseudopetiolate, oblong-obovate to lanceolate, margins slightly serrate. Inflorescence of solitary capitula or lax, few-headed corymbs. Capitula homogamous and discoid, few-flowered (5–12); involucre narrow, ecalyculate; phyllaries uniseriate, 4–7, connate at first, eventually suturing between one pair of adjacent phyllaries at achene maturity but remaining fused at base, glabrous or sparsely to densely pubescent; receptacle flat, epaleaceous, glabrous. Florets few to several (5–14), hermaphrodite, all fertile; corollas scarcely protruding from involucre, long-5-lobed, pink-purple or violet-blue; apical anther appendages triangular, about as long as wide, basal anther appendages ecalcarate, obtuse, filament collars balusterform; style base slightly bulbous, glabrous, style shaft glabrous, style arm apices truncate and tip penicillate. Achenes linear to narrow-cylindrical, glabrous, 5–10-ribbed, ribs pale; pappus rarely a single bifid or markedly lacerate scale, sometimes of several broad, often dissimilar scale-like elements fused at base, apices and margins of lacerate elements barbellate, off-white, or pappus absent.

A genus of 7 spp. of Angola, D.R. Congo and Zambia. Four spp. have been recorded from Zambia in the Flora area.

Jeffrey (Kew Bull. **41**(4): 875, 1986) commented that *Emiliella* may be united with *Emilia* but clearly had not done anything by 2005 in his joint account in the *Flora of Tropical East Africa*. The two genera are clearly distinct and easily separated, especially on the basis of the involucre, pappus, and the relative length of the achene to the pappus.

Mapaya & Cron (2021) have proposed a new phylogeny of just under half of the species of *Emilia* Cass., and within their new concept have proposed sinking both *Bafutia* C.D.Adams and *Emiliella*. Because both *Bafutia* and *Emiliella* 'nested within *Emilia*' they have seen fit to sink both genera, and all the species, within *Emilia*, and provided all the necessary combinations. Perhaps one of the more significant characters, the nature of the pappus, was not commented on. The standing of *Emiliella* remains as a separate genus in this Flora.

Moore (1918) originally described the capitula as 'homogama, disciformia', although clearly indicating only one type of floret; the capitula are discoid. Nordenstam (2006: 238) described the pappus as 'a single scale', an unusual state in the genus, since one or two spp. have obvious irregular scale-like elements (see Hind & Frisby 2014). Nothing like this is seen in *Emilia*, something not commented on by Mapaya & Cron (2021).

1. Plants diminutive herbs, 5–8 cm tall; pappus of a single excentric deeply laciniate scale; lower leaves distinctly petiolate, upper pseudopetiolate or sessile, margins entire to inconspicuously dentate; achenes c. 3 mm long . . . . . . . . . . . . . . . . . . . . . . . . . . .**1.** *exigua*
- Plants straggly, sometimes basally much-branched, herbs, 12–45 cm tall; pappus of basally connate scales, sometimes appearing deeply lacerate and with one deep sinus; all leaves sessile or pseudopetiolate, margins usually conspicuously lobed or coarsely toothed; achenes 6–9 mm long . . . . . . . . . . . . . . . . . . . . . . . . . . . . . . . . . . . . . . . . . . . . . . 2
2. Plants 12–18 cm tall; phyllaries sparsely pubescent . . . . . . . . . . . . . . . . . . .**2.** *drummondii*
- Plants 20–45 cm tall; phyllaries glabrous or densely pubescent . . . . . . . . . . . . . . . . . . . 3
3. Plants 30–35 cm tall; phyllaries 4, pubescent; florets 8; capitulum c. 12 mm tall . . . . . . . . . . . . . . . . . . . . . . . . . . . . . . . . . . . . . . . . . . . . . . . . . . . . . . . . . . .**3.** *zambiensis*
- Plants 40–45 cm tall; phyllaries 6–8, densely pubescent; florets 12–14; capitula 7–10 mm tall . . . . . . . . . . . . . . . . . . . . . . . . . . . . . . . . . . . . . . . . . . . . . . . . **4.** *luwiikae*

1. **Emiliella exigua** S.Moore in J. Bot **56**(668): 225 (1918). Type: 'Angola, along the Cubango in moist situations; *Gossweiler*, 2093.' (BM000924662 holotype).

> *Emilia exigua* (S.Moore) Mapaya & Cron in Taxon **70**-12417: 9 (e-published Dec. 2020); in Taxon **70**(1): 135 (2021).

Annual herb, 5–8 cm tall; roots fine and fibrous. Stems simple, or few-branched, finely ribbed, sparsely pubescent and densely pubescent in leaf axils, hairs multicellular, uniseriate and eglandular. Basal leaves opposite, simple, membranous, sessile to short-petiolate, ovate or elliptical to obovate, 3 × 1–1.5 mm wide, margins entire, apex obtuse, stem leaves alternate, simple, membranous, sessile, pseudopetiolate, pseudopetiole c. 4 mm long, narrowly winged, wings 0.25–0.5 mm wide, lamina obovate-spatulate, 12–15 × 3–5 mm, sparsely pubescent, hairs multicellular, uniseriate, eglandular; midrib prominent beneath, 3-veined from base, pinnate venation not prominent, margins entire or rarely with 2 pairs of shallow teeth, upper leaves lanceolate. Inflorescence of small solitary, terminal capitula, or of 2 or 3 capitula cymosely arranged, capitula pedicellate, pedicels 8–10 mm long (measured from uppermost bract to base of involucre), glabrous, bracteolate, bracteoles scale-like. Capitula 5–6 × 1–2 mm including corollas which protrude 1–2 mm; involucre ecalyculate; phyllaries 5, essentially glabrous, although usually sparsely pubescent at apex, green, apices usually tinged purple, margins scarious, apices acute, evidence of a rent opening between phyllaries with achene maturity. Florets 5–8, corollas pink, corolla tube cylindrical, c. 1.5 × c. 0.25 mm, glabrous throughout, throat campanulate, corolla lobes 5, c. 0.5 mm long, ascending, tapered towards apex, prominent orange central vein, thickened at lobe apex; anther cylinder transparent/whitish, included; apical anther appendages triangular, as long as wide, apex obtuse, basal anther appendages ecalcarate, shortly sagittate; filament collars conspicuous and balusterform; style arms squat, truncate, contained within anther cylinder, apices penicillate. Achenes narrowly cylindrical, 2.75–3 × 0.5–0.75 mm, body dark, 8–10-ribbed, ribs longitudinal, pale, glabrous throughout; carpopodium absent; pappus paleaceous, c. 1mm long, excentric, apices dissected into 4 or 5 laciniate lobes, abaxially barbellate, off-white.

**Zambia**. W: Northwestern Province, Mwinilunga Dist., Salujinga, near Jimbe source, 10°55'20"S, 24°7'25"E, 1311 m, 14.xi.2006, *Bingham* 13161 (K).

Also known from Angola and D.R. Congo. Label information indicates in 'suffrutex savanna', and the plants are usually found adjacent to water courses in damp ground or around ephemeral pools; 1311 m; flowering in October and November.

Conservation Status: Although apparently widely distributed the species is very infrequently collected. Population sizes may however be relatively moderate. Best recorded as DD (Data Deficient) until further fieldwork can be carried out throughout its range.

2. **Emiliella drummondii** Torre in Garcia de Orta, Sér. Bot. **2**(2): 86 (1975). Type: 'Zambia: Mongue Distr., edge of Bulosi [plain below Mongu], [9-]XI-1959, *Drummond & Cookson* 6270' (K000306841 holotype, LISC002787, PRE, SRGH).

> *Emilia drummondii* (Torre) Mapaya & Cron in Taxon **70**-12417: 9 (e-published Dec. 2020); in Taxon **70**(1): 135 (2021).

### Var. **drummondii**

Small annual herb, (5)12–18 cm tall. Stems erect, simple or branched, pubescent. Leaves sessile, lanceolate or spatulate or oblanceolate, margins remotely lobate or dentate, 1–4.5 × 0.2–0.7 cm, base auriculate, surfaces white pilose, apices obtuse or acute. Pedicels filiform, 0.6–1.5 cm long. Capitula solitary, terminal or axillary; involucre cylindrical, ecalyculate; phyllaries 6, sparsely pubescent, linear-oblong, c. 7 mm long. Florets 12, corollas pink; style arms truncate, penicillate. Achenes c. 6 mm long, ribbed, glabrous; pappus of basally connate scales, c. 2 mm long.

**Zambia**. B: Western Province, Sefula Dambo, Kasuka Village, 15°22'S, 23°13'E, 1050 m, 15.x.1993, *Bingham* 9747 (K).

This species was originally described with two varieties; var. *moxicoensis* possesses epappose achenes and probably deserve separate recognition, although not present in the Flora area.

The typical subspecies is only known from Zambia. Growing in water in a canal; 1050 m; flowering in October.

Conservation Status: A very poorly collected species, most probably overlooked, and quite probably mistaken for a narrow-headed *Emilia* sp.; DD (Data Deficient).

3. **Emiliella zambiensis** Torre in Garcia de Orta, Sér. Bot. **2**(2): 86 (1975). Type: 'Zambia: Chiusali Distr., Shiwa, Ngandu, 21-XII-1964, *Robinson* 6323' (K000306839 holotype, M0105130).

> *Emilia zambiensis* (Torre) Mapaya & Cron in Taxon **70**-12417: 9 (e-published Dec. 2020); in Taxon **70**(1): 135 (2021).

Annual herb to c. 35 cm tall. Stems erect or ascending-erect, branched at base, weakly sparsely pubescent, hairs long, uniseriate, multicellular and eglandular. Lower leaves pseudopetiolate, upper sessile, lanceolate, oblanceolate or spatulate, (1.5)2–4(4.5) × 0.2–1(1.5) cm, pubescent above and beneath, hairs long, multicellular, white, midrib prominent beneath, margins unequally lobed or dentate. Inflorescences of solitary terminal capitula, rarely subcorymbose and 2-headed, pedicels filiform, glabrescent or sparsely pubescent, (2)3–10(18) mm long. Capitula (7)11–13 mm tall; involucre ± cylindrical, c. 13 mm long, glabrous; phyllaries 4, pubescent, linear-lanceolate, apices obtuse or acute. Florets 8, corollas tubular, pink or purple, c. 3 mm long, 5-lobed, glabrous; style arm apices truncate, penicillate. Achenes c. 9 mm long, c. 10-ribbed, glabrous, narrowed slightly above; pappus of basally connate scales, c. 2.5 mm long, lacerate, apices barbellate, white.

Only known from the type collection from the Northern Province of Zambia. Permanently wet dambo; 1475 m; flowering between September and December.

Conservation Status: Most certainly undercollected; DD (Data Deficient).

4. **Emiliella luwiikae** D.J.N.Hind & Frisby in Kew Bull. **69**(4)-9550: 2 (2014). Type: 'Zambia, Western Province, Mongu, Bulozi floodplain, 8.xii.1995, *Bingham & Luwiika* 10718.' (K holotype, ?MRSC). FIGURE 6.4.**19**.

> *Emilia luwiikae* (D.J.N.Hind & Frisby) Mapaya & Cron in Taxon **70**-12417: 9 (e-published Dec. 2020); in Taxon **70**(1): 135 (2021).

Annual erect herb, 28–45 cm tall; roots fine, fibrous. Stems simple, finely ribbed, densely pubescent, hairs multicellular, uniseriate and eglandular. Leaves simple, alternate, membranous, sessile, lamina oblong-oblanceolate, 38–56 × 15–20 mm, pubescent, hairs multicellular, uniseriate, eglandular, midrib prominent beneath, pinnate venation not prominent, apex retuse, base often truncate, margins irregularly lobed at base and irregularly coarsely dentate. Inflorescence of solitary terminal capitula or a few-headed (1–2) cyme, capitula pedicellate, pedicels 12–15 mm long (measured from uppermost bract to base of involucres), densely pubescent, hairs multicellular, uniseriate, eglandular, bracteolate, bracteoles scale-like. Capitula 7–10 × 3–4 mm; involucres cylindrical; phyllaries 6–8, linear-oblong, densely pubescent, hairs multicellular, uniseriate, eglandular, green, ridged, scarious at margins, apices acute, papillose, connate at first eventually suturing between a pair of adjacent phyllaries at achene maturity. Florets 12–14, corollas pink, corolla tube cylindrical, glabrous, c. 1.5 × 0.25 mm, throat campanulate, corolla lobes 5, ascending, c. 0.5 mm long, tapered towards apex, central vein prominent, thickened, apex papillose; anther cylinder transparent and whitish, included within corolla, apical anther appendages triangular, twice as long as wide, apex obtuse; anther bases calcarate, filament collars conspicuous and balusterform; style shaft glabrous, style arms truncate, contained within anther cylinder, apices penicillate. Achenes narrowly cylindrical 6–7 × c. 0.5 mm, 8–10-ribbed, ribs longitudinal, prominent, body straw-coloured, glabrous throughout; carpopodium absent; pappus paleaceous, 1.6–2 mm long, of either 1 deeply laciniate scale with 1 prominent sinus to base (rarely with a small rudimentary scale at base of sinus), or of few connate scales except for 1 deep sinus, abaxially moderately to densely barbellate, free-cell ends adpressed, antrorse, surface lustrous, straw-coloured, apices laciniate.

**Zambia**. W: Mongu, Bulozi floodplain, 8.xii.1995, *Bingham & Luwiika* 10718 (K).

Moist areas on the edge of the Bulozi flood plain; c. 1050 m; probably flowering November-December.

Conservation Status: Currently only known from the type collection; DD (Data Deficient).

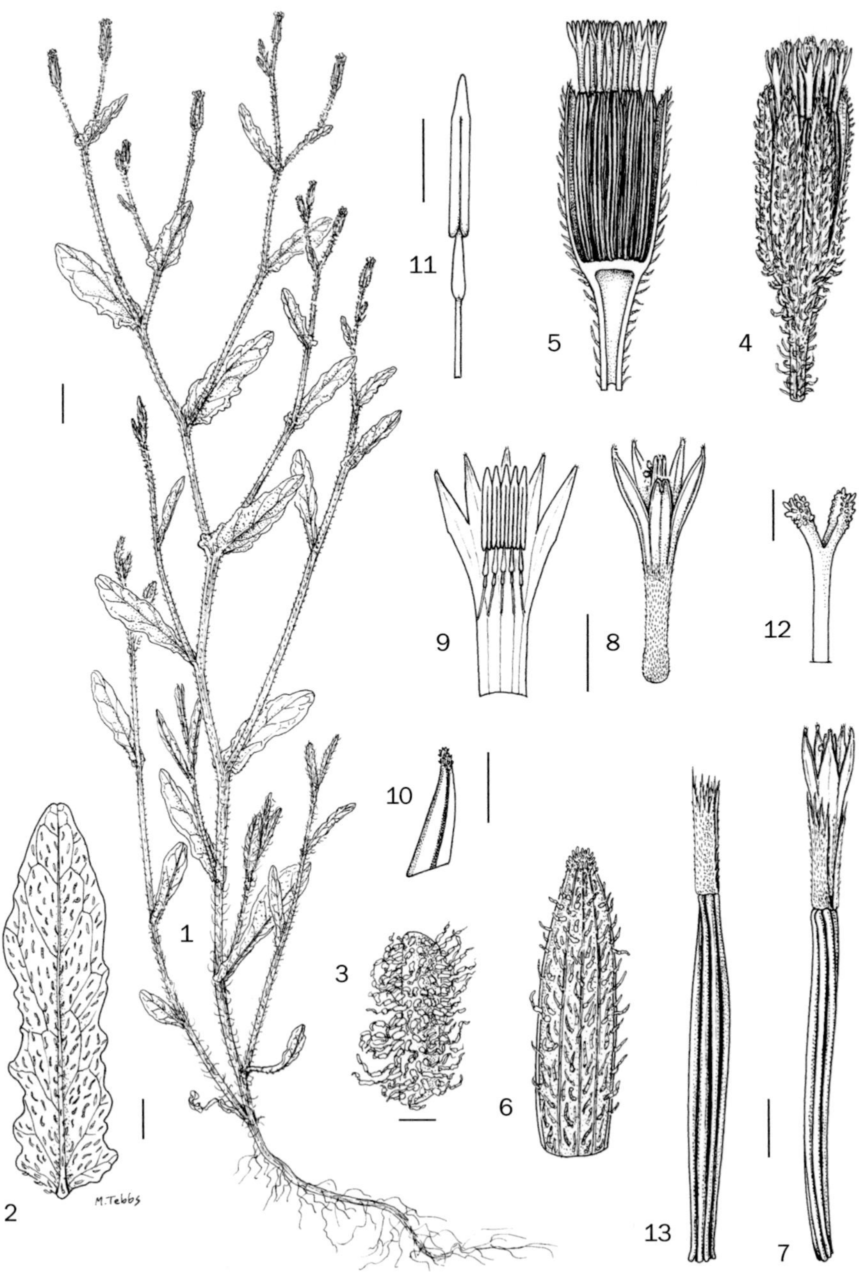

Fig. 6.4.**19**. EMILIELLA LUWIIKAE. 1, habit; 2, leaf; 3, stem detail; 4, late flowering-stage capitulum; 5, l.s. capitulum; 6, phyllary; 7, floret; 8, corolla (including anther cylinder and style); 9, corolla opened out; 10, detail of corolla lobe apex; 11, stamen; 12, style arms; 13, mature achene. All from *Bingham & Luwiika* 10718. Scale bars: 12 = 0.3 mm; 11 = 0.5 mm; 3–10 = 1 mm; 2 = 5 mm; 1 = 1 cm. Drawn by Margaret Tebbs. Reproduced from Kew Bulletin (2019) with permission of the artist.

## 110. **MIKANIOPSIS** Milne-Redh.

**Mikaniopsis** Milne-Redh. in Exell, Suppl. Cat. Vasc. Pl. S. Tomé: 27 (1956). —Jeffrey in Kew Bull. **41**(4): 878–879 (1986). —Bremer in Asterac. Cladist. Classif.: 505 (1994). — Lisowski, (Asterac. Fl. Afr. Cent. 2) Fragm. Flor. Geobot. **36** Suppl. 1: 440–448 (1991). — Malombe in F.T.E.A., Compositae **3**: 613–617 (2005).

Scandent perennial herbs or woody climbers. Stems simple or poorly branched, fluted, glabrous or arachnoid-pubescent. Leaves cauline, alternate, simple, glabrous or variously pubescent, margins entire, shallowly lobed or denticulate, 5–7-palmately veined, longly petiolate; petiole prehensile with basal thickening becoming woody and forming persistent hook which remains after leaf abscission. Inflorescence axillary, racemose, cymose or paniculate-corymbose, shorter than subtending leaves. Capitula homogamous and discoid or heterogamous and disciform, outer florets female, disc florets hermaphrodite; involucre calyculate, obconic, turbinate or narrowly campanulate; phyllaries uniseriate; receptacle flat or slightly convex, alveolate, glabrous. Florets 5–25, corollas with narrow cylindrical tube and campanulate 5-lobed upper limb, white, greenish or pale yellow; filament collars balusterform, anther bases sagittate, short-tailed; style arms long, slightly flattened, truncate, cone-shaped or rounded at apex, exappendiculate. Achenes oblong or cylindrical, 8–10-ribbed, glabrous; pappus setae 1–3-seriate, setose, whitish or stramineous.

A genus of about 15 spp. of tropical Africa, with the largest concentration of species in central Africa.

Leaves very broadly ovate to ovate-triangular; capitula heterogamous and disciform, phyllaries bases not swollen; corolla lobes without median resinous line . . . . . . . . . . . **1.** *tanganyikensis*
Leaves deltoid-ovate, sometimes rhomboid; capitula homogamous and discoid, phyllaries bases swollen; corolla lobes with median resinous line . . . . . . . . . . . . . . . . . . . . . . .**2.** *cissampelina*

1. **Mikaniopsis tanganyikensis** (R.E.Fr.) Milne-Redh. in Exell, Suppl. Cat. Vasc. Pl. S. Tomé: 30 (1956). —Jeffrey in Kew Bull. **41**(4): 879 (1986). —Lisowski (Asterac. Fl. Afr. Cent. 2) Fragm. Flor. Geobot. **36** Suppl. 1: 444 (1991). —Malombe in F.T.E.A., Compositae **3**: 616 (2005). Type: [Zambia:] 'Nordost-Rhodesia: Luvingo (zwischen dem Tanganyika- und Bangweolo-See), in einem auf feuchtem Boden wachsenden Wäldchen (fast verblüht und fruchttragend 24. Okt. [1911] – [*Fries*] n. 1103)' (UPS holotype, K – fragment drawings and note by Fries on same sheet & image).

    *Senecio tanganyikensis* R.E.Fr. in Wiss. Ergebn. Schwed. Rhodesia-Kongo-Exped. 1911–12, **1**: 345 (1916). —Brenan in T.T.C.L.: 160 (1949).

Woody or herbaceous climber to 6 m. Stems cylindrical, pale grey or greenish, corky, glabrous. Leaves with petiole 1.9–6 cm long, sometimes purple; leaves on sterile shoots very wide ovate or ovate-triangular, apex acuminate, base cordate, unfrequently truncate, margins coarsely sinuate-dentate or triangular-lobulate, sometimes entire; on fertile shoots ovate to very wide ovate, base cordate, apex acuminate, margins entire or with one or two teeth on each side near base; lamina 3.5–12 × 2.7–9 cm, almost glabrous beneath, young leaves tomentose, glabrous above, primary veins prominent beneath, secondary veins reticulate, conspicuous beneath, coriaceous to subcoriaceous. Inflorescence of 3–30 capitula in short dense axillary panicle or cymose corymb; peduncles of individual capitula 3.9–7.5 mm long, glabrescent or thinly arachnoid-floccose and glandular, bracts 3–4 mm long, lanceolate. Capitula heterogamous, disciform; involucre turbinate, 5–6.7 × 3–5.6 mm; calycular bracts 3, lanceolate, 1.6–3.5 mm long, sparsely hairy or glabrescent; phyllaries 8–11, oblong to oblanceolate, 4.5–6 × 1.1–1.6 mm, glabrous but puberulent at apex, margins whitish-scarious, apex obtuse to acute, base not swollen. Florets 8–13, outer female, with corollas often shorter than inner, inner hermaphrodite; corolla white, 5.5–7.5 mm long, swollen at base, lobes 0.5–1 mm long, without median resinous line; style arms truncate. Achenes subcylindrical, 2–4 mm long, glabrous; pappus setae 5–7.2 mm long, whitish.

**Zambia**. N: Chishimba Falls, 10.ix.1958, *Fanshawe* F4785 (K, NDO). W: Kitwe, 27.viii.1967, *Mutimushi* 2027 (K, NDO). E: 10 m N of Nyika, 17.ix.1958, *Lawton* 486 (K, NDO). **Malawi**. N: Mzimba Dist., 2.5 m SW of Chikangawa, 1700 m, 28.ix.1978, *Phillips* 4012 (K, MO).

Also known in D.R. Congo and Tanzania. Forest margins, swamp forest or gallery forest; 1200–1800 m.

Conservation Status: LC, widespread.

2. **Mikaniopsis cissampelina** (DC.) C.Jeffrey in Kew Bull. **41**(4): 879 (1986). Type: [South Africa:] 'in Africae Capensis territorio cesso legit cl. *Ecklon* pl. exs. n. 1367! [1835] ... (v. s.)' (G-DC holotype, HAL, M). FIGURE 6.4.**20**.

> *Cacalia? cissampelina* DC., Prodr. **6**: 331 (1838).
> *Senecio cissampelinus* (DC.) Sch.Bip. in Flora **28**: 499 (1845). —Hilliard & Burtt (1973: 376–378).

Branched climber to 5 m. Stems cylindrical, dark brownish, greyish or greenish, corky, arachnoid-pubescent, glabrescent, becoming woody. Leaves with petiole 2.7–5 cm long; lamina deltoid-ovate, sometimes rhomboid, 4.1–7.7 × 4–8 cm, apex acute to acuminate, base cordate, subcordate or truncate, margins blunt-3–5-angled or shallow-lobed, sinuses rounded, lobes deltoid, apices mucronate, sparsely setulose to almost glabrous beneath, glabrous above, primary veins prominent beneath, secondary veins inconspicuous, chartaceous to subcoriaceous. Inflorescence of 9–16 capitula in short corymbose panicle; peduncles of individual capitula 4.2–8.3 mm long, glabrescent or arachnoid-floccose, glandular, bracts 1.1–5.2 mm long, lanceolate or oblong. Capitula homogamous, discoid; involucre narrowly campanulate, 5.5–6.7 × 4.4–5.9 mm; calycular bracts 3–5, oblong to lanceolate, 1.7–3 mm long, sparsely ciliate at margins, hairy at apex; phyllaries 5–8, oblong to lanceolate, 4–5.2 × 1.2–1.5 mm, glabrous but hairy at apex, margins light greenish-scarious, apex acute, base swollen. Florets 8–11, hermaphrodite; corolla cream to pale yellow (sometimes described as 'creamy pink'), 6.2–7 mm long, slightly swollen at base, lobes c. 2 mm long, with median resinous line, strongly inrolled; style arms truncate, penicillate. Achenes broadly cylindrical, 2.5–3.7 mm long, glabrous; pappus setae 6–8 mm long, white.

**Zimbabwe**. E: Umtali, Mountain Home, 1525 m, 26.vii.1948, *Chase* 860 (K, SRGH).

Growing on forest trees, and in forest margins; 750–1525 m; flowering from (May)June to August.

Also in South Africa (Eastern and Western Cape, KwaZulu-Natal, Limpopo, Mpumalanga).

Conservation Status: LC in South Africa, only just extending into the Flora area.

The most southerly representative of the genus, appearing only in the Umtali area (according to herbarium records) in the Flora area. It is probably under-collected, as it normally flowers mid-winter (Hilliard 1977). Chase noted (*Chase* 860) that the florets were 'showy'; Grobbelaar (compiler) Rep. Agric. Res. Serv. (2000) reported that they were 'sweet almond scented'.

The two specimens from the Flora area differ slightly from South African ones by their cordate, subcordate or truncate leaf bases (vs. broad-cuneate to subtruncate, rarely cordate) and the creamy-whitish leaf venation (vs. purplish). The capitula seem also larger in South African specimens. More specimens are needed to understand the relevance of such variation.


# 111. **AUSTROSYNOTIS** C.Jeffrey

**Austrosynotis** C.Jeffrey in Kew Bull. **41**(4): 878 (1986). —Malombe in F.T.E.A., Compositae **3**: 611–613 (2005).

Scandent or climbing perennial herb. Leaves cauline, petiolate, petioles prehensile, base auriculate or exauriculate, lamina ovate-cordate, glabrous or arachnoid- or tomentose-pubescent, palmately veined, margins entire to dentate or denticulate. Inflorescence of many-headed paniculate corymbs, capitula pedicellate. Capitula radiate, heterogamous; involucre calyculate. Marginal florets few, radiate, female, ray limbs yellow; style arms truncate, penicillate. Disc florets hermaphrodite, several, corollas yellow; anther bases caudate; style arms truncate, apices penicillate. Achenes oblong, sparsely setuliferous; carpopodium a broad annulus concolorous with achene body, glabrous; pappus setae numerous, uniseriate, capillary, coarsely barbellate, white, caducous.

A monotypic genus known only from Malawi, Zambia and Tanzania.

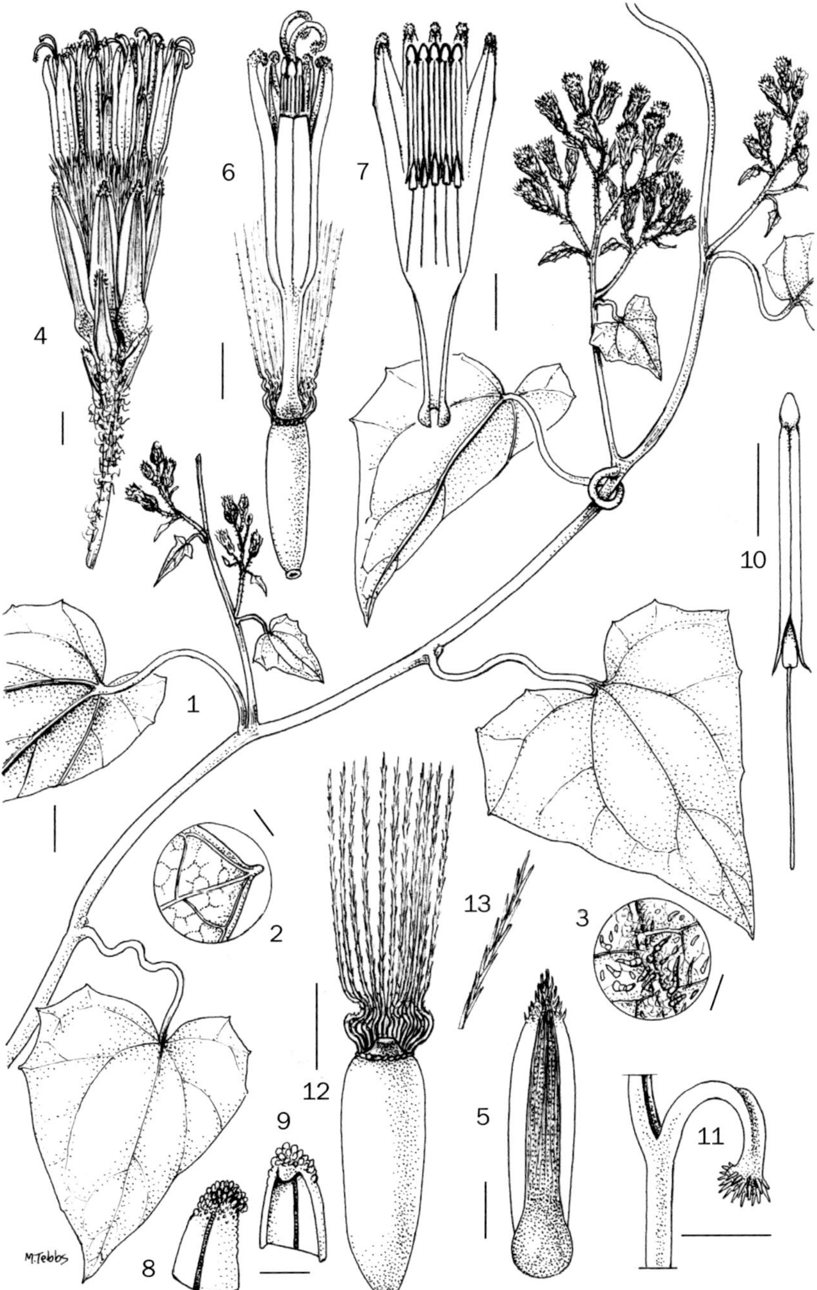

Fig. 6.4.**20**. MIKANIOPSIS CISSAMPELINA. 1, flowering stem; 2, detail of tooth on lower leaf margin; 3, detail of lower surface of leaf lamina; 4, capitulum; 5, phyllary; 6, floret; 7, corolla opened out to show attachment point of filaments; 8, detail of outer surface of corolla lobe; 9, detail of inner surface of corolla lobe; 10, stamen; 11, detail of style arm; 12, achene and basal portion of pappus setae; 13, detail of pappus seta. 1–3 from *Chase* 763; 4–12 from *Chase* 860. Scale bars: 1 = 10 mm, 2–12 = 1 mm. Drawn by Margaret Tebbs.

Fig. 6.4.**21**. AUSTROCYNOTIS RECTIRAMA. 1, habit (× ⅔); 2, capitulum (× 4); 3, ray floret (× 10); 4, disc floret (× 10); 5, achene, with one pappus seta remaining, others fallen (× 12). 1–4 from *Richards* 15048; 5 from *Salubeni* 1690. Drawn by Juliet Williamson. From Flora of Tropical East Africa.

**Austrosynotis rectirama** (Baker) C.Jeffrey in Kew Bull. **41**(4): 878 (1986). Type: [Malawi:] 'British Central Africa. Nyika plateau, alt. 6000–7000 ft., [July 1896] Whyte, 110.' (K000377631 holotype). FIGURE 6.4.**21**.

*Senecio rectiramus* Baker in Bull. Misc. Inform., Kew **1898**(139): 155 (1898).

Climber to 7 m; stems terete, floccose-tomentose, glabrescent, sometimes red, or purple-spotted. Leaves ovate, 3–8 × 2–6 cm, cordate, conspicuously sinuate-dentate, shortly acuminate, green, sparsely arachnoid and glabrescent above, densely grey-tomentose beneath; petioles 2.6–3.5 cm long, exauriculate or with small auricles at base, 3–5 × 5–7 mm, glabrous above, tomentose beneath, margins entire to dentate or denticulate, base amplexicaul, apex rounded. Inflorescences paniculate-corymbs, pedicels 2–10 mm long, bracteolate, bracteoles scale-like, linear, 1–2 mm long. Involucre obconic, 3–4 mm long; calycular bracts 5, lanceolate, 2.5–3 mm long, floccose, glabrescent, dark-tipped; phyllaries 13, oblong-lanceolate, 4 mm long, thinly floccose to tomentose, glabrescent, apex darker and bearded. Ray florets 5, ray limbs yellow, 4.5–7 × 1.2–1.5 mm, 4 or 5-veined. Disc floret c. 15, corollas yellow, 4.5–7 mm long, lobes 1.2–1.5 mm long; style arms rounded, unappendaged, fringed with short hairs. Achenes (immature) 1.5–2.3 mm long, 4-ribbed, body dark brown (often appearing blackish), sparsely finely setuliferous, setulae of twin-hairs, apices connate; pappus setae 5–8 mm long, white.

**Zambia**. Muchinga Province, Mpika Dist., Mutinondo Wilderness Area, Charlie's Rock to Quentins Rock, Chipundu Stream, 12°27'S, 31°18'E, 1450 m, 2.v.2015, *Bidgood et al.* 8539 (K). **Malawi**. N: Vitinithiza Hill slope, Nyika National Park, 10°45'21"S, 33°40'56"E, 1795 m, 4.viii.2008, *Mphamba* 867 (FRIM, K). C: Kota Kota District, Nchisi Mountain, 1400 m, 27.vii.1946, *Brass* 16987 (K, NY).

Also known from one collection in Tanzania. Dense riverine forest, *Brachystegia* woodland, rainforest margins; 1400–1800(2134) m; flowering April–August.

Conservation Status: With the exception of *Brass* 16987 the majority of the material in K (12 sheets) is from the North Province, Malawi, with a disjunction to the Zambian locality of *Bidgood et al.* 8539. This suggests that the species has a restricted distribution although perhaps not uncommon, and it should be looked for in suitable habitats in between. However, there is no indication of its relative abundance so the species is best recorded as DD (Data Deficient) until assessments can be made with further fieldwork.

# INDEX TO BOTANICAL NAMES

Accepted names in roman, synonyms in *italic*. Bold page numbers indicate main entries of accepted names (names with descriptions and in the keys), and illustrations.

# FAMILIES OF VASCULAR PLANTS REPRESENTED IN THE FLORA ZAMBESIACA AREA

## PTERIDOPHYTA

(Flora Zambesiaca families and family number.  Published 1970)

| | | | | | | | |
|---|---|---|---|---|---|---|---|
| Actiniopteridaceae | | Gleicheniaceae | 9 | Parkeriaceae | | | |
| see Adiantaceae | 18 | Grammitidaceae | 20 | see Adiantaceae | 18 | | |
| Adiantaceae | 18 | Hymenophyllaceae | 15 | Polypodiaceae | 21 | | |
| Aspidiaceae | 27 | Isoetaceae | 4 | Psilotaceae | 1 | | |
| Aspleniaceae | 23 | Lindsaeaceae | 19 | Pteridaceae | | | |
| Athyriaceae | 25 | Lomariopsidaceae | 26 | see Adiantaceae | 18 | | |
| Azollaceae | 13 | Lycopodiaceae | 2 | Salviniaceae | 12 | | |
| Blechnaceae | 28 | Marattiaceae | 7 | Schizaeaceae | 10 | | |
| Cyatheaceae | 14 | Marsileaceae | 11 | Selaginellaceae | 3 | | |
| Davalliaceae | 22 | Oleandraceae | | Thelypteridaceae | 24 | | |
| Dennstaedtiaceae | 16 | see Davalliaceae | 22 | Vittariaceae | 17 | | |
| Dryopteridaceae | | Ophioglossaceae | 6 | Woodsiaceae | | | |
| see Aspidiaceae | 27 | Osmundaceae | 8 | see Athyriaceae | 25 | | |
| Equisetaceae | 5 | | | | | | |

## GYMNOSPERMAE

(Flora Zambesiaca families and family number.  Volume 1(1) 1960)

| | | | | | |
|---|---|---|---|---|---|
| Cupressaceae | 3 | Cycadaceae | 1 | Podocarpaceae | 2 |

## ANGIOSPERMAE

(Flora Zambesiaca families, volume and part number and year of publication)

| | | | | | |
|---|---|---|---|---|---|
| Acanthaceae | | | Arecaceae | 13(2) | 2010 |
| tribes 1–5 | 8(5) | 2013 | Aristolochiaceae | 9(2) | 1997 |
| tribes 6–7 | 8(6) | 2015 | Asclepiadaceae | | |
| Agapanthaceae | 13(1) | 2008 | see Apocynaceae part 2 | 7(3) | 2020 |
| Agavaceae | 13(1) | 2008 | Asparagaceae | 13(1) | 2008 |
| Aizoaceae | 4 | 1978 | Asphodelaceae | 12(3) | 2001 |
| Alangiaceae | 4 | 1978 | Avicenniaceae | 8(7) | 2005 |
| Alismataceae | 12(2) | 2009 | Balanitaceae | 2(1) | 1963 |
| Alliaceae | 13(1) | 2008 | Balanophoraceae | 9(3) | 2006 |
| Aloaceae | 12(3) | 2001 | Balsaminaceae | 2(1) | 1963 |
| Amaranthaceae | 9(1) | 1988 | Barringtoniaceae | 4 | 1978 |
| Amaryllidaceae | 13(1) | 2008 | Basellaceae | 9(1) | 1988 |
| Anacardiaceae | 2(2) | 1966 | Begoniaceae | 4 | 1978 |
| Anisophylleaceae | | | Behniaceae | 13(1) | 2008 |
| see Rhizophoraceae | 4 | 1978 | Berberidaceae | 1(1) | 1960 |
| Annonaceae | 1(1) | 1960 | Bignoniaceae | 8(3) | 1988 |
| Anthericaceae | 13(1) | 2008 | Bixaceae | 1(1) | 1960 |
| Apocynaceae | | | Bombacaceae | 1(2) | 1961 |
| subfam. Apocynoideae | 7(2) | 1985 | Boraginaceae | 7(4) | 1990 |
| subfam. Asclepiadoideae | 7(3) | 2020 | Brexiaceae | 4 | 1978 |
| subfam. Periplocoideae | 7(3) | 2020 | Bromeliaceae | 13(2) | 2010 |
| subfam. Rauvolfioideae | 7(2) | 1985 | Buddlejaceae | | |
| subfam. Secamonoideae | 7(3) | 2020 | see Loganiaceae | 7(1) | 1983 |
| Aponogetonaceae | 12(2) | 2009 | Burmanniaceae | 12(2) | 2009 |
| Aquifoliaceae | 2(2) | 1966 | Burseraceae | 2(1) | 1963 |
| Araceae | 12(1) | 2011 | Buxaceae | 9(3) | 2006 |
| Araliaceae | 4 | 1978 | Cabombaceae | 1(1) | 1960 |

| Cactaceae | 4 | 1978 |
| Caesalpinioideae | | |
| see Leguminosae | 3(2) | 2006 |
| Campanulaceae | 7(1) | 1983 |
| Canellaceae | 7(4) | 1990 |
| Cannabaceae | 9(6) | 1991 |
| Cannaceae | 13(4) | 2010 |
| Capparaceae | 1(1) | 1960 |
| Caricaceae | 4 | 1978 |
| Caryophyllaceae | 1(2) | 1961 |
| Casuarinaceae | 9(6) | 1991 |
| Cecropiaceae | 9(6) | 1991 |
| Celastraceae | 2(2) | 1966 |
| Ceratophyllaceae | 9(6) | 1991 |
| Chenopodiaceae | 9(1) | 1988 |
| Chrysobalanaceae | 4 | 1978 |
| Colchicaceae | 12(2) | 2009 |
| Combretaceae | 4 | 1978 |
| Commelinaceae | - | - |
| Compositae | | |
| tribes 1–5 | 6(1) | 1992 |
| tribe 11 | 6(4) | 2026 |
| tribes 12–14 | 6(5) | 2025 |
| Connaraceae | 2(2) | 1966 |
| Convolvulaceae | 8(1) | 1987 |
| Cornaceae | 4 | 1978 |
| Costaceae | 13(4) | 2010 |
| Crassulaceae | 7(1) | 1983 |
| Cruciferae | 1(1) | 1960 |
| Cucurbitaceae | 4 | 1978 |
| Cuscutaceae | 8(1) | 1987 |
| Cymodoceaceae | 12(2) | 2009 |
| Cyperaceae | 14 | 2020 |
| Dichapetalaceae | 2(1) | 1963 |
| Dilleniaceae | 1(1) | 1960 |
| Dioscoreaceae | 12(2) | 2009 |
| Dipsacaceae | 7(1) | 1983 |
| Dipterocarpaceae | 1(2) | 1961 |
| Droseraceae | 4 | 1978 |
| Ebenaceae | 7(1) | 1983 |
| Elatinaceae | 1(2) | 1961 |
| Ericaceae | 7(1) | 1983 |
| Eriocaulaceae | 13(4) | 2010 |
| Eriospermaceae | 13(2) | 2010 |
| Erythroxylaceae | 2(1) | 1963 |
| Escalloniaceae | 7(1) | 1983 |
| Euphorbiaceae | 9(4) | 1996 |
| Euphorbiaceae | 9(5) | 2001 |
| Flacourtiaceae | 1(1) | 1960 |
| Flagellariaceae | 13(4) | 2010 |
| Fumariaceae | 1(1) | 1960 |
| Gentianaceae | 7(4) | 1990 |
| Geraniaceae | 2(1) | 1963 |
| Gesneriaceae | 8(3) | 1988 |
| Gisekiaceae | | |
| see Molluginaceae | 4 | 1978 |
| Goodeniaceae | 7(1) | 1983 |
| Gramineae | | |
| tribes 1–18 | 10(1) | 1971 |
| tribes 19–22 | 10(2) | 1999 |
| tribes 24–26 | 10(3) | 1989 |
| tribe 27 | 10(4) | 2002 |
| Guttiferae | 1(2) | 1961 |
| Haloragaceae | 4 | 1978 |
| Hamamelidaceae | 4 | 1978 |
| Hemerocallidaceae | 12(3) | 2001 |
| Hernandiaceae | 9(2) | 1997 |
| Heteropyxidaceae | 4 | 1978 |
| Hyacinthaceae | 13(3) | 2023 |
| Hydnoraceae | 9(2) | 1997 |
| Hydrocharitaceae | 12(2) | 2009 |
| Hydrophyllaceae | 7(4) | 1990 |
| Hydrostachyaceae | 9(2) | 1997 |
| Hypericaceae | | |
| see Guttiferae | 1(2) | 1961 |
| Hypoxidaceae | 12(3) | 2001 |
| Icacinaceae | 2(1) | 1963 |
| Illecebraceae | 1(2) | 1961 |
| Iridaceae | 12(4) | 1993 |
| Irvingiaceae | 2(1) | 1963 |
| Ixonanthaceae | 2(1) | 1963 |
| Juncaceae | 13(4) | 2010 |
| Juncaginaceae | 12(2) | 2009 |
| Labiatae | | |
| see Lamiaceae, Verbenaceae | | |
| Lamiaceae | | |
| Viticoideae, Pingoideae | 8(7) | 2005 |
| Scutellaroideae- | | |
| Nepetoideae | 8(8) | 2013 |
| Lauraceae | 9(2) | 1997 |
| Lecythidaceae | | |
| see Barringtoniaceae | 4 | 1978 |
| Leeaceae | 2(2) | 1966 |
| Leguminosae, | | |
| Caesalpinioideae | 3(2) | 2007 |
| Mimosoideae | 3(1) | 1970 |
| Papilionoideae | 3(3) | 2007 |
| Papilionoideae | 3(4) | 2012 |
| Papilionoideae | 3(5) | 2001 |
| Papilionoideae | 3(6) | 2000 |
| Papilionoideae | 3(7) | 2002 |
| Lemnaceae | | |
| see Araceae | 12(1) | 2011 |
| Lentibulariaceae | 8(3) | 1988 |
| Liliaceae sensu stricto | 12(2) | 2009 |
| Limnocharitaceae | 12(2) | 2009 |
| Linaceae | 2(1) | 1963 |
| Lobeliaceae | 7(1) | 1983 |
| Loganiaceae | 7(1) | 1983 |
| Loranthaceae | 9(3) | 2006 |
| Lythraceae | 4 | 1978 |
| Malpighiaceae | 2(1) | 1963 |
| Malvaceae | 1(2) | 1961 |
| Marantaceae | 13(4) | 2010 |
| Mayacaceae | 13(2) | 2010 |
| Melastomataceae | 4 | 1978 |

| Meliaceae | 2(1) | 1963 |
| Melianthaceae | 2(2) | 1966 |
| Menispermaceae | 1(1) | 1960 |
| Menyanthaceae | 7(4) | 1990 |
| Mesembryanthemaceae | 4 | 1978 |
| Mimosoideae | | |
|   see Leguminosae | 3(1) | 1970 |
| Molluginaceae | 4 | 1978 |
| Monimiaceae | 9(2) | 1997 |
| Montiniaceae | 4 | 1978 |
| Moraceae | 9(6) | 1991 |
| Musaceae | 13(4) | 2010 |
| Myristicaceae | 9(2) | 1997 |
| Myricaceae | 9(3) | 2006 |
| Myrothamnaceae | 4 | 1978 |
| Myrsinaceae | 7(1) | 1983 |
| Myrtaceae | 4 | 1978 |
| Najadaceae | 12(2) | 2009 |
| Nesogenaceae | 8(7) | 2005 |
| Nyctaginaceae | 9(1) | 1988 |
| Nymphaeaceae | 1(1) | 1960 |
| Ochnaceae | 2(1) | 1963 |
| Olacaceae | 2(1) | 1963 |
| Oleaceae | 7(1) | 1983 |
| Oliniaceae | 4 | 1978 |
| Onagraceae | 4 | 1978 |
| Opiliaceae | 2(1) | 1963 |
| Orchidaceae | 11(1) | 1995 |
| Orchidaceae | 11(2) | 1998 |
| Orobanchaceae | | |
|   see Scrophulariaceae | 8(2) | 1990 |
| Oxalidaceae | 2(1) | 1963 |
| Palmae | | |
|   see Arecaceae | 13(2) | 2010 |
| Pandanaceae | 12(2) | 2009 |
| Papaveraceae | 1(1) | 1960 |
| Papilionoideae | | |
|   see Leguminosae | - | - |
| Passifloraceae | 4 | 1978 |
| Pedaliaceae | 8(3) | 1988 |
| Periplocaceae | | |
|   see Apocynaceae part 2 | 7(3) | 2020 |
| Philesiaceae | | |
|   see Behniaceae | 13(1) | 2008 |
| Phormiaceae | | |
|   see Hemerocallidaceae | 12(3) | 2001 |
| Phytolaccaceae | 9(1) | 1988 |
| Piperaceae | 9(2) | 1997 |
| Pittosporaceae | 1(1) | 1960 |
| Plantaginaceae | 9(1) | 1988 |
| Plumbaginaceae | 7(1) | 1983 |
| Podostemaceae | 9(2) | 1997 |
| Polygalaceae | 1(1) | 1960 |
| Polygonaceae | 9(3) | 2006 |
| Pontederiaceae | 13(2) | 2010 |
| Portulacaceae | 1(2) | 1961 |
| Potamogetonaceae | 12(2) | 2009 |
| Primulaceae | 7(1) | 1983 |
| Proteaceae | 9(3) | 2006 |
| Ptaeroxylaceae | 2(2) | 1966 |
| Rafflesiaceae | 9(2) | 1997 |
| Ranunculaceae | 1(1) | 1960 |
| Resedaceae | 1(1) | 1960 |
| Restionaceae | 13(4) | 2010 |
| Rhamnaceae | 2(2) | 1966 |
| Rhizophoraceae | 4 | 1978 |
| Rosaceae | 4 | 1978 |
| Rubiaceae | | |
|   subfam. Rubioideae | 5(1) | 1989 |
|   tribe Vanguerieae | 5(2) | 1998 |
|   subfam.Cinchonoideae | 5(3) | 2003 |
| Rutaceae | 2(1) | 1963 |
| Salicaceae | 9(6) | 1991 |
| Salvadoraceae | 7(1) | 1983 |
| Santalaceae | 9(3) | 2006 |
| Sapindaceae | 2(2) | 1966 |
| Sapotaceae | 7(1) | 1983 |
| Scrophulariaceae | 8(2) | 1990 |
| Selaginaceae | | |
|   see Scrophulariaceae | 8(2) | 1990 |
| Simaroubaceae | 2(1) | 1963 |
| Smilacaceae | 12(2) | 2009 |
| Solanaceae | 8(4) | 2005 |
| Sonneratiaceae | 4 | 1978 |
| Sphenocleaceae | 7(1) | 1983 |
| Sterculiaceae | 1(2) | 1961 |
| Strelitziaceae | 13(4) | 2010 |
| Taccaceae | | |
|   see Dioscoreaceae | 12(2) | 2009 |
| Tecophilaeaceae | 12(3) | 2001 |
| Tetragoniaceae | 4 | 1978 |
| Theaceae | 1(2) | 1961 |
| Thymelaeaceae | 9(3) | 2006 |
| Tiliaceae | 2(1) | 1963 |
| Trapaceae | 4 | 1978 |
| Turneraceae | 4 | 1978 |
| Typhaceae | 13(4) | 2010 |
| Ulmaceae | 9(6) | 1991 |
| Umbelliferae | 4 | 1978 |
| Urticaceae | 9(6) | 1991 |
| Vacciniaceae | | |
|   see Ericaceae | 7(1) | 1983 |
| Vahliaceae | 4 | 1978 |
| Valerianaceae | 7(1) | 1983 |
| Velloziaceae | 12(2) | 2009 |
| Verbenaceae | 8(7) | 2005 |
| Violaceae | 1(1) | 1960 |
| Viscaceae | 9(3) | 2006 |
| Vitaceae | 2(2) | 1966 |
| Xyridaceae | 13(4) | 2010 |
| Zannichelliaceae | 12(2) | 2009 |
| Zingiberaceae | 13(4) | 2010 |
| Zosteraceae | 12(2) | 2009 |
| Zygophyllaceae | 2(1) | 1963 |